Praise for *Software Architecture: The Hard Parts*

"This book provides the missing manual around building microservices and analyzing the nuances of architectural decisions throughout the whole tech stack. In this book, you get a catalog of architectural decisions you can make when building your distributed system and what are the pros and cons associated with each decision. This book is a must for every architect that is building modern distributed systems."

—*Aleksandar Serafimoski, Lead Consultant, Thoughtworks*

"It's a must-read for technologists who are passionate about architecture. Great articulation of patterns."

—*Vanya Seth, Head Of Tech, Thoughtworks India*

"Whether you're an aspiring architect or an experienced one leading a team, no handwaving, this book will guide you through the specifics of how to succeed in your journey to create enterprise applications and microservices."

—*Dr. Venkat Subramaniam,*
Award-winning Author and Founder of Agile Developer, Inc.

"*Software Architecture: The Hard Parts* provides the reader with valuable insight, practices, and real-world examples on pulling apart highly coupled systems and building them back up again. By gaining effective trade-off analysis skills, you will start to make better architecture decisions."

—*Joost van Wenen,*
Managing Partner & Cofounder, Infuze Consulting

"I loved reading this comprehensive body of work on distributed architectures! A great mix of solid discussions on fundamental concepts, together with tons of practical advice."

—David Kloet, Independent Software Architect

"Splitting a big ball of mud is no easy work. Starting from the code and getting to the data, this book will help you see the services that should be extracted and the services that should remain together."

—Rubén Díaz-Martínez, Software Developer at Codesai

"This book will equip you with the theoretical background and with a practical framework to help answer the most difficult questions faced in modern software architecture."

—James Lewis, Technical Director, Thoughtworks

软件架构难点解惑（影印版）

Software Architecture: The Hard Parts

[美] 尼尔·福特（Neal Ford）
[美] 马克·理查兹（Mark Richards）
[美] 普拉莫德·塞得拉吉（Pramod Sadalage）
[美] 扎马克·德加尼（Zhamak Dehghani）著

Beijing · Boston · Farnham · Sebastopol · Tokyo 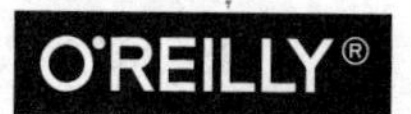

O'Reilly Media, Inc.授权东南大学出版社出版

南京 东南大学出版社

图书在版编目(CIP)数据

软件架构难点解惑 = Software Architecture: The Hard Parts：影印版：英文 /（美）尼尔·福特（Neal Ford）等著. —南京：东南大学出版社，2023.3

ISBN 978-7-5766-0591-4

Ⅰ. ①软… Ⅱ. ①尼… Ⅲ. ①软件设计—英文 Ⅳ. ①TP311.1

中国国家版本馆 CIP 数据核字(2023)第 001812 号

图字：10-2022-475 号

软件架构难点解惑（影印版）

著　　者：[美]尼尔·福特(Neal Ford)，[美]马克·理查兹(Mark Richards)，[美]普拉莫德·塞得拉吉(Pramod Sadalage)，[美]扎马克·德加尼(Zhamak Dehghani)
责任编辑：张　烨　　封面设计：Karen Montgomery，张　健　　责任印制：周荣虎
出版发行：东南大学出版社
社　　址：南京四牌楼 2 号　　邮编：210096　　电话：025-83793330
网　　址：http://www.seupress.com
电子邮件：press@ seupress.com
经　　销：全国各地新华书店
印　　刷：常州市武进第三印刷有限公司
开　　本：787mm×1000mm　1/16
印　　张：28.75
字　　数：563 千
版　　次：2023 年 3 月第 1 版
印　　次：2023 年 3 月第 1 次印刷
书　　号：ISBN 978-7-5766-0591-4
定　　价：138.00 元

本社图书若有印装质量问题，请直接与营销部联系。电话(传真)：025-83791830

Table of Contents

Part II. Putting Things Back Together

Preface

When two of your authors, Neal and Mark, were writing the book *Fundamentals of Software Architecture*, we kept coming across complex examples in architecture that we wanted to cover but that were too difficult. Each one offered no easy solutions but rather a collection of messy trade-offs. We set those examples aside into a pile we called "The Hard Parts." Once that book was finished, we looked at the now gigantic pile of hard parts and tried to figure out: *why are these problems so difficult to solve in modern architectures*?

We took all the examples and worked through them like architects, applying trade-off analysis for each situation, but also paying attention to the process we used to arrive at the trade-offs. One of our early revelations was the increasing importance of data in architecture decisions: who can/should access data, who can/should write to it, and how to manage the separation of analytical and operational data. To that end, we asked experts in those fields to join us, which allows this book to fully incorporate decision making from both angles: architecture to data and data to architecture.

The result is this book: a collection of difficult problems in modern software architecture, the trade-offs that make the decisions hard, and ultimately an illustrated guide to show you how to apply the same trade-off analysis to your own unique problems.

Conventions Used in This Book

The following typographical conventions are used in this book:

Italic
: Indicates new terms, URLs, email addresses, filenames, and file paths.

`Constant width`
: Used for program listings, as well as within paragraphs to refer to program elements such as variable or function names, databases, data types, environment variables, statements, and keywords.

`Constant width bold`
: Shows commands or other text that should be typed literally by the user.

`Constant width italic`
: Shows text that should be replaced with user-supplied values or by values determined by context.

This element signifies a tip or suggestion.

Using Code Examples

Supplemental material (code examples, exercises, etc.) is available for download at *http://architecturethehardparts.com*.

If you have a technical question or a problem using the code examples, please send email to *bookquestions@oreilly.com*.

This book is here to help you get your job done. In general, if example code is offered with this book, you may use it in your programs and documentation. You do not need to contact us for permission unless you're reproducing a significant portion of the code. For example, writing a program that uses several chunks of code from this book does not require permission. Selling or distributing examples from O'Reilly books does require permission. Answering a question by citing this book and quoting example code does not require permission. Incorporating a significant amount of example code from this book into your product's documentation does require permission.

We appreciate, but generally do not require, attribution. An attribution usually includes the title, author, publisher, and ISBN. For example: "*Software Architecture: The Hard Parts* by Neal Ford, Mark Richards, Pramod Sadalage, and Zhamak Dehghani (O'Reilly). Copyright 2022 Neal Ford, Mark Richards, Pramod Sadalage, and Zhamak Dehghani, 978-1-492-08689-5."

If you feel your use of code examples falls outside fair use or the permission given above, feel free to contact us at *permissions@oreilly.com*.

O'Reilly Online Learning

O'REILLY®

For more than 40 years, *O'Reilly Media* has provided technology and business training, knowledge, and insight to help companies succeed.

Our unique network of experts and innovators share their knowledge and expertise through books, articles, and our online learning platform. O'Reilly's online learning platform gives you on-demand access to live training courses, in-depth learning paths, interactive coding environments, and a vast collection of text and video from O'Reilly and 200+ other publishers. For more information, visit *http://oreilly.com*.

How to Contact Us

Please address comments and questions concerning this book to the publisher:

O'Reilly Media, Inc.
1005 Gravenstein Highway North
Sebastopol, CA 95472
800-998-9938 (in the United States or Canada)
707-829-0515 (international or local)
707-829-0104 (fax)

We have a web page for this book, where we list errata, examples, and any additional information. You can access this page at *https://oreil.ly/sa-the-hard-parts*.

Email *bookquestions@oreilly.com* to comment or ask technical questions about this book.

For news and information about our books and courses, visit *http://oreilly.com*.

Find us on Facebook: *http://facebook.com/oreilly*

Follow us on Twitter: *http://twitter.com/oreillymedia*

Watch us on YouTube: *http://youtube.com/oreillymedia*

Acknowledgments

Mark and Neal would like to thank all the people who attended our (almost exclusively online) classes, workshops, conference sessions, and user group meetings, as well as all the other people who listened to versions of this material and provided invaluable feedback. Iterating on new material is especially tough when we can't do it live, so we appreciate those who commented on the many iterations. We thank the publishing team at O'Reilly, who made this as painless an experience as writing a book can be. We also thank a few random oases of sanity-preserving and idea-sparking groups that have names like Pasty Geeks and the Hacker B&B.

Thanks to those who did the technical review of our book—Vanya Seth, Venkat Subramanian, Joost van Weenen, Grady Booch, Ruben Diaz, David Kloet, Matt Stein, Danilo Sato, James Lewis, and Sam Newman. Your valuable insights and feedback helped validate our technical content and make this a better book.

We especially want to acknowledge the many workers and families impacted by the unexpected global pandemic. As knowledge workers, we faced inconveniences that pale in comparison to the massive disruption and devastation wrought on so many of our friends and colleagues across all walks of life. Our sympathies and appreciation especially go out to health care workers, many of whom never expected to be on the front line of a terrible global tragedy yet handled it nobly. Our collective thanks can never be adequately expressed.

Acknowledgments from Mark Richards

In addition to the preceding acknowledgments, I once again thank my lovely wife, Rebecca, for putting up with me through yet another book project. Your unending support and advice helped make this book happen, even when it meant taking time away from working on your own novel. You mean the world to me, Rebecca. I also thank my good friend and coauthor Neal Ford. Collaborating with you on the materials for this book (as well as our last one) was truly a valuable and rewarding experience. You are, and always will be, my friend.

Acknowledgments from Neal Ford

I would like to thank my extended family, Thoughtworks as a collective, and Rebecca Parsons and Martin Fowler as individual parts of it. Thoughtworks is an extraordinary group of people who manage to produce value for customers while keeping a keen eye toward why things work so that we can improve them. Thoughtworks supported this book in many ways and continues to grow Thoughtworkers who challenge and inspire me every day. I also thank our neighborhood cocktail club for a regular escape from routine, including the weekly outside, socially distanced versions that helped us all survive the odd time we just lived through. I thank my long-time friend Norman Zapien, who never ceases to provide enjoyable conversation. Lastly, I thank my wife, Candy, who continues to support this lifestyle that has me staring at things like book writing rather than our cats too much.

Acknowledgments from Pramod Sadalage

I thank my wife, Rupali, for all the support and understanding, and my lovely girls, Arula and Arhana, for the encouragement; daddy loves you both. All the work I do would not be possible without the clients I work with and various conferences that have helped me iterate on the concepts and content. I thank AvidXchange, the latest client I am working with, for its support and providing great space to iterate on new concepts. I also thank Thoughtworks for its continued support in my life, and Neal Ford, Rebecca Parsons, and Martin Fowler for being amazing mentors; you all make me a better person. Lastly, thank you to my parents, especially my mother, Shobha, whom I miss every day. *I miss you, MOM.*

Acknowledgments from Zhamak Dehghani

I thank Mark and Neal for their open invitation to contribute to this amazing body of work. My contribution to this book would not have been possible without the continuous support of my husband, Adrian, and patience of my daughter, Arianna. I love you both.

CHAPTER 1

What Happens When There Are No "Best Practices"?

Why does a technologist like a software architect present at a conference or write a book? Because they have discovered what is colloquially known as a "best practice," a term so overused that those who speak it increasingly experience backlash. Regardless of the term, technologists write books when they have figured out a novel solution to a general problem and want to broadcast it to a wider audience.

But what happens for that vast set of problems that have no good solutions? Entire classes of problems exist in software architecture that have no general good solutions, but rather present one messy set of trade-offs cast against an (almost) equally messy set.

Software developers build outstanding skills in searching online for solutions to a current problem. For example, if they need to figure out how to configure a particular tool in their environment, expert use of Google finds the answer.

But that's not true for architects.

For architects, many problems present unique challenges because they conflate the exact environment and circumstances of your organization—what are the chances that someone has encountered exactly this scenario *and* blogged it or posted it on Stack Overflow?

Architects may have wondered why so few books exist about architecture compared to technical topics like frameworks, APIs, and so on. Architects rarely experience common problems but constantly struggle with decision making in novel situations. For architects, every problem is a snowflake. In many cases, the problem is novel not just within a particular organization but rather throughout the world. No books or conference sessions exist for those problems!

Architects shouldn't constantly seek out silver-bullet solutions to their problems; they are as rare now as in 1986, when Fred Brooks coined the term:

> There is no single development, in either technology or management technique, which by itself promises even one order of magnitude [tenfold] improvement within a decade in productivity, in reliability, in simplicity.
>
> —Fred Brooks from "No Silver Bullet"

Because virtually every problem presents novel challenges, the real job of an architect lies in their ability to objectively determine and assess the set of trade-offs on either side of a consequential decision to resolve it as well as possible. The authors don't talk about "best solutions" (in this book or in the real world) because "best" implies that an architect has managed to maximize all the possible competing factors within the design. Instead, our tongue-in-cheek advice is as follows:

Don't try to find the *best* design in software architecture; instead, strive for the *least worst* combination of trade-offs.

Often, the best design an architect can create is the least worst collection of trade-offs —no single architecture characteristics excels as it would alone, but the balance of all the competing architecture characteristics promote project success.

Which begs the question: "How can an architect *find* the least worst combination of trade-offs (and document them effectively)?" This book is primarily about decision making, enabling architects to make better decisions when confronted with novel situations.

Why "The Hard Parts"?

Why did we call this book *Software Architecture: The Hard Parts?* Actually, the "hard" in the title performs double duty. First, *hard* connotes *difficult*, and architects constantly face difficult problems that literally (and figuratively) no one has faced before, involving numerous technology decisions with long-term implications layered on top of the interpersonal and political environment where the decision must take place.

Second, *hard* connotes *solidity*—just as in the separation of *hardware* and *software*, the *hard* one should change much less because it provides the foundation for the *soft* stuff. Similarly, architects discuss the distinction between *architecture* and *design*, where the former is structural and the latter is more easily changed. Thus, in this book, we talk about the foundational parts of architecture.

The definition of software architecture itself has provided many hours of non-productive conversation among its practitioners. One favorite tongue-in-cheek definition is that "software architecture is the *stuff* that's hard to change later." That *stuff* is what our book is about.

Giving Timeless Advice About Software Architecture

The software development ecosystem constantly and chaotically shifts and grows. Topics that were all the rage a few years ago have either been subsumed by the ecosystem and disappeared or replaced by something different/better. For example, 10 years ago, the predominant architecture style for large enterprises was orchestration-driven, service-oriented architecture. Now, virtually no one builds in that architecture style anymore (for reasons we'll uncover along the way); the current favored style for many distributed systems is microservices. How and why did that transition happen?

When architects look at a particular style (especially a historical one), they must consider the constraints in place that lead to that architecture becoming dominant. At the time, many companies were merging to become *enterprises*, with all the attendant integration woes that come with that transition. Additionally, open source wasn't a viable option (often for political rather than technical reasons) for large companies. Thus, architects emphasized shared resources and centralized orchestration as a solution.

However, in the intervening years, open source and Linux became viable alternatives, making operating systems *commercially* free. However, the real tipping point occurred when Linux became *operationally* free with the advent of tools like Puppet and Chef, which allowed development teams to programmatically spin up their environments as part of an automated build. Once that capability arrived, it fostered an architectural revolution with microservices and the quickly emerging infrastructure of containers and orchestration tools like Kubernetes.

This illustrates that the software development ecosystem expands and evolves in completely unexpected ways. One new capability leads to another one, which unexpectedly creates new capabilities. Over the course of time, the ecosystem completely replaces itself, one piece at a time.

This presents an age-old problem for authors of books about technology generally and software architecture specifically—how can we write something that isn't old immediately?

We don't focus on technology or other implementation details in this book. Rather, we focus on *how* architects make decisions, and how to objectively weigh trade-offs when presented with novel situations. We use contemporaneous scenarios and examples to provide details and context, but the underlying principles focus on trade-off analysis and decision making when faced with new problems.

The Importance of Data in Architecture

> Data is a precious thing and will last longer than the systems themselves.
>
> —Tim Berners-Lee

For many in architecture, data is everything. Every enterprise building any system must deal with data, as it tends to live much longer than systems or architecture, requiring diligent thought and design. However, many of the instincts of data architects to build tightly coupled systems create conflicts within modern distributed architectures. For example, architects and DBAs must ensure that business data survives the breaking apart of monolith systems and that the business can still derive value from its data regardless of architecture undulations.

It has been said that *data is the most important asset in a company*. Businesses want to extract value from the data that they have and are finding new ways to deploy data in decision making. Every part of the enterprise is now data driven, from servicing existing customers, to acquiring new customers, increasing customer retention, improving products, predicting sales, and other trends. This reliance on data means that all software architecture is in the service of data, ensuring the right data is available and usable by all parts of the enterprise.

The authors built many distributed systems a few decades ago when they first became popular, yet decision making in modern microservices seems more difficult, and we wanted to figure out why. We eventually realized that, back in the early days of distributed architecture, we mostly still persisted data in a single relational database. However, in microservices and the philosophical adherence to a *bounded context* from Domain-Driven Design (*https://oreil.ly/bW8CH*), as a way of limiting the scope of implementation detail coupling, data has moved to an architectural concern, along with transactionality. Many of the hard parts of modern architecture derive from tensions between data and architecture concerns, which we untangle in both Part I and Part II.

One important distinction that we cover in a variety of chapters is the separation between *operational* versus *analytical* data:

Operational data
: Data used for the operation of the business, including sales, transactional data, inventory, and so on. This data is what the company runs on—if something interrupts this data, the organization cannot function for very long. This type of data is defined as *Online Transactional Processing* (OLTP), which typically involves inserting, updating, and deleting data in a database.

Analytical data

Data used by data scientists and other business analysts for predictions, trending, and other business intelligence. This data is typically not transactional and often not relational—it may be in a graph database or snapshots in a different format than its original transactional form. This data isn't critical for the day-to-day operation but rather for the long-term strategic direction and decisions.

We cover the impact of both operational and analytical data throughout the book.

Architectural Decision Records

One of the most effective ways of documenting architecture decisions is through *Architectural Decision Records* (ADRs (*https://adr.github.io*)). ADRs were first evangelized by Michael Nygard in a blog post (*https://oreil.ly/yDcU2*) and later marked as "adopt" in the Thoughtworks Technology Radar (*https://oreil.ly/0nwHw*). An ADR consists of a short text file (usually one to two pages long) describing a specific architecture decision. While ADRs can be written using plain text, they are usually written in some sort of text document format like AsciiDoc (*http://asciidoc.org*) or Markdown (*https://www.markdownguide.org*). Alternatively, an ADR can also be written using a wiki page template. We devoted an entire chapter to ADRs in our previous book, *Fundamentals of Software Architecture* (O'Reilly).

We will be leveraging ADRs as a way of documenting various architecture decisions made throughout the book. For each architecture decision, we will be using the following ADR format with the assumption that each ADR is approved:

ADR: A short noun phrase containing the architecture decision

Context

In this section of the ADR we will add a short one- or two-sentence description of the problem, and list the alternative solutions.

Decision

In this section we will state the architecture decision and provide a detailed justification of the decision.

Consequences

In this section of the ADR we will describe any consequences after the decision is applied, and also discuss the trade-offs that were considered.

A list of all the Architectural Decision Records created in this book can be found in Appendix B.

Documenting a decision is important for an architect, but governing the proper use of the decision is a separate topic. Fortunately, modern engineering practices allow automating many common governance concerns by using architecture fitness functions.

Architecture Fitness Functions

Once an architect has identified the relationship between components and codified that into a design, how can they make sure that the implementers will adhere to that design? More broadly, how can architects ensure that the design principles they define become reality if they aren't the ones to implement them?

These questions fall under the heading of *architecture governance*, which applies to any organized oversight of one or more aspects of software development. As this book primarily covers architecture structure, we cover how to automate design and quality principles via fitness functions in many places.

Software development has slowly evolved over time to adapt unique engineering practices. In the early days of software development, a manufacturing metaphor was commonly applied to software practices, both in the large (like the Waterfall development process) and small (integration practices on projects). In the early 1990s, a rethinking of software development engineering practices, lead by Kent Beck and the other engineers on the C3 project, called eXtreme Programming (XP), illustrated the importance of incremental feedback and automation as key enablers of software development productivity. In the early 2000s, the same lessons were applied to the intersection of software development and operations, spawning the new role of DevOps and automating many formerly manual operational chores. Just as before, automation allows teams to go faster because they don't have to worry about things breaking without good feedback. Thus, *automation* and *feedback* have become central tenets for effective software development.

Consider the environments and situations that lead to breakthroughs in automation. In the era before continuous integration, most software projects included a lengthy integration phase. Each developer was expected to work in some level of isolation from others, then integrate all the code at the end into an integration phase. Vestiges of this practice still linger in version control tools that force branching and prevent continuous integration. Not surprisingly, a strong correlation existed between project size and the pain of the integration phase. By pioneering continuous integration, the XP team illustrated the value of rapid, continuous feedback.

The DevOps revolution followed a similar course. As Linux and other open source software became "good enough" for enterprises, combined with the advent of tools that allowed programmatic definition of (eventually) virtual machines, operations personnel realized they could automate machine definitions and many other repetitive tasks.

In both cases, advances in technology and insights led to automating a recurring job that was handled by an expensive role—which describes the current state of architecture governance in most organizations. For example, if an architect chooses a particular architecture style or communication medium, how can they make sure that a developer implements it correctly? When done manually, architects perform code reviews or perhaps hold architecture review boards to assess the state of governance. However, just as in manually configuring computers in operations, important details can easily fall through superficial reviews.

Using Fitness Functions

In the 2017 book *Building Evolutionary Architectures* (O'Reilly), the authors (Neal Ford, Rebecca Parsons, and Patrick Kua) defined the concept of an *architectural fitness function*: any mechanism that performs an objective integrity assessment of some architecture characteristic or combination of architecture characteristics. Here is a point-by-point breakdown of that definition:

Any mechanism

Architects can use a wide variety of tools to implement fitness functions; we will show numerous examples throughout the book. For example, dedicated testing libraries exist to test architecture structure, architects can use monitors to test operational architecture characteristics such as performance or scalability, and chaos engineering frameworks test reliability and resiliency.

Objective integrity assessment

One key enabler for automated governance lies with objective definitions for architecture characteristics. For example, an architect can't specify that they want a "high performance" website; they must provide an object value that can be measured by a test, monitor, or other fitness function.

Architects must watch out for *composite architecture characteristics*—ones that aren't objectively measurable but are really composites of other measurable things. For example, "agility" isn't measurable, but if an architect starts pulling the broad term *agility* apart, the goal is for teams to be able to respond quickly and confidently to change, either in ecosystem or domain. Thus, an architect can find measurable characteristics that contribute to agility: deployability, testability, cycle time, and so on. Often, the lack of ability to measure an architecture characteristic indicates too vague a definition. If architects strive toward measurable properties, it allows them to automate fitness function application.

Some architecture characteristic or combination of architecture characteristics

This characteristic describes the two scopes for fitness functions:

Atomic

These fitness functions handle a single architecture characteristic in isolation. For example, a fitness function that checks for component cycles within a codebase is atomic in scope.

Holistic

Holistic fitness functions validate a combination of architecture characteristics. A complicating feature of architecture characteristics is the synergy they sometimes exhibit with other architecture characteristics. For example, if an architect wants to improve security, a good chance exists that it will affect performance. Similarly, scalability and elasticity are sometimes at odds—supporting a large number of concurrent users can make handling sudden bursts more difficult. Holistic fitness functions exercise a combination of interlocking architecture characteristics to ensure that the combined effect won't negatively affect the architecture.

An architect implements fitness functions to build protections around unexpected change in architecture characteristics. In the Agile software development world, developers implement unit, functional, and user acceptance tests to validate different dimensions of the *domain* design. However, until now, no similar mechanism existed to validate the *architecture characteristics* part of the design. In fact, the separation between fitness functions and unit tests provides a good scoping guideline for architects. Fitness functions validate architecture characteristics, not domain criteria; unit tests are the opposite. Thus, an architect can decide whether a fitness function or unit test is needed by asking the question: "Is any domain knowledge required to execute this test?" If the answer is "yes," then a unit/function/user acceptance test is appropriate; if "no," then a fitness function is needed.

For example, when architects talk about *elasticity*, it's the ability of the application to withstand a sudden burst of users. Notice that the architect doesn't need to know any details about the domain—this could be an ecommerce site, an online game, or something else. Thus, *elasticity* is an architectural concern and within the scope of a fitness function. If on the other hand the architect wanted to validate the proper parts of a mailing address, that is covered via a traditional test. Of course, this separation isn't purely binary—some fitness functions will touch on the domain and vice versa, but the differing goals provide a good way to mentally separate them.

Here are a couple of examples to make the concept less abstract.

One common architect goal is to maintain good internal structural integrity in the codebase. However, malevolent forces work against the architect's good intentions on many platforms. For example, when coding in any popular Java or .NET development environment, as soon as a developer references a class not already imported, the IDE helpfully presents a dialog asking the developer if they would like to auto-import the reference. This occurs so often that most programmers develop the habit of swatting the auto-import dialog away like a reflex action.

However, arbitrarily importing classes or components among one another spells disaster for modularity. For example, Figure 1-1 illustrates a particularly damaging anti-pattern that architects aspire to avoid.

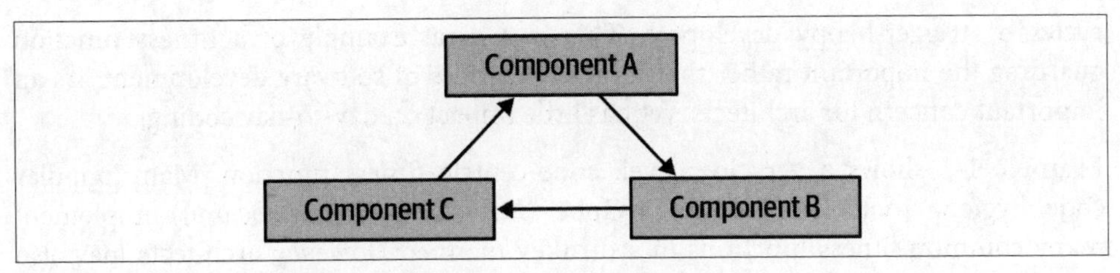

Figure 1-1. Cyclic dependencies between components

In this anti-pattern, each component references something in the others. Having a network of components such as this damages modularity because a developer cannot reuse a single component without also bringing the others along. And, of course, if the other components are coupled to other components, the architecture tends more and more toward the Big Ball of Mud (*https://oreil.ly/usx7p*) anti-pattern. How can architects govern this behavior without constantly looking over the shoulders of trigger-happy developers? Code reviews help but happen too late in the development cycle to be effective. If an architect allows a development team to rampantly import across the codebase for a week until the code review, serious damage has already occurred in the codebase.

The solution to this problem is to write a fitness function to avoid component cycles, as shown in Example 1-1.

Example 1-1. Fitness function to detect component cycles

```
public class CycleTest {
    private JDepend jdepend;

    @BeforeEach
    void init() {
          jdepend = new JDepend();
          jdepend.addDirectory("/path/to/project/persistence/classes");
          jdepend.addDirectory("/path/to/project/web/classes");
          jdepend.addDirectory("/path/to/project/thirdpartyjars");
    }

    @Test
    void testAllPackages() {
          Collection packages = jdepend.analyze();
          assertEquals("Cycles exist", false, jdepend.containsCycles());
    }
}
```

In the code, an architect uses the metrics tool JDepend (*https://oreil.ly/ozzzk*) to check the dependencies between packages. The tool understands the structure of Java packages and fails the test if any cycles exist. An architect can wire this test into the continuous build on a project and stop worrying about the accidental introduction of cycles by trigger-happy developers. This is a great example of a fitness function guarding the important rather than urgent practices of software development: it's an important concern for architects, yet has little impact on day-to-day coding.

Example 1-1 shows a very low-level, code-centric fitness function. Many popular code hygiene tools (such as SonarQube (*https://www.sonarqube.org*)) implement many common fitness functions in a turnkey manner. However, architects may also want to validate the macro structure of the architecture as well as the micro. When designing a layered architecture such as the one in Figure 1-2, the architect defines the layers to ensure separation of concerns.

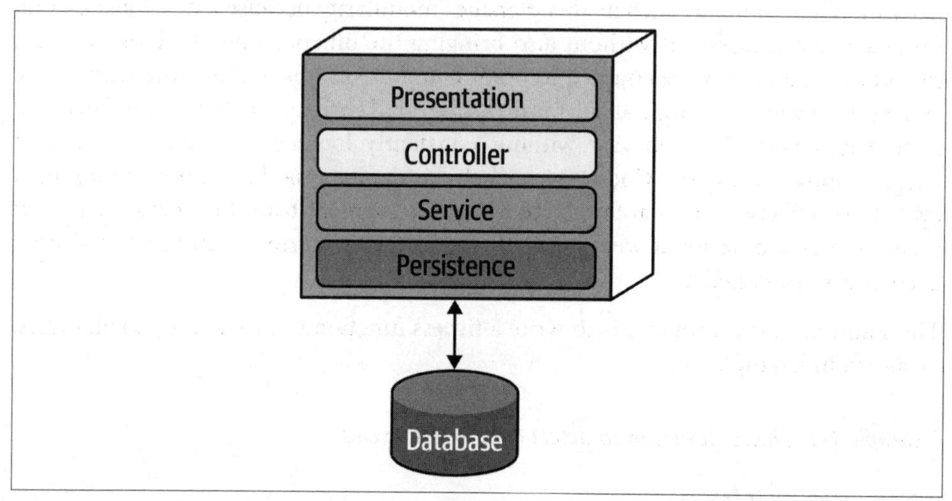

Figure 1-2. Traditional layered architecture

However, how can the architect ensure that developers will respect these layers? Some developers may not understand the importance of the patterns, while others may adopt a "better to ask forgiveness than permission" attitude because of some overriding local concern, such as performance. But allowing implementers to erode the reasons for the architecture hurts the long-term health of the architecture.

ArchUnit (*https://www.archunit.org*) allows architects to address this problem via a fitness function, shown in Example 1-2.

Example 1-2. ArchUnit fitness function to govern layers

```
layeredArchitecture()
    .layer("Controller").definedBy("..controller..")
    .layer("Service").definedBy("..service..")
    .layer("Persistence").definedBy("..persistence..")

    .whereLayer("Controller").mayNotBeAccessedByAnyLayer()
```

```
    .whereLayer("Service").mayOnlyBeAccessedByLayers("Controller")
    .whereLayer("Persistence").mayOnlyBeAccessedByLayers("Service")
```

In Example 1-2, the architect defines the desirable relationship between layers and writes a verification fitness function to govern it. This allows an architect to establish architecture principles outside the diagrams and other informational artifacts, and verify them on an ongoing basis.

A similar tool in the .NET space, NetArchTest (*https://oreil.ly/EMXpv*), allows similar tests for that platform. A layer verification in C# appears in Example 1-3.

Example 1-3. NetArchTest for layer dependencies

```
// Classes in the presentation should not directly reference repositories
var result = Types.InCurrentDomain()
    .That()
    .ResideInNamespace("NetArchTest.SampleLibrary.Presentation")
    .ShouldNot()
    .HaveDependencyOn("NetArchTest.SampleLibrary.Data")
    .GetResult()
    .IsSuccessful;
```

Tools continue to appear in this space with increasing degrees of sophistication. We will continue to highlight many of these techniques as we illustrate fitness functions alongside many of our solutions.

Finding an objective outcome for a fitness function is critical. However, *objective* doesn't imply *static*. Some fitness functions will have noncontextual return values, such as true/false or a numeric value such as a performance threshold. However, other fitness functions (deemed *dynamic*) return a value based on some context. For example, when measuring *scalability*, architects measure the number of concurrent users and also generally measure the performance for each user. Often, architects design systems so that as the number of users goes up, performance per user declines slightly—but doesn't fall off a cliff. Thus, for these systems, architects design performance fitness functions that take into account the number of concurrent users. As long as the measure of an architecture characteristic is objective, architects can test it.

While most fitness functions should be automated and run continually, some will necessarily be manual. A manual fitness function requires a person to handle the validation. For example, for systems with sensitive legal information, a lawyer may need to review changes to critical parts to ensure legality, which cannot be automated. Most deployment pipelines support manual stages, allowing teams to accommodate manual fitness functions. Ideally, these are run as often as reasonably possible—a validation that doesn't run can't validate anything. Teams execute fitness functions either on demand (rarely) or as part of a continuous integration work stream (most common). To fully achieve the benefit of validations such as fitness functions, they should be run continually.

Continuity is important, as illustrated in this example of enterprise-level governance using fitness functions. Consider the following scenario: what does a company do when a zero-day exploit is discovered in one of the development frameworks or libraries the enterprise uses? If it's like most companies, security experts scour projects to find the offending version of the framework and make sure it's updated, but that process is rarely automated, relying on many manual steps. This isn't an abstract question; this exact scenario affected a major financial institution described in The Equifax Data Breach. Like the architecture governance described previously, manual processes are error prone and allow details to escape.

The Equifax Data Breach

On September 7, 2017, Equifax, a major credit scoring agency in the US, announced that a data breach had occurred. Ultimately, the problem was traced to a hacking exploit of the popular Struts web framework in the Java ecosystem (Apache Struts vCVE-2017-5638). The foundation issued a statement announcing the vulnerability and released a patch on March 7, 2017. The Department of Homeland Security contacted Equifax and similar companies the next day, warning them of this problem, and they ran scans on March 15, 2017, which didn't reveal all of the affected systems. Thus, the critical patch wasn't applied to many older systems until July 29, 2017, when Equifax's security experts identified the hacking behavior that lead to the data breach.

Imagine an alternative world in which every project runs a deployment pipeline, and the security team has a "slot" in each team's deployment pipeline where they can deploy fitness functions. Most of the time, these will be mundane checks for safeguards like preventing developers from storing passwords in databases and similar regular governance chores. However, when a zero-day exploit appears, having the same mechanism in place everywhere allows the security team to insert a test in every project that checks for a certain framework and version number; if it finds the dangerous version, it fails the build and notifies the security team. Teams configure deployment pipelines to awaken for any change to the ecosystem: code, database schema, deployment configuration, and fitness functions. This allows enterprises to universally automate important governance tasks.

Fitness functions provide many benefits for architects, not the least of which is the chance to do some coding again! One of the universal complaints among architects is that they don't get to code much anymore—but fitness functions are often code! By building an executable specification of the architecture, which anyone can validate anytime by running the project's build, architects must understand the system and its ongoing evolution well, which overlaps with the core goal of keeping up with the code of the project as it grows.

However powerful fitness functions are, architects should avoid overusing them. Architects should not form a cabal and retreat to an ivory tower to build an impossibly complex, interlocking set of fitness functions that merely frustrate developers and teams. Instead, it's a way for architects to build an executable checklist of *important* but not *urgent* principles on software projects. Many projects drown in urgency, allowing some important principles to slip by the side. This is the frequent cause of technical debt: "We know this is bad, but we'll come back to fix it later"—and later never comes. By codifying rules about code quality, structure, and other safeguards against decay into fitness functions that run continually, architects build a quality checklist that developers can't skip.

A few years ago, the excellent book *The Checklist Manifesto* by Atul Gawande (Picador) highlighted the use of checklists by professionals like surgeons, airline pilots, and those other fields who commonly use (sometimes by force of law) checklists as part of their job. It isn't because they don't know their job or are particularly forgetful; when professionals perform the same task over and over, it becomes easy to fool themselves when it's accidentally skipped, and checklists prevent that. Fitness functions represent a checklist of important principles defined by architects and run as part of the build to make sure developers don't accidentally (or purposefully, because of external forces like schedule pressure) skip them.

We utilize fitness functions throughout the book when an opportunity arises to illustrate governing an architectural solution as well as the initial design.

Architecture Versus Design: Keeping Definitions Simple

A constant area of struggle for architects is keeping *architecture* and *design* as separate but related activities. While we don't want to wade into the never-ending argument about this distinction, we strive in this book to stay firmly on the *architecture* side of that spectrum for several reasons.

First, architects must understand underlying architecture principles to make effective decisions. For example, the decision between *synchronous* versus *asynchronous* communication has a number of trade-offs before architects layer in implementation details. In the book *Fundamentals of Software Architecture*, the authors coined the second law of software architecture: *why* is more important than *how*. While ultimately architects must understand how to implement solutions, they must first understand why one choice has better trade-offs than another.

Second, by focusing on architecture concepts, we can avoid the numerous implementations of those concepts. Architects can implement asynchronous communication in a variety of ways; we focus on why an architect would choose asynchronous communication and leave the implementation details to another place.

Third, if we start down the path of implementing all the varieties of options we show, this would be the longest book ever written. Focus on architecture principles allows us to keep things as generic as they can be.

To keep subjects as grounded in architecture as possible, we use the simplest definitions possible for key concepts. For example, *coupling* in architecture can fill entire books (and it has). To that end, we use the following simple, verging on simplistic, definitions:

Service
: In colloquial terms, a *service* is a cohesive collection of functionality deployed as an independent executable. Most of the concepts we discuss with regard to services apply broadly to distributed architectures, and specifically microservices architectures.

  In the terms we define in Chapter 2, a *service* is part of an architecture quantum, which includes further definitions of both static and dynamic coupling between services and other quanta.

Coupling
: Two artifacts (including services) are coupled if a change in one might require a change in the other to maintain proper functionality.

Component
: An architectural building block of the application that does some sort of business or infrastructure function, usually manifested through a package structure (Java), namespace (C#), or a physical grouping of source code files within some sort of directory structure. For example, the component Order History might be implemented through a set of class files located in the namespace `app.business.order.history`.

Synchronous communication
: Two artifacts communicate synchronously if the caller must wait for the response before proceeding.

Asynchronous communication
: Two artifacts communicate asynchronously if the caller does not wait for the response before proceeding. Optionally, the caller can be notified by the receiver through a separate channel when the request has completed.

Orchestrated coordination
: A workflow is orchestrated if it includes a service whose primary responsibility is to coordinate the workflow.

Choreographed coordination
A workflow is choreographed when it lacks an orchestrator; rather, the services in the workflow share the coordination responsibilities of the workflow.

Atomicty
A workflow is *atomic* if all parts of the workflow maintain a consistent state at all times; the opposite is represented by the spectrum of *eventual consistency*, covered in Chapter 6.

Contract
We use the term *contract* broadly to define the interface between two software parts, which may encompass method or function calls, integration architecture remote calls, dependencies, and so on. Anywhere two pieces of software join, a contract is involved.

Software architecture is by its nature abstract: we cannot know what unique combination of platforms, technologies, commercial software, and the other dizzying array of possibilities our readers might have, except that no two are exactly alike. We cover many abstract ideas, but must ground them with some implementation details to make them concrete. To that end, we need a problem to illustrate architecture concepts against—which leads us to the Sysops Squad.

Introducing the Sysops Squad Saga

> *saga*
> A long story of heroic achievement.
>
> —Oxford English Dictionary

We discuss a number of sagas in this book, both literal and figurative. Architects have co-opted the term *saga* to describe transactional behavior in distributed architectures (which we cover in detail in Chapter 12). However, discussions about architecture tend to become abstract, especially when considering abstract problems such as the hard parts of architecture. To help solve this problem and provide some real-world context for the solutions we discuss, we kick off a literal saga about the *Sysops Squad.*

We use the Sysops Squad saga within each chapter to illustrate the techniques and trade-offs described in this book. While many books on software architecture cover new development efforts, many real-world problems exist within existing systems. Therefore, our story starts with the existing Sysops Squad architecture highlighted here.

Penultimate Electronics is a large electronics giant that has numerous retail stores throughout the country. When customers buy computers, TVs, stereos, and other electronic equipment, they can choose to purchase a support plan. When problems occur, customer-facing technology experts (the Sysops Squad) come to the customer's residence (or work office) to fix problems with the electronic device.

The four main users of the Sysops Squad ticketing application are as follows:

Administrator
: The administrator maintains the internal users of the system, including the list of experts and their corresponding skill set, location, and availability. The administrator also manages all of the billing processing for customers using the system, and maintains static reference data (such as supported products, name-value pairs in the system, and so on).

Customer
: The customer registers for the Sysops Squad service and maintains their customer profile, support contracts, and billing information. Customers enter problem tickets into the system, and also fill out surveys after the work has been completed.

Sysops Squad expert
: Experts are assigned problem tickets and fix problems based on the ticket. They also interact with the knowledge base to search for solutions to customer problems and enter notes about repairs.

Manager
: The manager keeps track of problem ticket operations and receives operational and analytical reports about the overall Sysops Squad problem ticket system.

Nonticketing Workflow

The nonticketing workflows include those actions that administrators, managers, and customers perform that do not relate to a problem ticket. These workflows are outlined as follows:

1. Sysops Squad experts are added and maintained in the system through an administrator, who enters in their locale, availability, and skills.
2. Customers register with the Sysops Squad system and have multiple support plans based on the products they purchased.
3. Customers are automatically billed monthly based on credit card information contained in their profile. Customers can view billing history and statements through the system.
4. Managers request and receive various operational and analytical reports, including financial reports, expert performance reports, and ticketing reports.

Ticketing Workflow

The ticketing workflow starts when a customer enters a problem ticket into the system, and ends when the customer completes the survey after the repair is done. This workflow is outlined as follows:

1. Customers who have purchased the support plan enter a problem ticket by using the Sysops Squad website.
2. Once a problem ticket is entered in the system, the system then determines which Sysops Squad expert would be the best fit for the job based on skills, current location, service area, and availability.
3. Once assigned, the problem ticket is uploaded to a dedicated custom mobile app on the Sysops Squad expert's mobile device. The expert is also notified via a text message that they have a new problem ticket.
4. The customer is notified through an SMS text message or email (based on their profile preference) that the expert is on their way.
5. The expert uses the custom mobile application on their phone to retrieve the ticket information and location. The Sysops Squad expert can also access a knowledge base through the mobile app to find out what has been done in the past to fix the problem.
6. Once the expert fixes the problem, they mark the ticket as "complete." The sysops squad expert can then add information about the problem and repair the knowledge base.
7. After the system receives notification that the ticket is complete, it sends an email to the customer with a link to a survey, which the customer then fills out.
8. The system receives the completed survey from the customer and records the survey information.

A Bad Scenario

Things have not been good with the Sysops Squad problem ticket application lately. The current trouble ticket system is a large monolithic application that was developed many years ago. Customers are complaining that consultants are never showing up because of lost tickets, and often the wrong consultant shows up to fix something they know nothing about. Customers have also been complaining that the system is not always available to enter new problem tickets.

Change is also difficult and risky in this large monolith. Whenever a change is made, it usually takes too long and something else usually breaks. Because of reliability issues, the Sysops Squad system frequently "freezes up," or crashes, resulting in all application functionality not being available anywhere from five minutes to two hours while the problem is identified and the application restarted.

If something isn't done soon, Penultimate Electronics will be forced to abandon the very lucrative support contract business line and lay off all the Sysops Squad administrators, experts, managers, and IT development staff—including the architects.

Sysops Squad Architectural Components

The monolithic system for the Sysops Squad application handles ticket management, operational reporting, customer registration, and billing, as well as general administrative functions such as user maintenance, login, and expert skills and profile maintenance. Figure 1-3 and the corresponding Table 1-1 illustrate and describe the components of the existing monolithic application (the `ss.` part of the namespace specifies the Sysops Squad application context).

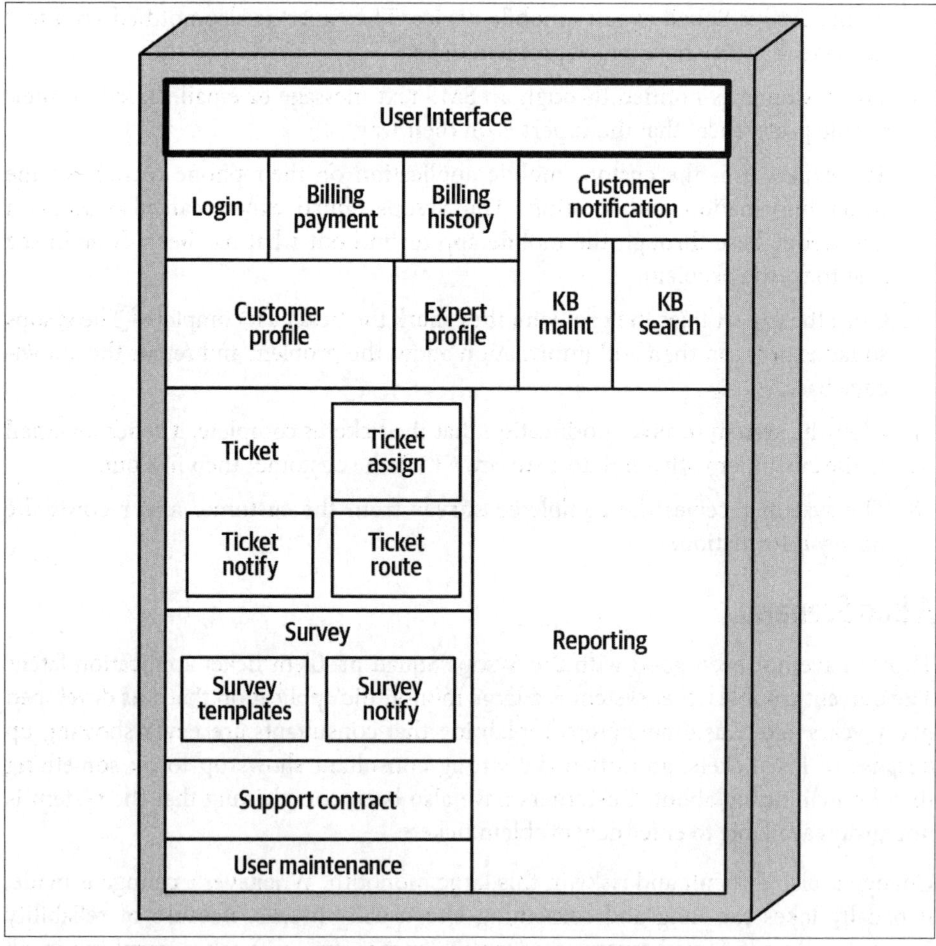

Figure 1-3. Components within the existing Sysops Squad application

Table 1-1. Existing Sysops Squad components

Component	Namespace	Responsibility
Login	`ss.login`	Internal user and customer login and security logic
Billing payment	`ss.billing.payment`	Customer monthly billing and customer credit card info
Billing history	`ss.billing.history`	Payment history and prior billing statements
Customer notification	`ss.customer.notification`	Notify customer of billing, general info
Customer profile	`ss.customer.profile`	Maintain customer profile, customer registration
Expert profile	`ss.expert.profile`	Maintain expert profile (name, location, skills, etc.)
KB maint	`ss.kb.maintenance`	Maintain and view items in the knowledge base
KB search	`ss.kb.search`	Query engine for searching the knowledge base
Reporting	`ss.reporting`	All reporting (experts, tickets, financial)
Ticket	`ss.ticket`	Ticket creation, maintenance, completion, common code
Ticket assign	`ss.ticket.assign`	Find an expert and assign the ticket
Ticket notify	`ss.ticket.notify`	Notify customer that the expert is on their way
Ticket route	`ss.ticket.route`	Send the ticket to the expert's mobile device app
Support contract	`ss.supportcontract`	Support contracts for customers, products in the plan
Survey	`ss.survey`	Maintain surveys, capture and record survey results
Survey notify	`ss.survey.notify`	Send survey email to customer
Survey templates	`ss.survey.templates`	Maintain various surveys based on type of service
User maintenance	`ss.users`	Maintain internal users and roles

These components will be used in subsequent chapters to illustrate various techniques and trade-offs when dealing with breaking applications into distributed architectures.

Sysops Squad Data Model

The Sysops Squad application with its various components listed in Table 1-1 uses a single schema in the database to host all its tables and related database code. The database is used to persist customers, users, contracts, billing, payments, knowledge base, and customer surveys; the tables are listed in Table 1-2, and the ER model is illustrated in Figure 1-4.

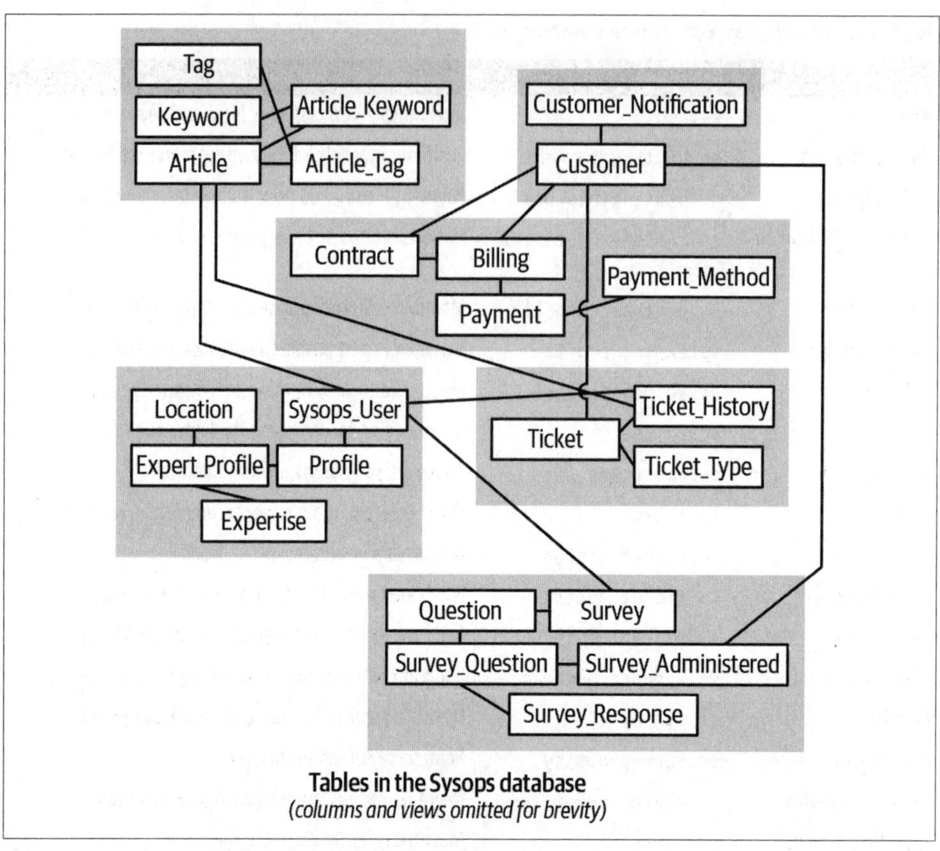

Figure 1-4. Data model within the existing Sysops Squad application

Table 1-2. Existing Sysops Squad database tables

Table	Responsibility
Customer	Entities needing Sysops support
Customer_Notification	Notification preferences for customers
Survey	A survey for after-support customer satisfaction
Question	Questions in a survey
Survey_Question	A question is assigned to the survey
Survey_Administered	Survey question is assigned to customer
Survey_Response	A customer's response to the survey
Billing	Billing information for support contract
Contract	A contract between an entity and Sysops for support
Payment_Method	Payment methods supported for making payment
Payment	Payments processed for billings

Table	Responsibility
SysOps_User	The various users in Sysops
Profile	Profile information for Sysops users
Expert_Profile	Profiles of experts
Expertise	Various expertise within Sysops
Location	Locations served by the expert
Article	Articles for the knowledge base
Tag	Tags on articles
Keyword	Keyword for an article
Article_Tag	Tags associated to articles
Article_Keyword	Join table for keywords and articles
Ticket	Support tickets raised by customers
Ticket_Type	Different types of tickets
Ticket_History	The history of support tickets

The Sysops data model is a standard third normal form data model with only a few stored procedures or triggers. However, a fair number of views exist that are mainly used by the Reporting component. As the architecture team tries to break up the application and move toward distributed architecture, it will have to work with the database team to accomplish the tasks at the database level. This setup of database tables and views will be used throughout the book to discuss various techniques and trade-offs to accomplish the task of breaking apart the database.

PART I
Pulling Things Apart

As many of us discovered when we were children, a great way to understand how something fits together is to first pull it apart. To understand complex subjects (such as trade-offs in distributed architectures), an architect must figure out where to start untangling.

In the book *What Every Programmer Should Know About Object-Oriented Design* (*https://oreil.ly/bLPm4*) (Dorset House), Meilir Page-Jones made the astute observation that coupling in architecture may be split into static and dynamic coupling. *Static* coupling refers to the way architectural parts (classes, components, services, and so on) are *wired* together: dependencies, coupling degree, connection points, and so on. An architect can often measure static coupling at compile time as it represents the static dependencies within the architecture.

Dynamic coupling refers to how architecture parts *call* one another: what kind of communication, what information is passed, strictness of contracts, and so on.

Our goal is to investigate how to do trade-off analysis in distributed architectures; to do that, we must pull the moving pieces apart so that we can discuss them in isolation to understand them fully before putting them back together.

Part I primarily deals with *architectural structure*, how things are statically coupled together. In Chapter 2, we tackle the problem of defining the scope of static and dynamic coupling in architectures, and present the entire picture that we must pull apart to understand. Chapter 3 begins that process, defining modularity and separation in architecture. Chapter 4 provides tools to evaluate and deconstruct codebases, and Chapter 5 supplies patterns to assist the process.

Data and transactions have become increasingly important in architecture, driving many trade-off decisions by architects and DBAs. Chapter 6 addresses the architectural impacts of data, including how to reconcile service and data boundaries. Finally, Chapter 7 ties together architecture coupling with data concerns to define *integrators* and *disintegrators*—forces that encourage a larger or smaller service size and boundary.

CHAPTER 2

Discerning Coupling in Software Architecture

`Wednesday, November 3, 13:00`

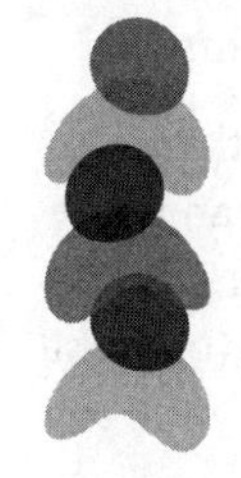

Logan, the lead architect for Penultimate Electronics, interrupted a small group of architects in the cafeteria, discussing distributed architectures. "Austen, are you wearing a cast *again*?"

"No, it's just a splint," replied Austen. "I sprained my wrist playing extreme disc golf over the weekend—it's almost healed."

"What is...never mind. What is this impassioned conversation I barged in on?"

"Why wouldn't someone always choose the *saga pattern* in microservices to wire together transactions?" asked Austen. "That way, architects can make the services as small as they want."

"But don't you have to use *orchestration* with sagas?" asked Addison. "What about times when we need asynchronous communication? And, how complex will the transactions get? If we break things down too much, can we really guarantee data fidelity?"

"You know," said Austen, "if we use an *enterprise service bus*, we can get it to manage most of that stuff for us."

"I thought no one used ESBs anymore—shouldn't we use Kafka for stuff like that?"

"They aren't even the same thing!" said Austen.

Logan interrupted the increasingly heated conversation. "It is an apples-to-oranges comparison, but none of these tools or approaches is a silver bullet. Distributed architectures like microservices are difficult, especially if architects cannot untangle all the forces at play. What we need is an approach or framework that helps us figure out the hard problems in our architecture."

"Well," said Addison, "whatever we do, it has to be as decoupled as possible—everything I've read says that architects should embrace decoupling as much as possible."

"If you follow that advice," said Logan, "Everything will be so decoupled that nothing can communicate with anything else—it's hard to build software that way! Like a lot of things, coupling isn't inherently bad; architects just have to know how to apply it appropriately. In fact, I remember a famous quote about that from a Greek philosopher...."

> All things are poison, and nothing is without poison; the dosage alone makes it so a thing is not a poison.
>
> —Paracelsus

One of the most difficult tasks an architect will face is untangling the various forces and trade-offs at play in distibuted architectures. People who provide advice constantly extol the benefits of "loosely coupled" systems, but how can architects design systems where nothing connects to anything else? Architects design fine-grained microservices to achieve decoupling, but then orchestration, transactionality, and asynchronicity become huge problems. Generic advice says "decouple," but provides no guidelines for *how* to achieve that goal while still constructing useful systems.

Architects struggle with granularity and communication decisions because there are no clear universal guides for making decisions—no best practices exist that can apply to real-world complex systems. Until now, architects lacked the correct perspective and terminology to allow a careful analysis that could determine the best (or least worst) set of trade-offs on a case-by-case basis.

Why have architects struggled with decisions in distributed architectures? After all, we've been building distributed systems since the last century, using many of the same mechanisms (message queues, events, and so on). Why has the complexity ramped up so much with microservices?

The answer lies with the fundamental philosophy of microservices, inspired by the idea of a *bounded context*. Building services that model bounded contexts required a subtle but important change to the way architects designed distributed systems because now transactionality is a first-class architectural concern. In many of the distributed systems architects designed prior to microservices, event handlers typically connected to a single relational database, allowing it to handle details such as integrity and transactions. Moving the database within the service boundary moves data concerns into architecture concerns.

As we've said before, *"Software architecture" is the stuff you can't Google answers for*. A skill that modern architects must build is the ability to do trade-off analysis. While several frameworks have existed for decades (such as Architecture Trade-off Analysis

Method, or ATAM (*https://oreil.ly/okbuO*)), they lack focus on real problems architects face on a daily basis.

This book focuses on how architects can perform trade-off analysis for any number of scenarios unique to their situation. As in many things in architecture, the advice is simple; the hard parts lie in the details, particularly how difficult parts become entangled, making it difficult to see and understand the individual parts, as illustrated in Figure 2-1.

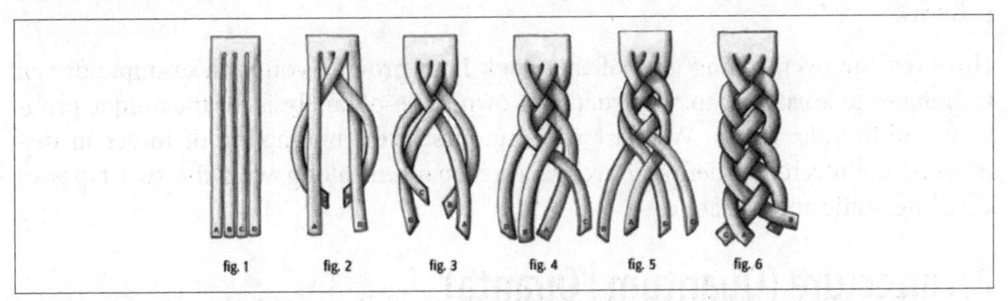

Figure 2-1. A braid entangles hair, making the individual strands hard to identify

When architects look at entangled problems, they struggle with performing trade-off analysis because of the difficulties separating the concerns, so that they may consider them independently. Thus, the first step in trade-off analysis is untangle the dimensions of the problem, analyzing what parts are coupled to one another and what impact that coupling has on change. For this purpose, we use the simplest definition of the word *coupling*:

Coupling
: Two parts of a software system are coupled if a change in one might cause a change in the other.

Often, software architecture creates multidimensional problems, where multiple forces all interact in interdependent ways. To analyze trade-offs, an architect must first determine what forces need to trade off with each other.

Thus, here's our advice for modern trade-off analysis in software architecture:

1. Find what parts are entangled together.
2. Analyze how they are coupled to one another.
3. Assess trade-offs by determining the impact of change on interdependent systems.

While the steps are simple, the hard parts lurk in the details. Thus, to illustrate this framework in practice, we take one of the most difficult (and probably the closest to generic) problems in distributed architectures, which is related to microservices:

How do architects determine the size and communication styles for microservices?
Determining the proper size for microservices seems a pervasive problem—too-small services create transactional and orchestration issues, and too-large services create scale and distribution issues.

To that end, the remainder of this book untangles the many aspects to consider when answering the preceding question. We provide new terminology to differentiate similar but distinct patterns and show practical examples of applying these and other patterns.

However, the overarching goal of this book is to provide you with example-driven techniques to learn how to construct your own trade-off analysis for the unique problems within your realm. We start with our first great untangling of forces in distributed architectures: defining architecture quantum along with the two types of coupling, static and dynamic.

Architecture (Quantum | Quanta)

The term *quantum* is, of course, used heavily in the field of physics known as *quantum mechanics*. However, the authors chose the word for the same reasons physicists did. *Quantum* originated from the Latin word *quantus*, meaning "how great" or "how many." Before physics co-opted it, the legal profession used it to represent the "required or allowed amount" (for example, in damages paid). The term also appears in the mathematics field of topology, concerning the properties of families of shapes. Because of its Latin roots, the singular is *quantum*, and the plural is *quanta*, similar to the datum/data symmetry.

An architecture quantum measures several aspects of both topology and behavior in software architecture related to how parts connect and communicate with one another:

Architecture quantum
An architecture quantum is an independently deployable artifact with high functional cohesion, high static coupling, and synchronous dynamic coupling. A common example of an architecture quantum is a well-formed microservice within a workflow.

Static coupling
Represents how static dependencies resolve within the architecture via contracts. These dependencies include operating system, frameworks, and/or libraries delivered via transitive dependency management, and any other operational requirement to allow the quantum to operate.

Dynamic coupling

Represents how quanta communicate at runtime, either synchronously or asynchronously. Thus, fitness functions for these characteristics must be *continuous*, typically utilizing monitors.

Even though both static and dynamic coupling seem similar, architects must distinguish two important differences. An easy way to think about the difference is that *static coupling* describes how services are *wired* together, whereas *dynamic coupling* describes how services *call* one another at runtime. For example, in a microservices architecture, a service must contain dependent components such as a database, representing static coupling—the service isn't operational without the necessary data. That service may call other services during the course of a workflow, which represents dynamic coupling. Neither service requires the other to be present to function, except for this runtime workflow. Thus, static coupling analyzes operational dependencies, and dynamic coupling analyzes communication dependencies.

These definitions include important characteristics; let's cover each in detail as they inform most of the examples in the book.

Independently Deployable

Independently deployable implies several aspects of an architecture quantum—each quantum represents a separate deployable unit within a particular architecture. Thus, a monolithic architecture—one that is deployed as a single unit—is by definition a single architecture quantum. Within a distributed architecture such as microservices, developers tend toward the ability to deploy services independently, often in a highly automated way. Thus, from an independently deployable standpoint, a service within a microservices architecture represents an architecture quantum (contingent on coupling—as discussed next).

Making each architecture quantum represent a deployable asset within the architecture serves several useful purposes. First, the boundary represented by an architecture quantum serves as a useful common language among architects, developers, and operations. Each understands the common scope under question: architects understand the coupling characteristics, developers understand the scope of behavior, and the operations team understands the deployable characteristics.

Second, the architecture quantum represents one of the forces (static coupling) architects must consider when striving for proper granularity of services within a distributed architecture. Often, in microservices architectures, developers face the difficult question of what service granularity offers the optimum set of trade-offs. Some of those trade-offs revolve around deployability: what release cadence does this service require, what other services might be affected, what engineering practices are involved, and so on. Architects benefit from a firm understanding of exactly where deployment boundaries lie in distributed architectures. We discuss service granularity and its attendant trade-offs in Chapter 7.

Third, *independent deployability* forces the architecture quantum to include common coupling points such as databases. Most discussions about architecture conveniently ignore issues such as databases and user interfaces, but real-world systems must commonly deal with those problems. Thus, any system that uses a shared database fails the architecture quantum criteria for independent deployment unless the database deployment is in lockstep with the application. Many distributed systems that would otherwise qualify for multiple quanta fail the independently deployable part if they share a common database that has its own deployment cadence. Thus, merely considering the deployment boundaries doesn't solely provide a useful measure. Architects should also consider the second criteria for an architecture quantum, high functional cohesion, to limit the architecture quantum to a useful scope.

High Functional Cohesion

High functional cohesion refers structurally to the proximity of related elements: classes, components, services, and so on. Throughout history, computer scientists defined a variety of cohesion types, scoped in this case to the generic *module*, which may be represented as *classes* or *components*, depending on platform. From a domain standpoint, the technical definition of *high functional cohesion* overlaps with the goals of the *bounded context* in domain-driven design: behavior and data that implements a particular domain workflow.

From a purely independent deployability standpoint, a giant monolithic architecture qualifies as an architecture quantum. However, it almost certainly isn't highly functionally cohesive, but rather includes the functionality of the entire system. The larger the monolith, the less likely it is singularly functionally cohesive.

Ideally, in a microservices architecture, each service models a single domain or workflow, and therefore exhibits high functional cohesion. Cohesion in this context isn't about how services interact to perform work, but rather how independent and coupled one service is to another service.

High Static Coupling

High static coupling implies that the elements inside the architecture quantum are tightly wired together, which is really an aspect of contracts. Architects recognize things like REST or SOAP as contract formats, but method signatures and operational dependencies (via coupling points such as IP addresses or URLs) also represent contracts. Thus, contracts are an *architecture hard part*; we cover coupling issues involving all types of contracts, including how to choose appropriate ones, in Chapter 13.

An architecture quantum is, in part, a measure of static coupling, and the measure is quite simple for most architecture topologies. For example, the following diagrams show the architecture styles featured in *Fundamentals of Software Architecture*, with the architecture quantum static coupling illustrated.

Any of the monolithic architecture styles will necessarily have a quantum of one, as illustrated in Figure 2-2.

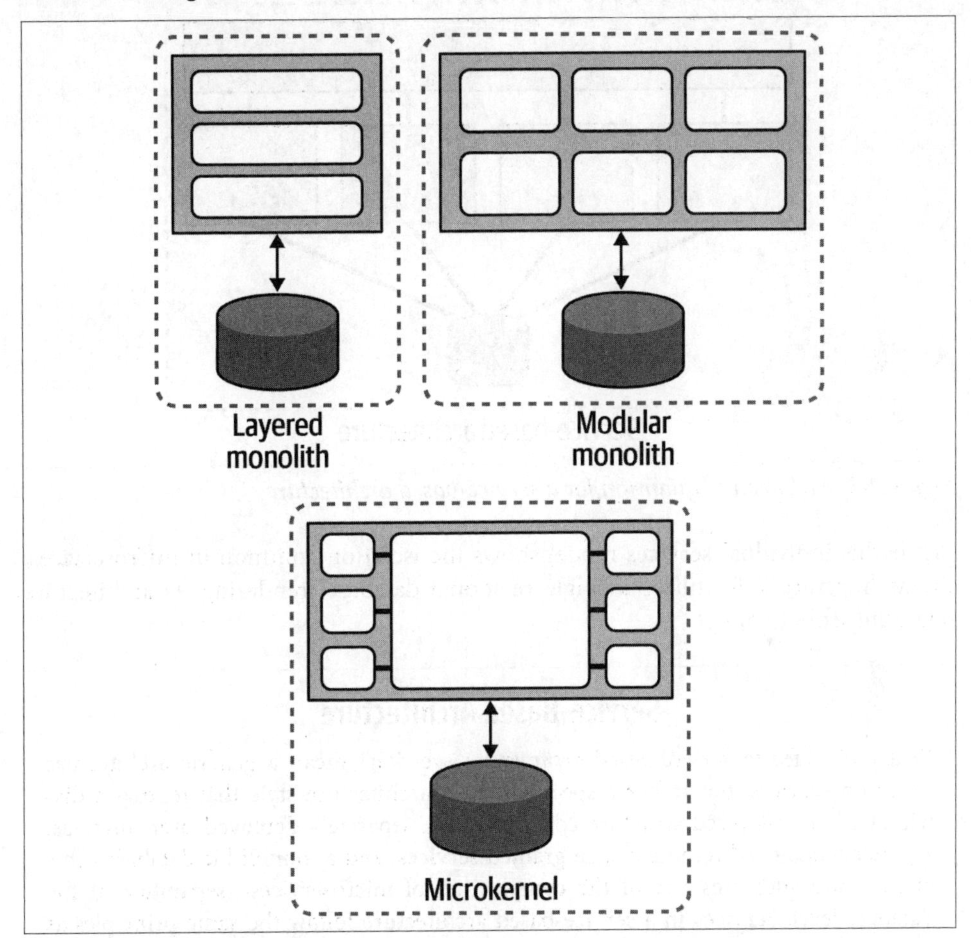

Figure 2-2. Monolithic architectures always have a quantum of one

As you can see, any architecture that deploys as a single unit and utilizes a single database will always have a single quantum. The architecture quantum measure of static coupling includes the database, and a system that relies on a single database cannot have more than a single quantum. Thus, the static coupling measure of an architecture quantum helps identify coupling points in architecture, not just within the software components under development. Most monolithic architectures contain a single coupling point (typically, a database) that makes its quantum measure one.

Distributed architectures often feature decoupling at the component level; consider the next set of architecture styles, starting with the service-based architecture shown in Figure 2-3.

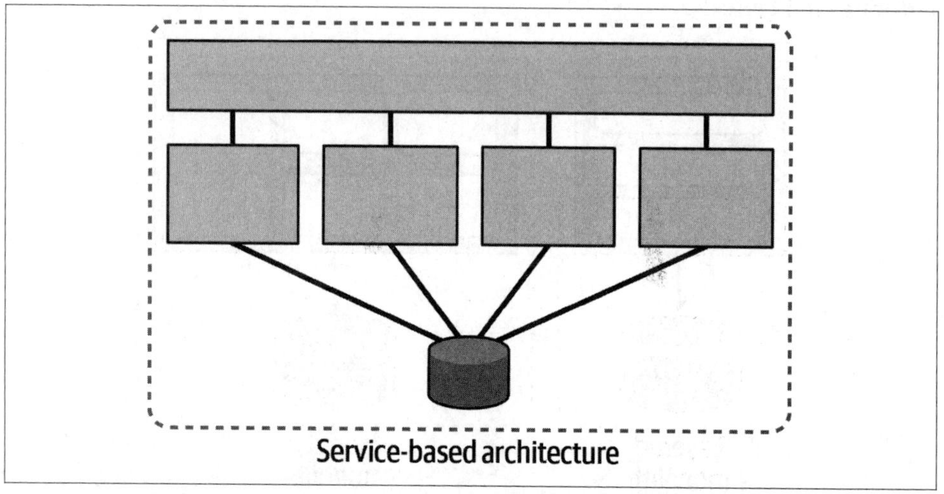

Figure 2-3. Architecture quantum for a service-based architecture

While this individual services model shows the isolation common in microservices, the architecture still utilizes a single relational database, rendering its architecture quantum score to one.

Service-Based Architecture

When we refer to *service-based architecture*, we don't mean a generic architecture based on services, but rather a specific hybrid architecture style that follows a distributed macro-layered structure consisting of a separately deployed user interface, separately deployed remote coarse-grained services, and a monolithic database. This architecture addresses one of the complexities of microservices—separation at the database level. Services in a service-based architecture follow the same principles as microservices (based on domain-driven design's bounded context) but rely on a single relational database because the architects didn't see value in separation (or saw too many negative trade-offs).

Service-based architectures are common targets when restructuring monolithic architectures, allowing for decomposition without disrupting existing database schemas and integration points. We cover decomposition patterns in Chapter 5.

So far, the static coupling measurement of architecture quantum has evaluated all the topologies to *one*. However, distributed architectures create the possibility of multiple quanta but don't necessarily guarantee it. For example, the mediator style of event-driven architecture will always be evaluated to an single architecture quantum, as illustrated in Figure 2-4.

Even though this style represents a distributed architecture, two coupling points push it toward a single architecture quantum: the database, as common with the previous monolithic architectures, but also the `Request Orchestrator` itself—any holistic coupling point necessary for the architecture to function forms an architecture quantum around it.

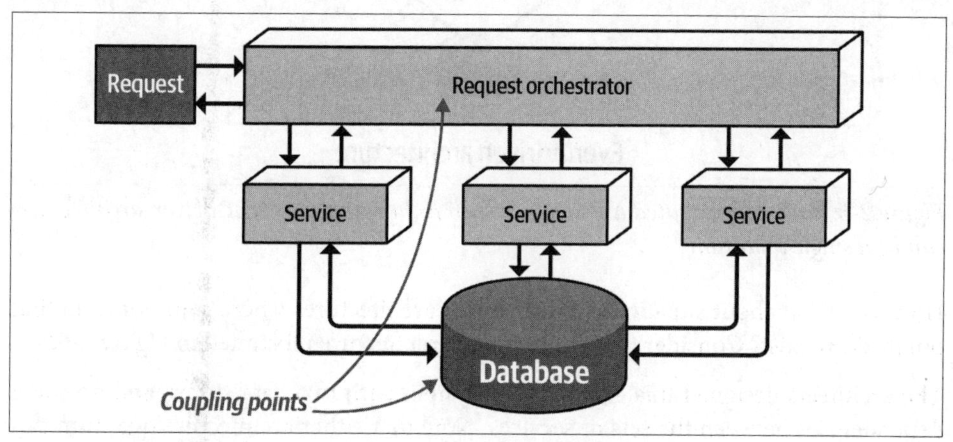

Figure 2-4. A mediated EDA has a single architecture quantum

Broker event-driven architectures (without a central mediator) are less coupled, but that doesn't guarantee complete decoupling. Consider the event-driven architecture illustrated in Figure 2-5.

This broker-style event driven architecture (without a central mediator) is nevertheless a single architecture quantum because all the services utilize a single relational database, which acts as a common coupling point. The question answered by the static analysis for an architecture quantum is, "Is this dependent of the architecture necessary to bootstrap this service?" Even in the case of an event-driven architecture where some of the services don't access the database, if they rely on services that *do* access the database, then they become part of the static coupling of the architecture quantum.

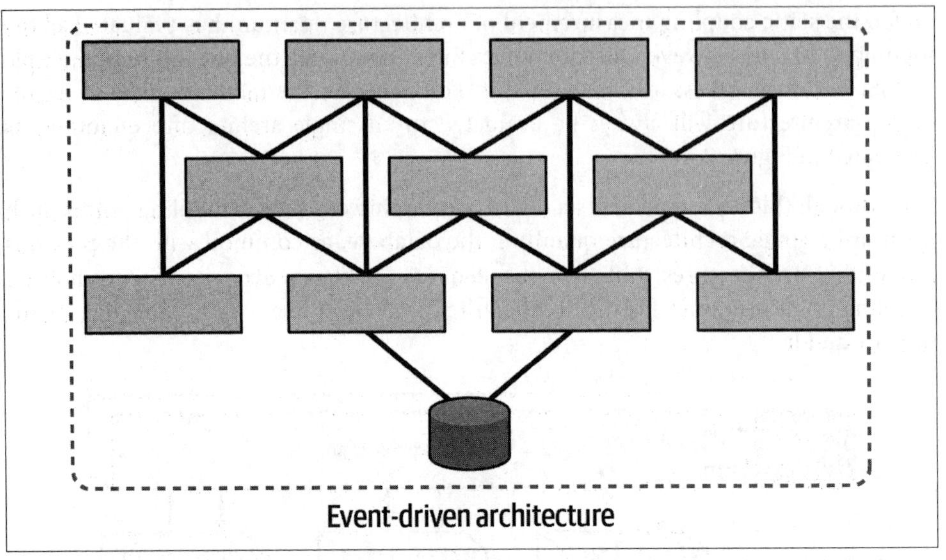

Figure 2-5. Even a distributed architecture such as broker-style event-driven architecture can be a single quantum

However, what about situations in distributed architectures where common coupling points don't exist? Consider the event-driven architecture illustrated in Figure 2-6.

The architects designed this event-driven system with two data stores, and no static dependencies between the sets of services. Note that either architecture quantum can run in a production-like ecosystem. It may not be able to participate in all workflows required by the system, but it runs successfully and operates—sends requests and receives them within the architecture.

The static coupling measure of an architecture quantum assesses the coupling dependencies between architectural and operational components. Thus, the operating system, data store, message broker, container orchestration, and all other operational dependencies form the static coupling points of an architecture quantum, using the strictest possible contracts, operational dependencies (more about the role of contracts in architecture quanta in Chapter 13).

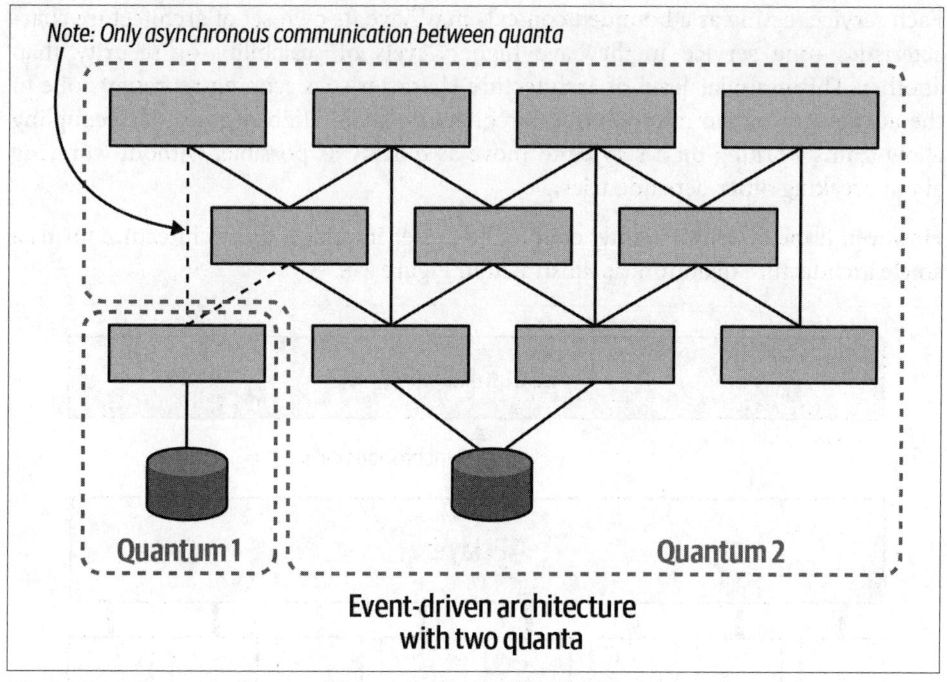

Figure 2-6. An event-driven architecture with multiple quanta

The microservices architecture style features highly decoupled services, including data dependencies. Architects in these architectures favor high degrees of decoupling and take care not to create coupling points between services, allowing each individual service to each form its own quanta, as shown in Figure 2-7.

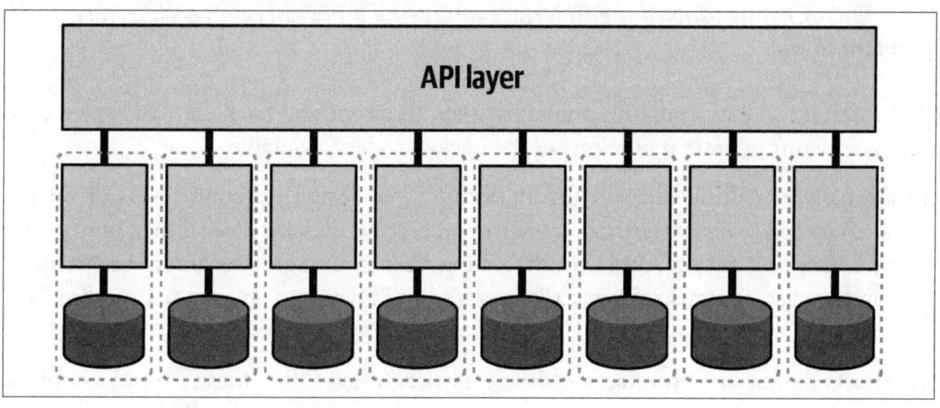

Figure 2-7. Microservices may form their own quanta

Each service (acting as a bounded context) may have its own set of architecture characteristics—one service might have higher levels of scalability or security than another. This granular level of architecture characteristics scoping represents one of the advantages of the microservices architecture style. High degrees of decoupling allow teams working on a service to move as quickly as possible, without worrying about breaking other dependencies.

However, if the system is tightly coupled to a user interface, the architecture forms a single architecture quantum, as illustrated in Figure 2-8.

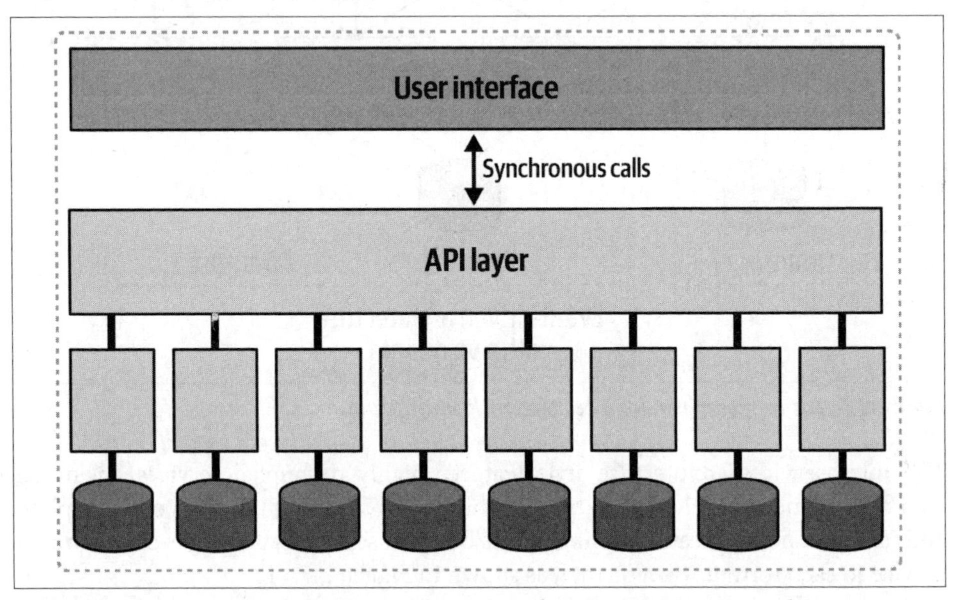

Figure 2-8. A tightly coupled user interface can reduce a microservices architecture quantum to one

User interfaces create coupling points between the front and back end, and most user interfaces won't operate if portions of the backend aren't available.

Additionally, it will be difficult for an architect to design different levels of operational architecture characteristics (performance, scale, elasticity, reliability, and so on) for each service if they all must cooperate together in a single user interface (particularly in the case of synchronous calls, covered in "Dynamic Quantum Coupling" on page 38).

Architects design user interfaces utilizing asynchronicity that doesn't create coupling between front and back. A trend on many microservices projects is to use a *micro frontend* framework for user interface elements in a microservices architecture. In such an architecture, the user interface elements that interact on behalf of the services are emitted from the services themselves. The user interface surface acts as a canvas where the user interface elements can appear, and also facilitates loosely coupled communication between components, typically using events. Such an architecture is illustrated in Figure 2-9.

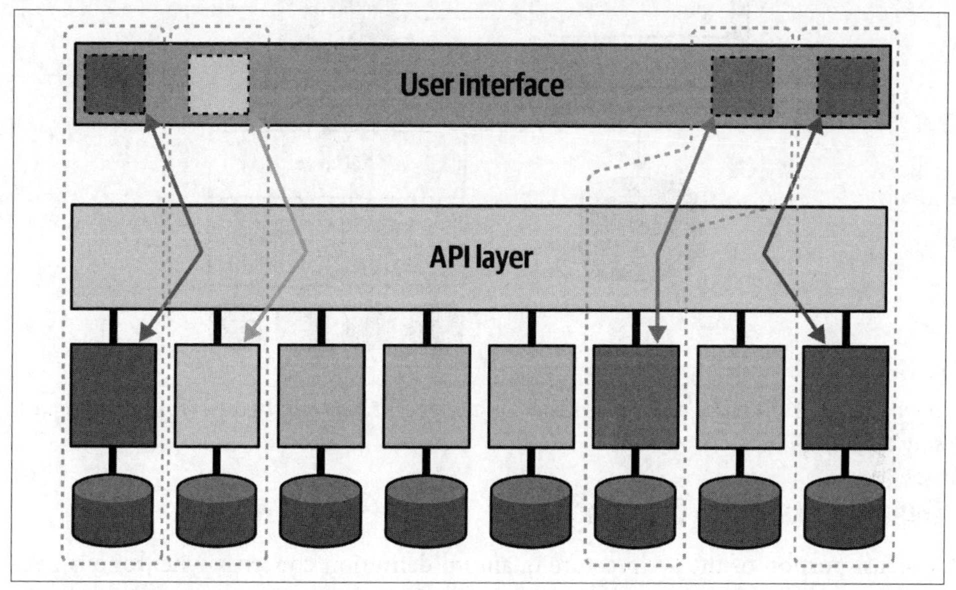

Figure 2-9. In a micro-frontend architecture, each service + user interface component forms an architecture quantum

In this example, the four tinted services along with their corresponding micro-frontends form architecture quanta: each of these services may have different architecture characteristics.

Any coupling point in an architecture can create static coupling points from a quantum standpoint. Consider the impact of a shared database between two systems, as illustrated in Figure 2-10.

The static coupling of a system provides valuable insight, even in complex systems involving integration architecture. Increasingly, a common architect technique for understanding legacy architecture involves creating a static quantum diagram of how things are "wired" together, which helps determine what systems will be impacted by change and offers a way of understanding (and potentially decoupling) the architecture.

Static coupling is only one-half of the forces at play in distributed architectures. The other is dynamic coupling.

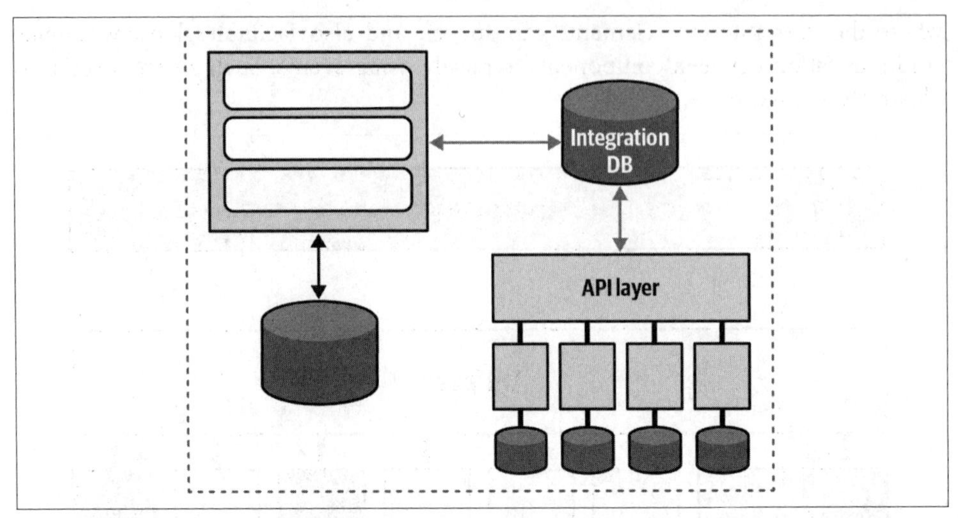

Figure 2-10. A shared database forms a coupling point between two systems, creating a single quantum

Dynamic Quantum Coupling

The last portion of the architecture quantum definition concerns synchronous coupling at runtime—in other words, the behavior of architecture quanta as they interact with one another to form workflows within a distributed architecture.

The nature of *how* services call one another creates difficult trade-off decisions because it represents a multidimensional decision space, influenced by three interlocking forces:

Communication
: Refers to the type of connection synchronicity used: synchronous or asynchronous.

Consistency
: Describes whether the workflow communication requires atomicity or can utilize eventual consistency.

Coordination
: Describes whether the workflow utilizes an orchestrator or whether the services communicate via choreography.

Communication

When two services communicate with each other, one of the fundamental questions for an architect is whether that communication should be synchronous or asynchronous.

Synchronous communication requires the requestor to wait for the response from the receiver, as shown in Figure 2-11.

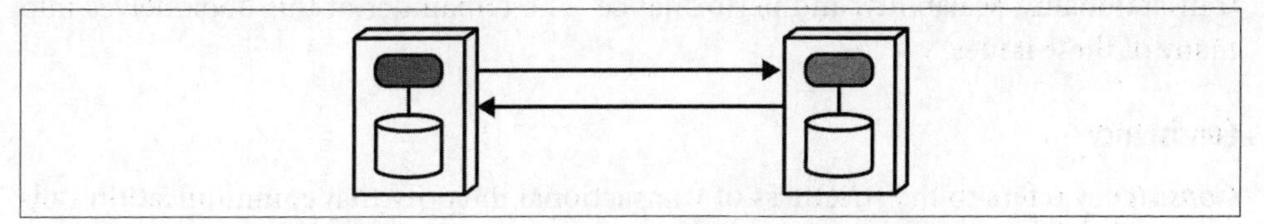

Figure 2-11. A synchronous call waits for a result from the receiver

The calling service makes a call (using one of a number of protocols that support synchronous calls, such as gRPC) and *blocks* (does no further processing) until the receiver returns a value (or status indicating a state change or error condition).

Asynchronous communication occurs between two services when the caller posts a message to the receiver (usually via a mechanism such as a message queue) and, once the caller gets acknowledgment that the message will be processed, it returns to work. If the request required a response value, the receiver can use a reply queue to (asynchronously) notify the caller of the result, which is illustrated in Figure 2-12.

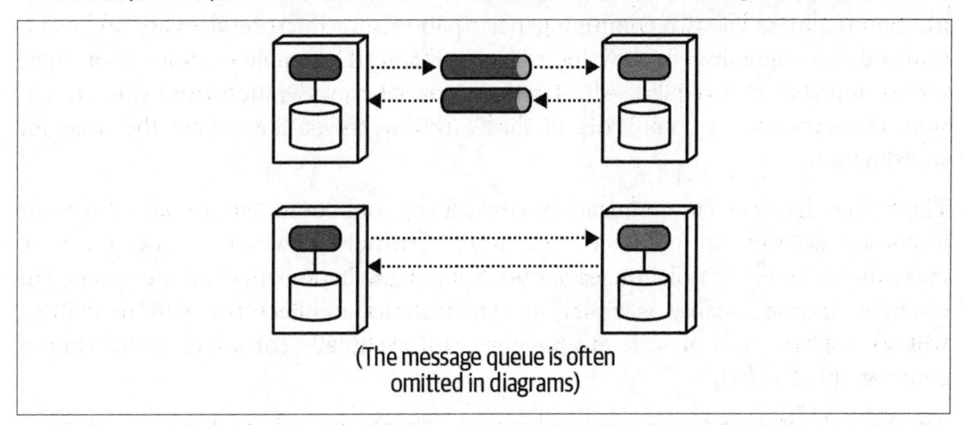

Figure 2-12. Asynchronous communication allows parallel processing

The caller posts a message to a message queue and continues processing until notified by the receiver that the requested information is available via return call. Generally, architects use message queues (illustrated via the gray cylindrical tube in the top diagram in Figure 2-12) to implement asynchronous communication, but queues are common and create noise on diagrams, so many architects leave them off, as shown in the lower diagram. And, of course, architects can implement asynchronous communication without message queues by using a variety of libraries or frameworks. Each diagram variety implies asynchronous messaging; the second provides visual shorthand and less implementation detail.

Architects must consider significant trade-offs when choosing how services will communicate. Decisions around communication affect synchronization, error handling, transactionality, scalability, and performance. The remainder of this book delves into many of these issues.

Consistency

Consistency refers to the strictness of transactional integrity that communication calls must adhere to. Atomic transactions (all-or-nothing transactions requiring consistency *during* the processing of a request) lie on one side of the spectrum, whereas different degrees of eventual consistency lie on the other side.

Transactionality—having several services participate in an all-or-nothing transaction—is one of the most difficult problems to model in distibuted architectures, resulting in the general advice to try to avoid cross-service transactions. We discuss consistency and the intersection of data and architecture in Chapters 6, 9, 10, and 12.

Coordination

Coordination refers to how much coordination the workflow modeled by the communication requires. The two common generic patterns for microservices are orchestration and choreography, which we describe in Chapter 11. Simple workflows—a single service replying to a request—don't require special consideration from this dimension. However, as the complexity of the workflow grows, the greater the need for coordination.

These three factors—communication, consistency, and coordination—all inform the important decision an architect must make. Critically, however, architects cannot make these choices in isolation; each option has a gravitation effect on the others. For example, transactionality is easier in synchronous architectures with mediation, whereas higher levels of scale are possible with eventually consistent asynchronous choreographed systems.

Thinking about these forces as related to each other forms a three-dimensional space, illustrated in Figure 2-13.

Each force in play during service communication appears as a dimension. For a particular decision, an architect could graph the position in space representing the strength of these forces.

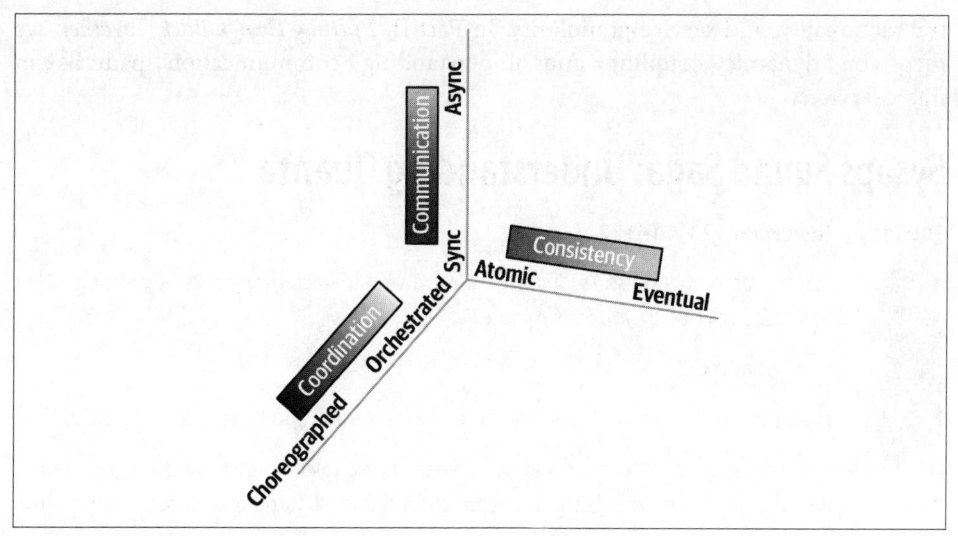

Figure 2-13. The dimensions of dynamic quantum coupling

When an architect can build a clear understanding of forces at play within a given situation, it creates criteria for trade-off analysis. In the case of dynamic coupling, Table 2-1 shows a framework for identifying fundamental pattern names based on the eight possible combinations.

Table 2-1. The matrix of dimensional intersections for distributed architectures

Pattern name	Communication	Consistency	Coordination	Coupling
Epic Saga[(sao)]	synchronous	atomic	orchestrated	very high
Phone Tag Saga[(sac)]	synchronous	atomic	choreographed	high
Fairy Tale Saga[(seo)]	synchronous	eventual	orchestrated	high
Time Travel Saga[(sec)]	synchronous	eventual	choreographed	medium
Fantasy Fiction Saga[(aao)]	asynchronous	atomic	orchestrated	high
Horror Story[(aac)]	asynchronous	atomic	choreographed	medium
Parallel Saga[(aeo)]	asynchronous	eventual	orchestrated	low
Anthology Saga[(aec)]	asynchronous	eventual	choreographed	very low

To fully understand this matrix, we must first investigate each of the dimensions individually. Therefore, the following chapters help you build context to understand the individual trade-offs for communication, consistency, and coordination, then entangle them back together in Chapter 12.

In the remaining chapters in Part I, we focus on static coupling and understanding the various dimensions at play in distributed architectures, including data ownership, transactionality, and service granularity. In Part II, *Putting Things Back Together*, we focus on dynamic coupling and understanding communication patterns in microservices.

Sysops Squad Saga: Understanding Quanta

`Tuesday, November 23, 14:32`

Austen came to Addison's office wearing an uncharacteristic cross expression. "Hey, Addison, can I bother you for a minute?"

"Sure, what's up?"

"I've been reading about this architecture quantum stuff, and I just...don't...get...it!"

Addison laughed, "I know what you mean. I struggled with it when it was purely abstract, but when you ground it in practical things, it turns out to be a useful set of perspectives."

"What do you mean?"

"Well," said Addison, "the architecture quantum basically defines a DDD bounded context in architectural terms."

"Why not just use bounded context, then?" asked Austen.

"Bounded context has a specific definition in DDD, and overloading it with stuff about architecture just makes people constantly have to differentiate. They are similar, but not the same thing. The first part about *functional cohesion* and *independent deployment* certainly matches a service based on bounded context. But the architecture quantum definition goes further by identifying types of coupling—that's where the *static* and *dynamic* stuff comes in."

"What is that all about? Isn't *coupling* just *coupling*? Why make the distinction?"

"It turns out that a bunch of different concerns revolve around the different types," said Addison. "Let's take the *static* one first, which I like to think of as how things are wired together. Another way to think about it: consider one of the services we're building in our target architecture. What is all the wiring required to bootstrap that service?"

"Well, it's written in Java, using a Postgres database, and running in Docker—that's it, right?"

"You're missing a lot." said Addison. "What if you had to build that service from scratch, assuming we had nothing in place? It's Java, but also using SpringBoot and, what, about 15 or 20 different frameworks and libraries?"

"That's right, we can look in the Maven POM file to figure out all those dependencies. What else?"

"The idea behind *static quantum coupling* is the wiring required to function. We're using events to communicate between services—what about the event broker?"

"But isn't that the dynamic part?"

"Not the *presence* of the broker. If the service (or, more broadly, architecture quantum) I want to bootstrap utilizes a message broker to function, the broker must be present. When the service *calls* another service via the broker, we get into the dynamic side."

"OK, that makes sense," said Austen. "If I think about what it would take to bootstrap it from scratch, that's the static quantum coupling."

"That's right. And just that information is super useful. We recently built a diagram of the static quantum coupling for each of our services defensively."

Austen laughed. "Defensively? What do you..."

"We were performing a reliability analysis to determine if I change this *thing*, what might break, where *thing* could be anything in our architecture or operations. They're trying to do risk mitigation —if we change a service, they want to know what must be tested."

"I see—that's the static quantum coupling. I can see how that's a useful view. It also shows how teams might impact one another. That seems really useful. Is there a tool we can download that figures that out for us?"

"Wouldn't that be nice!" laughed Addison. "Unfortunately, no one with our unique mix of architecture has built and open sourced exactly the tool we want. However, some of the platform team is working on a tool to automate it, necessarily customized to our architecture. They're using the container manifests, POM files, NPM dependencies, and other dependency tools to build and maintain a list of build dependencies. We have also instituted observability for all our services, so we now have consistent log files about what systems call each other, when, and how often. They're using that to build a call graph to see how things are connected."

"OK, so static coupling is how things are wired together. What about *dynamic* coupling?"

"Dynamic coupling concerns how quanta *communicate* with each other, particularly synchronous versus asynchronous calls and their impact on operational architecture characteristics—things like performance, scale, elasticity, reliability, and so on. Consider *elasticity* for a moment—remember the difference between scalability and elasticity?"

Austen smirked. "I didn't know there was going to be a test. Let's see...*scalability* is the ability to support a large number of concurrent users; *elasticity* is the ability to support a burst of user requests in a short time frame."

"Correct! Gold star for you. OK, let's think about *elasticity*. Suppose in our future state architecture we have two services like Ticketing and Assignment, and the two types of calls. We carefully designed our services to be highly statically decoupled from one another, so that they can be independently elastic. That's the other side effect of static coupling, by the way—it identifies the scope of things like operational architecture characteristics. Let's say that Ticketing is operating at 10 times the elastic scale of Assignment, and we need to make a call between them. If we make a synchronous call, the whole workflow will bog down, as the caller waits for the slower service to process and return. If on the other hand we make an asynchronous call, using the message queue as a buffer, we can allow the two services to execute operationally independently, allowing the caller to add messages to the queue and continue working, receiving notification when the workflow is complete."

"Oh, I see, I see! The architecture quantum defines the scope of architecture characteristics—it's obvious how the static coupling can affect that. But I see now that, depending on the type of call you make, you might temporarily couple two services together."

"That's right," said Addison. "The architecture quanta can entangle one another temporarily, during the course of a call, if the nature of the call ties things like performance, responsiveness, scale, and a bunch of others."

"OK, I think I understand what an architecture quantum is, and how the coupling definitions work. But I'm never going to get that quantum/quanta thing straight!"

"Same for datum/data, but no one ever uses datum!" laughed Addison. "You'll see a lot more of the impact of dynamic coupling on workflows and transactional sagas as you keep digging into our architecture."

"I can't wait!"

CHAPTER 3

Architectural Modularity

`Tuesday, September 21 09:33`

It was the same conference room they had been in a hundred times before, but today the atmosphere was different. Very different. As people gathered, no small talk was exchanged. Only silence. The sort of dead silence that you could cut with a knife. Yes, that was indeed an appropriate cliche given the topic of the meeting.

The business leaders and sponsors of the failing Sysops Squad ticketing application met with the application architects, Addison and Austen, with the purpose of voicing their concern and frustration about the inability of the IT department to fix the never-ending issues associated with the trouble ticket application. "Without a working application," they had said, "we cannot possibly continue to support this business line."

As the tense meeting ended, the business sponsors quietly filed out one by one, leaving Addison and Austen alone in the conference room.

"That was a bad meeting," said Addison. "I can't believe they're actually blaming *us* for all the issues we're currently facing with the trouble ticket application. This is a really bad situation."

"Yeah, I know," said Austen. "Especially the part about possibly closing down the product support business line. We'll be assigned to other projects, or worse, maybe even let go. Although I'd rather be spending all of my time on the soccer field or on the slopes skiing in the winter, I really can't afford to lose this job."

"Neither can I," said Addison. "Besides, I really like the development team we have in place, and I'd hate to see it broken up."

"Me too," said Austen. "I still think breaking apart the application would solve most of these issues."

"I agree with you," said Addison, "but how do we convince the business to spend more money and time to refactor the architecture? You saw how they complained in the meeting about the amount of money we've already spent applying patches here and there, only to create additional issues in the process."

"You're right," Austen said. "They would never agree to an expensive and time-consuming architecture migration effort at this point."

"But if we both agree that we need to break apart the application to keep it alive, how in the world are we going to convince the business and get the funding and time we need to completely restructure the Sysops Squad application?" asked Addison.

"Beats me," said Austen. "Let's see if Logan is available to discuss this problem with us."

Addison looked online and saw that Logan, the lead architect for Penultimate Electronics, was available. Addison sent a message explaining that they wanted to break apart the existing monolithic application, but weren't sure how to convince the business that this approach would work. Addison explained in the message that they were in a real bind and could use some advice. Logan agreed to meet with them and joined them in the conference room.

"What makes you so sure that breaking apart the Sysops Squad application will solve all of the issues?" asked Logan.

"Because," said Austen, "we've tried patching the code over and over, and it doesn't seem to be working. We still have way too many issues."

"You're completely missing my point," said Logan. "Let me ask you the question a different way. What assurances do you have that breaking apart the system will accomplish anything more than just spending more money and wasting more valuable time?"

"Well," said Austen, "actually, we don't."

"Then how do you know breaking apart the application is the right approach?" asked Logan.

"We already told you," said Austen, "because nothing else we try seems to work!"

"Sorry," said Logan, "but you know as well as I do that's not a reasonable justification for the business. You'll never get the funding you need with that kind of reason."

"So, what would be a good business justification?" asked Addison. "How do we sell this approach to the business and get the additional funding approved?"

"Well," said Logan, "to build a good business case for something of this magnitude, you first need to understand the benefits of architectural modularity, match those benefits to the issues you are facing with the current system, and finally analyze and document the trade-offs involved with breaking apart the application."

Businesses today face a torrent of change; market evolution seems to keep accelerating at a blistering pace. Business drivers (such as mergers and acquisitions), increased competition in the marketplace, increased consumer demand, and increased innovation (such as automation through machine learning and artificial intelligence) necessarily require changes to underlying computer systems. In many cases, those changes in computer systems consequently necessitate changes to the underlying architectures supporting them.

However, it's not only business that's undergoing constant and rapid change—it's also the technical environment in which those computer systems reside. Containerization, the move to cloud-based infrastructure, the adoption of DevOps, and even new advancements in continuous delivery pipelines all impact the underlying architecture of those computer systems.

It's difficult in today's world to manage all of this constant and rapid change with respect to software architecture. Software architecture is the foundational structure of a system, and is therefore generally thought of as something that should remain stable and not undergo frequent change, similar to the underlying structural aspects of a large building or skyscraper. However, unlike the structural architecture of a building, software architecture must constantly change and adapt to meet the new demands of today's business and technology environment.

Consider the increased number of mergers and acquisitions happening in today's marketplace. When one company acquires another, not only does it acquire the physical aspects of a company (such as people, buildings, inventory, and so on) but also more customers. Can the existing systems in either company scale to meet the increase in user volume as a result of the merger or acquisition? Scalability is a big part of mergers and acquisitions, as is agility and extensibility, all of which are *architectural concerns*.

Large monolithic (single deployment) systems generally do not provide the level of scalability, agility, and extensibility required to support most mergers and acquisitions. The capacity for additional machine resources (threads, memory, and CPU) fills up very quickly. To illustrate this point, consider the water glass shown in Figure 3-1. The glass represents the server (or virtual machine), and the water represents the application. As monolithic applications grow to handle increased consumer demand and user load (whether from mergers, acquisitions, or company growth), they begin to consume more and more resources. As more water is added to the glass (representing the growing monolithic application), the glass begins to fill up. Adding another glass (represented as another server or virtual machine) does nothing, because the new glass would contain the same amount of water as the first one.

Figure 3-1. A full glass representing a large monolithic application close to capacity

One aspect of architectural modularity is breaking large monolithic applications into separate and smaller parts to provide more capacity for further scalability and growth, while at the same time facilitating constant and rapid change. In turn, these capabilities can help achieve a company's strategic goals.

By adding another empty glass to our water glass example and breaking the water (application) into two separate parts, half the water can now be poured into the new empty glass, providing 50% more capacity, as shown in Figure 3-2. The water glass analogy is a great way of explaining architectural modularity (the breaking up of monolithic applications) to business stakeholders and C-level executives, who will inevitably be paying for the architecture-refactoring effort.

Figure 3-2. Two half-full glasses representing an application broken apart with plenty of capacity for growth

Increased scalability is only one benefit of architectural modularity. Another important benefit is *agility*, the ability to respond quickly to change. An article from *Forbes* (*https://oreil.ly/2im3v*) in January 2020 by David Benjamin and David Komlos stated the following:

> There is one thing that will separate the pack into winners and losers: the on-demand capability to make bold and decisive course-corrections that are executed effectively and with urgency.

Businesses must be agile in order to survive in today's world. However, while business stakeholders may be able to make quick decisions and change direction quickly, the company's technology staff may not be able to implement those new directives fast enough to make a difference. Enabling technology to move as fast as the business (or, conversely, preventing technology from slowing the business) requires a certain level of architectural agility.

Modularity Drivers

Architects shouldn't break a system into smaller parts unless clear business drivers exist. The primary business drivers for breaking applications into smaller parts include *speed-to-market* (sometimes called time-to-market) and achieving a level of *competitive advantage* in the marketplace.

Speed-to-market is achieved through architectural agility—the ability to respond quickly to change. Agility is a compound architectural characteristic made up of many other architecture characteristics, including maintainability, testability, and deployability.

Competitive advantage is achieved through speed-to-market combined with scalability and overall application availability and fault tolerance. The better a company does, the more it grows, hence the need for more scalability to support increased user activity. *Fault tolerance*, the ability of an application to fail and continue to operate, is necessary to ensure that as parts of the application fail, other parts are still able to function as normal, minimizing the overall impact to the end user. Figure 3-3 illustrates the relationship between the technical drivers and the resulting business drivers for modularity (enclosed within boxes).

Businesses must be agile to survive in today's fast-paced and ever-changing volatile market, meaning the underlying architectures must be agile as well. As illustrated in Figure 3-3, the five key architectural characteristics to support agility, speed-to-market, and, ultimately, competitive advantage in today's marketplace are availability (fault tolerance), scalability, deployability, testability, maintainability.

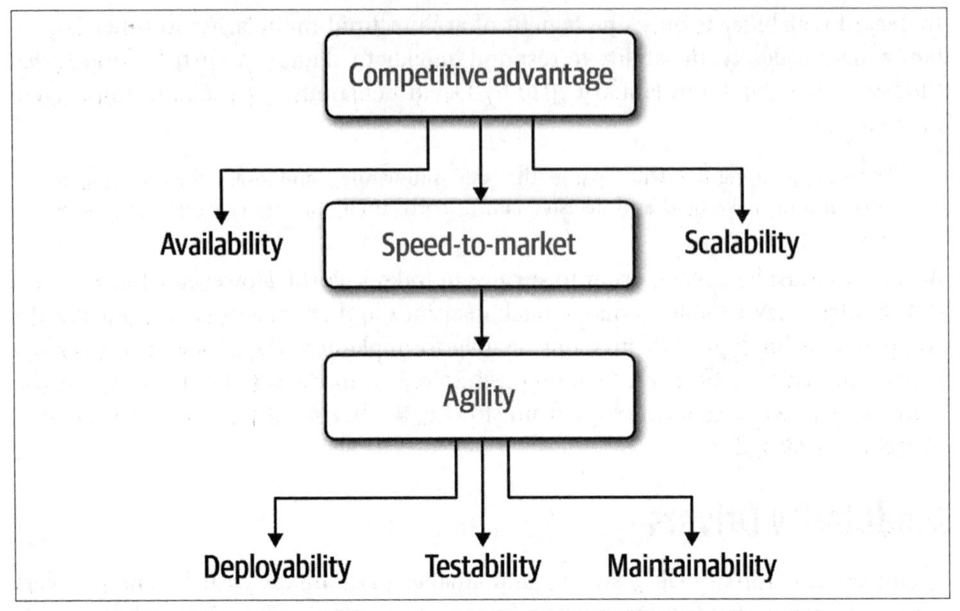

Figure 3-3. The drivers for modularity and the relationships among them

Note that *architectural modularity* does not always have to translate to a distributed architecture. Maintainability, testability, and deployability (defined in the following sections) can also be achieved through monolithic architectures such as a modular monolith or even a microkernel architecture (see Appendix B for a list of references providing more information about these architecture styles). Both of these architecture styles offer a level of architectural modularity based on the way the components are *structured*. For example, with a modular monolith, components are grouped into well-formed domains, providing for what is known as a *domain partitioned architecture* (see *Fundamentals of Software Architecture*, Chapter 8, page 103). With the microkernel architecture, functionality is partitioned into separate plug-in components, allowing for a much smaller testing and deployment scope.

Maintainability

Maintainability is about the ease of adding, changing, or removing features, as well as applying internal changes such as maintenance patches, framework upgrades, third-party upgrades, and so on. As with most composite architecture characteristics, maintainability is hard to define objectively. Alexander von Zitzewitz, software architect and founder of hello2morrow (*http://www.hello2morrow.com*), wrote an article (*https://oreil.ly/TbFjN*) about a new metric for objectively defining the maintainability level of an application. While von Zitzewitz's maintainability metric is fairly complicated and involves lots of factors, its initial form is as follows:

$$ML = 100 * \sum_{i=1}^{k} c_i$$

where ML is the maintainability level of the overall system (percentage from 0% to 100%), k is the total number of logical components in the system, and c_i is the coupling level for any given component, with a special focus on incoming coupling levels. This equation basically demonstrates that the higher the incoming coupling level between components, the lower the overall maintainability level of the codebase.

Putting aside complicated mathematics, some of the typical metrics used for determining the relative maintainability of an application based on components (the architectural building blocks of an application) include the following:

Component coupling
: The degree and manner to which components know about one another

Component cohesion
: The degree and manner to which the operations of a component interrelate

Cyclomatic complexity
: The overall level of indirection and nesting within a component

Component size
: The number of aggregated statements of code within a component

Technical versus domain partitioning
: Components aligned by technical usage or by domain purpose—see Appendix A

Within the context of architecture, we are defining a *component* as an architectural building block of the application that does some sort of business or infrastructure function, usually manifested through a package structure (Java), namespace (C#), or physical grouping of files (classes) within some sort of directory structure. For example, the component Order History might be implemented through a set of class files located in the namespace `app.business.order.history`.

Large monolithic architectures generally have low levels of maintainability due to the technical partitioning of functionality into layers, the tight coupling between components, and weak component cohesion from a domain perspective. For example, consider a new requirement within a traditional monolithic layered architecture to add an expiration date to items contained in a customer's wish list (items in a list to maybe purchase at a later time). Notice in Figure 3-4 that the change scope of the new requirement is at an *application level* since the change is propagated to all of the layers within the application.

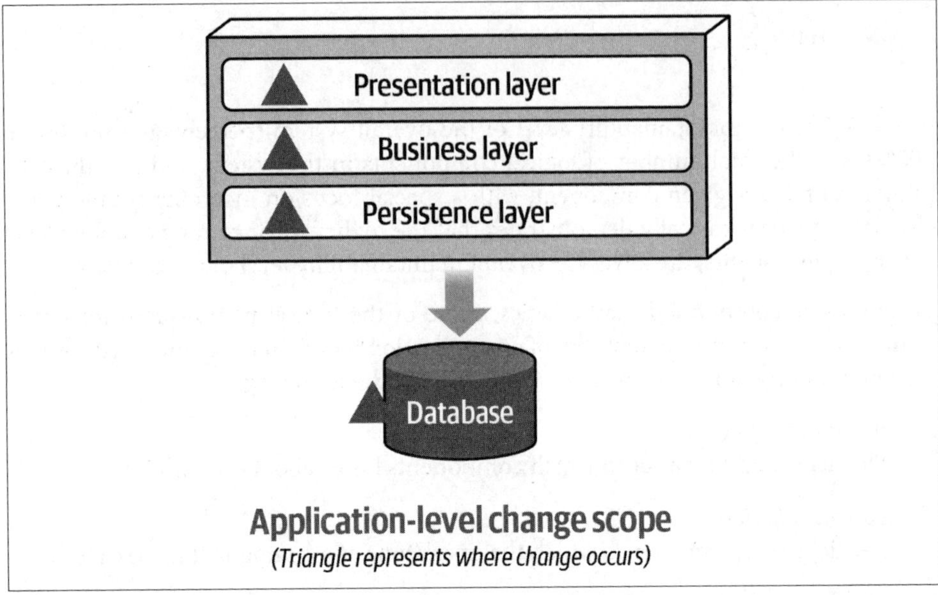

Figure 3-4. With monolithic layered architectures, change is at an application level

Depending on the team structure, implementing this simple change to add an expiration date to wish list items in a monolithic layered architecture could possibly require the coordination of at least three teams:

- A member from the user interface team would be needed to add the new expiry field to the screen.
- A member from the backend team would be needed to add business rules associated with the expiry date and change contracts to add the new expiry field.
- A member from the database team would be needed to change the table schema to add the new expiry column in the Wishlist table.

Since the Wishlist domain is spread throughout the entire architecture, it becomes harder to maintain a particular domain or subdomain (such as Wishlist). Modular architectures, on the other hand, partition domains and subdomains into smaller, separately deployed units of software, thereby making it easier to modify a domain or subdomain. Notice that with a distributed service-based architecture, as shown in Figure 3-5, the change scope of the new requirement is at a *domain level* within a particular domain service, making it easier to isolate the specific deployment unit requiring the change.

Moving to even more architectural modularity such as a microservices architecture, as illustrated in Figure 3-6, places the new requirement at a *function-level* change scope, isolating the change to a specific service responsible for the wish list functionality.

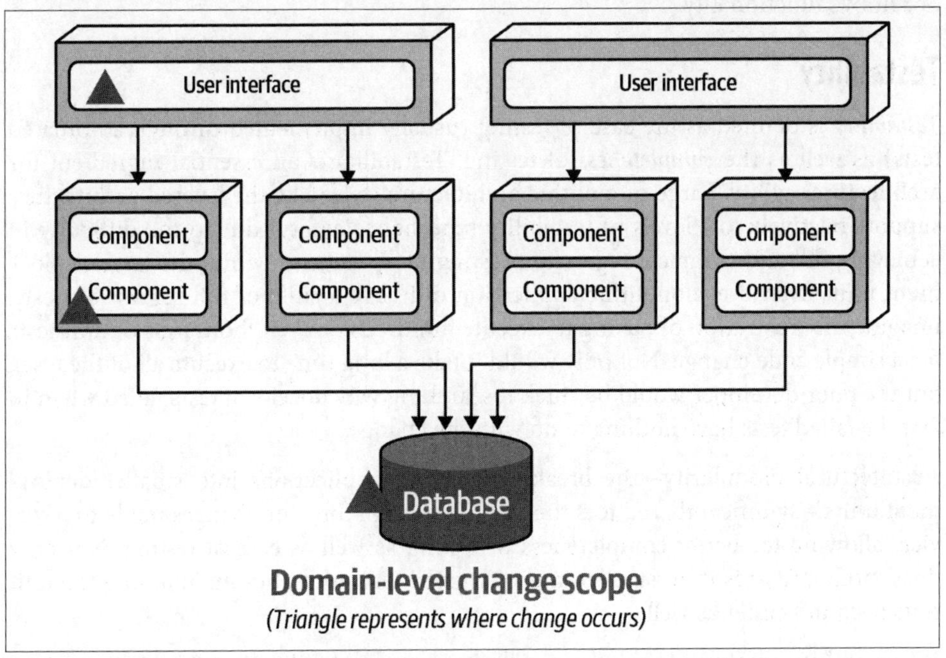

Figure 3-5. With service-based architectures, change is at a domain level

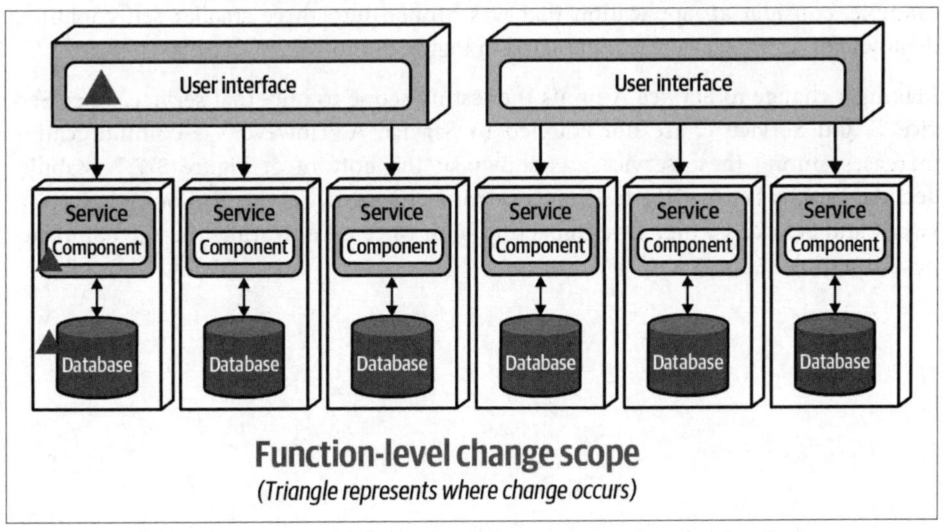

Figure 3-6. With microservices architectures, change is at a function level

These three progressions toward modularity demonstrate that as the level of architectural modularity increases, so does maintainability, making it easier to add, change, or remove functionality.

Testability

Testability is defined as the ease of testing (usually implemented through automated tests) as well as the *completeness* of testing. Testability is an essential ingredient for architectural agility. Large monolithic architecture styles like the layered architecture support relatively low levels of testability (and hence agility) due to the difficulty in achieving full and complete regression testing of all features within the large deployment unit. Even if a monolithic application did have a suite of full regression tests, imagine the frustration of having to execute hundreds or even thousands of unit tests for a simple code change. Not only would it take a long time to execute all of the tests, but the poor developer would be stuck researching why dozens of tests failed when in fact the failed tests have nothing to do with the change.

Architectural modularity—the breaking apart of applications into smaller deployment units—significantly reduces the overall testing scope for changes made to a service, allowing for better completeness of testing as well as ease of testing. Not only does modularity result in smaller, more targeted test suites, but maintaining the unit tests becomes easier as well.

While architectural modularity generally improves testability, it can sometimes lead to the same problems that exist with monolithic, single-deployment applications. For example, consider an application that was broken into three smaller self-contained deployment units (services), as depicted in Figure 3-7.

Making a change to Service A limits the testing scope to only that service, since Service B and Service C are not coupled to Service A. However, as communication increases among these services, as shown at the bottom of Figure 3-7, testability declines rapidly because the testing scope for a change to Service A now includes Service B and Service C, therefore impacting both the ease of testing and the completeness of testing.

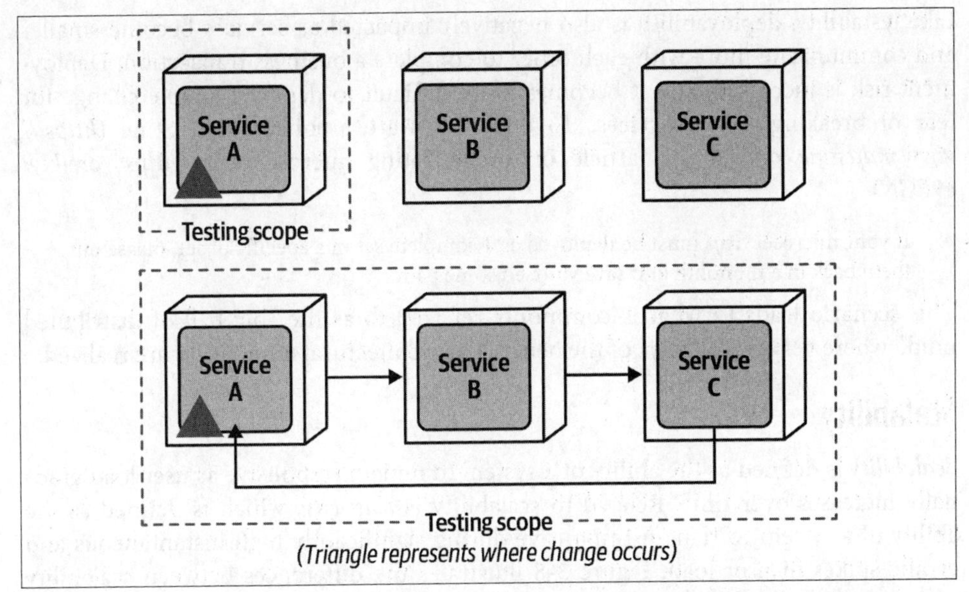

Figure 3-7. Testing scope is increased as services communicate with one another

Deployability

Deployability is not only about the ease of deployment—it is also about the frequency of deployment and the overall risk of deployment. To support agility and respond quickly to change, applications must support all three of these factors. Deploying software every two weeks (or more) not only increases the overall risk of deployment (due to grouping multiple changes together), but in most cases unnecessarily delays new features or bug fixes that are ready to be pushed out to customers. Of course, deployment frequency must be balanced with the customer's (or end user's) ability to be able to absorb changes quickly.

Monolithic architectures generally support low levels of deployability due to the amount of ceremony involved in deploying the application (such as code freezes, mock deployments, and so on), the increased risk that something else might break once new features or bug fixes are deployed, and a long time frame between deployments (weeks to months). Applications having a certain level of architectural modularity in terms of separately deployed units of software have less deployment ceremony, less risk of deployment, and can be deployed more frequently than a large, single monolithic application.

Like testability, deployability is also negatively impacted as services become smaller and communicate more with each other to complete a business transaction. Deployment risk is increased, and it becomes more difficult to deploy a simple change for fear of breaking other services. To quote software architect Matt Stine (*https://www.mattstine.com*) in his article on orchestrating microservices (*https://oreil.ly/e9EGN*):

> If your microservices must be deployed as a complete set in a specific order, please put them back in a monolith and save yourself some pain.

This scenario leads to what is commonly referred to as the "big ball of distributed mud," where very few (if any) of the benefits of architectural modularity are realized.

Scalability

Scalability is defined as the ability of a system to remain responsive as user load gradually increases over time. Related to scalability is *elasticity*, which is defined as the ability of a system to remain responsive during significantly high instantaneous and erratic spikes in user load. Figure 3-8 illustrates the differences between scalability and elasticity.

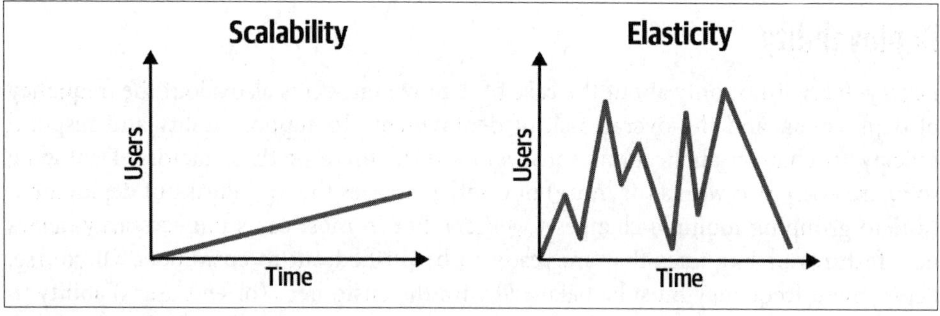

Figure 3-8. Scalability is different from elasticity

While both of these architectural characteristics include responsiveness as a function of the number of concurrent requests (or users in the system), they are handled differently from an architectural and implementation standpoint. Scalability generally occurs over a longer period of time as a function of normal company growth, whereas elasticity is the immediate response to a spike in user load.

A great example to further illustrate the difference is that of a concert-ticketing system. Between major concert events, there is usually a fairly light concurrent user load. However, the minute tickets go on sale for a popular concert, concurrent user load significantly spikes. The system may go from 20 concurrent users to 3,000 concurrent users in a matter of seconds. To maintain responsiveness, the system must have the capacity to handle the high peaks in user load, and also have the ability to instantaneously start up additional services to handle the spike in traffic. Elasticity relies on services having a very small *mean time to startup* (MTTS), which is achieved *architecturally* by having very small, fine-grained services. With an appropriate architectural solution in place, MTTS (and hence elasticity) can then be further managed through design-time techniques such as small lightweight platforms and runtime environments.

Although both scalability and elasticity improve with finer-grained services, elasticity is more a function of granularity (the size of a deployment unit), whereas scalability is more a function of modularity (the breaking apart of applications into separate deployment units). Consider the traditional layered architecture, service-based architecture, and microservices architecture styles and their corresponding star ratings for scalability and elasticity, as illustrated in Figure 3-9 (the details of these architecture styles and their corresponding star ratings can be found in our previous book, *Fundamentals of Software Architecture*. Note that one star means that the capability is not well supported by the architecture style, whereas five stars means that capability is a major feature of the architecture style and is well supported.

Notice that scalability and elasticity rate relatively low with the monolithic layered architecture. Large monolithic layered architectures are both difficult and expensive to scale because all of the application functionality must scale to the same degree (application-level scalability and poor MTTS). This can become particularly costly in cloud-based infrastructures. However, with service-based architecture, notice that scalability improves, but not as much as elasticity. This is because domain services in a service-based architecture are coarse grained and usually contain the entire domain in one deployment unit (such as order processing or warehouse management), and generally have too long of a mean time to startup (MTTS) to respond fast enough to immediate demand for elasticity due to their large size (domain-level scalability and fair MTTS). Notice that with microservices, both scalability and elasticity are maximized because of the small, single-purpose, fine-grained nature of each separately deployed service (function-level scalability and excellent MTTS).

Like testability and deployability, the more services communicate with one other to complete a single business transaction, the greater the negative impact on scalability and elasticity. For this reason, it is important to keep synchronous communication among services to a minimum when requiring high levels of scalability and elasticity.

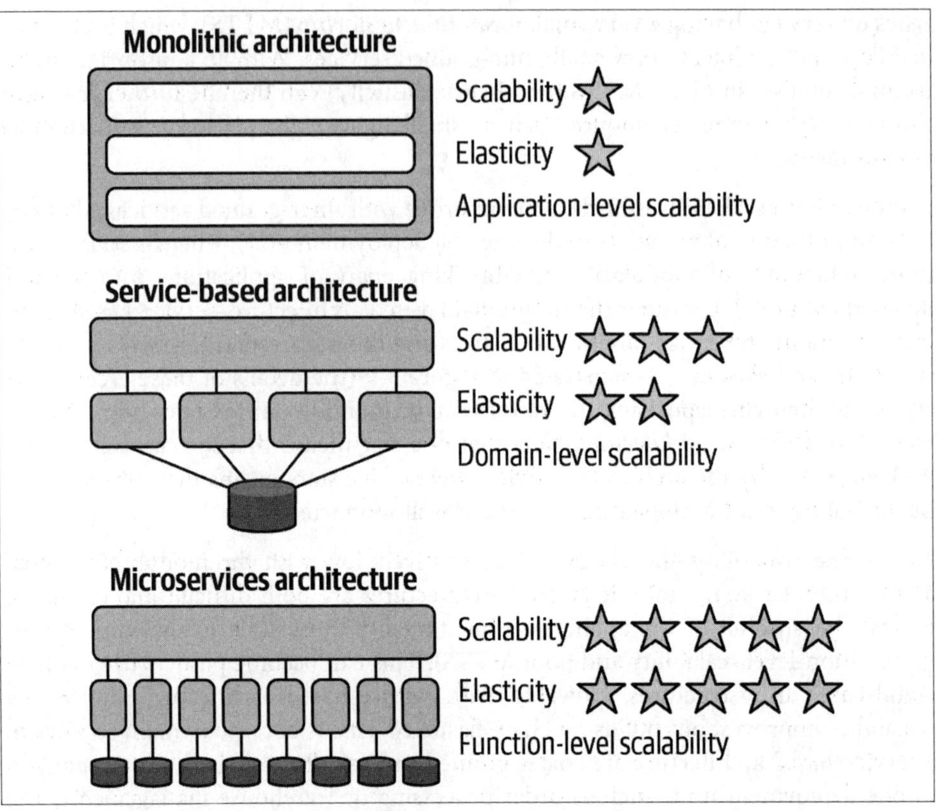

Figure 3-9. Scalability and elasticity improve with modularity

Availability/Fault Tolerance

Like many architecture characteristics, *fault tolerance* has varying definitions. Within the context of architectural modularity, we define fault tolerance as the ability for some parts of the system to remain responsive and available as other parts of the system fail. For example, if a fatal error (such as an out-of-memory condition) in the payment-processing portion of a retail application occurs, the users of the system should still be able to search for items and place orders, even though the payment processing is unavailable.

All monolithic systems suffer from low levels of fault tolerance. While fault tolerance can be somewhat mitigated in a monolithic system by having multiple instances of the entire application load balanced, this technique is both expensive and ineffective. If the fault is due to a programming bug, that bug will exist in both instances, therefore potentially bringing down both instances.

Architectural modularity is essential to achieving domain-level and function-level fault tolerance in a system. By breaking apart the system into multiple deployment units, catastrophic failure is isolated to only that deployment unit, thereby allowing the rest of the system to function normally. There is a caveat to this, however: if other services are synchronously dependent on a service that is failing, fault tolerance is not achieved. This is one of the reasons asynchronous communication between services is essential for maintaining a good level of fault tolerance in a distributed system.

Sysops Squad Saga: Creating a Business Case

`Thursday, September 30, 12:01`

Armed with a better understanding of what is meant by *architectural modularity* and the corresponding drivers for breaking apart a system, Addison and Austen met to discuss the Sysops Squad issues and try to match them to modularity drivers in order to build a solid business justification to present to the business sponsors.

"Let's take each of the issues we are facing and see if we can match them to some of the modularity drivers," said Addison. "That way, we can demonstrate to the business that breaking apart the application will in fact address the issues we are facing."

"Good idea," said Austen. "Let's start with the first issue they talked about in the meeting—change. We cannot seem to effectively apply changes to the existing monolithic system without something else breaking. Also, changes take way too long, and testing the changes is a real pain."

"And the developers are constantly complaining that the codebase is too large, and it's difficult to find the right place to apply changes to new features or bug fixes," said Addison.

"OK," said Austen, "so clearly, overall maintainability is a key issue here."

"Right," said Addison. "So, by breaking apart the application, it would not only decouple the code, but it would isolate and partition the functionality into separately deployed services, making it easier for developers to apply changes."

"Testability is another key characteristic related to this problem, but we have that covered already because of all our automated unit tests," said Austen.

"Actually, it's not," replied Addison. "Take a look at this."

Addison showed Austen that over 30% of the test cases are commented out or obsolete, and there are missing test cases for some of the critical workflow parts of the system. Addison also explained that the developers were continually complaining that the entire unit test suite had to be run for any change (big or small), which not only took a long time, but developers were faced with having to fix issues not related to their change. This was one of the reasons it was taking so long to apply even the simplest of changes.

"Testability is about the ease of testing, but also the completeness of testing," said Addison. "We have neither. By breaking apart the application, we can significantly reduce the scope of testing for changes made to the application, group relevant automated unit tests together, and get better completeness of testing—hence fewer bugs."

"The same is true with deployability," continued Addison. "Because we have a monolithic application, we have to deploy the entire system, even for a small bug fix. Because our deployment risk is so high, Parker insists on doing production releases on a monthly basis. What Parker doesn't understand is that by doing so, we pile multiple changes onto every release, some of which haven't even been tested in conjunction with each other."

"I agree," said Austen, "and besides, the mock deployments and code freezes we do for each release take up valuable time—time we don't have. However, what we're talking about here is not an architecture issue, but purely a deployment pipeline issue."

"I disagree," said Addison. "It's definitely architecture related as well. Think about it for a minute, Austen. If we broke the system into separately deployed services, then a change for any given service would be scoped to that service only. For example, let's say we make yet another change to the ticket assignment process. If that process was a separate service, not only would the testing scope be reduced, but we would significantly reduce the deployment risk. That means we could deploy more frequently with much less ceremony, as well as significantly reduce the number of bugs."

"I see what you mean," said Austen, "and while I agree with you, I still maintain that at some point we will have to modify our current deployment pipeline as well."

Satisfied that breaking apart the Sysops Squad application and moving to a distributed architecture would address the change issues, Addison and Austen moved on to the other business sponsor concerns.

"OK," said Addison, "the other big thing the business sponsors complained about in the meeting was overall customer satisfaction. Sometimes the system isn't available, the system seems to crash at certain times during the day, and we've experienced too many lost tickets and ticket routing issues. It's no wonder customers are starting to cancel their support plans."

"Hold on," said Austen. "I have some latest metrics here that show it's not the core ticketing functionality that keeps bringing the system down, but the customer survey functionality and reporting."

"This is excellent news," said Addison. "So by breaking apart that functionality of the system into separate services, we can isolate those faults, keeping the core ticketing functionality operational. That's a good justification in and of itself!"

"Exactly," said Austen. "So, we are in agreement then that overall availability through fault tolerance will address the application not always being available for the customers since they only interact with the ticketing portion of the system."

"But what about the system freezing up?" asked Addison. "How do we justify that part with breaking up the application?"

"It just so happens I asked Sydney from the Sysops Squad development team to run some analysis for me regarding exactly that issue," said Austen. "It turns out that it is a combination of two things. First, whenever we have more than 25 customers creating tickets at the same time, the system freezes. But, check this out—whenever they run the operational reports during the day when customers are entering problem tickets, the system also freezes up."

"So," said Addison, "it appears we have both a scalability and a database load issue here."

"Exactly!" Austen said. "And get this—by breaking up the application *and* the monolithic database, we can segregate reporting into its own system and also provide the added scalability for the customer-facing ticketing functionality."

Satisfied that they had a good business case to present to the business sponsors and confident that this was the right approach for saving this business line, Addison created an Architecture Decision Record (ADR) for the decision to break apart the system and create a corresponding business case presentation for the business sponsors.

> *ADR: Migrate Sysops Squad Application to a Distributed Architecture*
>
> *Context*
> The Sysops Squad is currently a monolithic problem ticket application that supports many different business functions related to problem tickets, including customer registration, problem ticket entry and processing, operations and analytical reporting, billing and payment processing, and various administrative maintenance functions. The current application has numerous issues involving scalability, availability, and maintainability.
>
> *Decision*
> We will migrate the existing monolithic Sysops Squad application to a distributed architecture. Moving to a distributed architecture will accomplish the following:
>
> - Make the core ticketing functionality more available for our external customers, therefore providing better fault tolerance
> - Provide better scalability for customer growth and ticket creation, resolving the frequent application freeze-ups we've been experiencing
> - Separate the reporting functionality and reporting load on the database, resolving the frequent application freeze-ups we've been experiencing
> - Allow teams to implement new features and fix bugs much faster than with the current monolithic application, therefore providing for better overall agility
> - Reduce the amount of bugs introduced into the system when changes occur, therefore providing better testability
> - Allow us to deploy new features and bug fixes at a much faster rate (weekly or even daily), therefore providing better deployability

Consequences

The migration effort will cause delays for new features being introduced since most of the developers will be needed for the architecture migration.

The migration effort will incur additional cost (cost estimates to be determined).

Until the existing deployment pipeline is modified, release engineers will have to manage the release and monitoring of multiple deployment units.

The migration effort will require us to break apart the monolithic database.

Addison and Austen met with the business sponsors for the Sysops Squad problem ticketing system and presented their case in a clear and concise manner. The business sponsors were pleased with the presentation and agreed with the approach, informing Addison and Austen to move forward with the migration.

CHAPTER 4

Architectural Decomposition

`Monday, October 4, 10:04`

Now that Addison and Austen had the go-ahead to move to a distributed architecture and break apart the monolithic Sysops Squad application, they needed to determine the best approach for how to get started.

"The application is so big I don't even know where to start. It's as big as an elephant!" exclaimed Addison.

"Well," said Austen. "How do you eat an elephant?"

"Ha, I've heard that joke before, Austen. One bite at a time, of course!" laughed Addison.

"Exactly. So let's use the same principle with the Sysops Squad application," said Austen. "Why don't we just start breaking it apart, one bite at a time? Remember how I said reporting was one of the things causing the application to freeze up? Maybe we should start there."

"That might be a good start," said Addison, "but what about the data? Just making reporting a separate service doesn't solve the problem. We'd need to break apart the data as well, or even create a separate reporting database with data pumps to feed it. I think that's too big of a bite to take starting out."

"You're right," said Austen. "Hey, what about the knowledge base functionality? That's fairly standalone and might be easier to extract."

"That's true. And what about the survey functionality? That should be easy to separate out as well," said Addison. "The problem is, I can't help feeling like we should be tackling this with more of a methodical approach rather than just eating the elephant bite by bite."

"Maybe Logan can give us some advice," said Austen.

Addison and Austen met with Logan to discuss some of the approaches they were considering for how to break apart the application. They explained to Logan that they wanted to start with the knowledge base and survey functionality but weren't sure what to do after that.

"The approach you're suggesting," said Logan, "is what is known as the Elephant Migration Anti-Pattern. Eating the elephant one bite at a time may seem like a good approach at the start, but in most cases it leads to an unstructured approach that results in a big ball of distributed mud, what some people also call a distributed monolith. I would not recommend that approach."

"So, what other approaches exist? Are there patterns we can use to break apart the application?" asked Addison.

"You need to take a holistic view of the application and apply either tactical forking or component-based decomposition," said Logan. "Those are the two most effective approaches I know of."

Addison and Austen looked at Logan. "But how do we know which one to use?"

Whereas architectural modularity describes the *why* for breaking apart a monolithic application, architectural decomposition describes the *how*. Breaking apart large, complex monolithic applications can be a complex and time-consuming undertaking, and it's important to know whether it is even feasible to begin such an effort and how to approach it.

Component-based decomposition and tactical forking are two common approaches for breaking apart monolithic applications. *Component-based decomposition* is an extraction approach that applies various refactoring patterns for refining and extracting components (the logical building blocks of an application) to form a distributed architecture in an incremental and controlled fashion. The *tactical forking* approach involves making replicas of an application and chipping away at the unwanted parts to form services, similar to the way a sculptor creates a beautiful work of art from a block of granite or marble.

Which approach is most effective? The answer to this question is, of course, *it depends*. One of the main factors in selecting a decomposition approach is how well the existing monolithic application code is structured. Do clear components and component boundaries exist within the codebase, or is the codebase largely an unstructured big ball of mud?

As the flowchart in Figure 4-1 illustrates, the first step in an architecture decomposition effort is to first determine whether the codebase is even decomposable. We cover this topic in detail in the next section. If the codebase is decomposable, the next step is to determine if the source code is largely an unstructured mess with no clearly definable components. If that's the case, then tactical forking (see "Tactical Forking" on page 73) is probably the right approach. However, if the source code files are structured in a way that combines like functionality within well-defined (or even loosely defined) components, then a component-based decomposition approach (see "Component-Based Decomposition" on page 71) is the way to go.

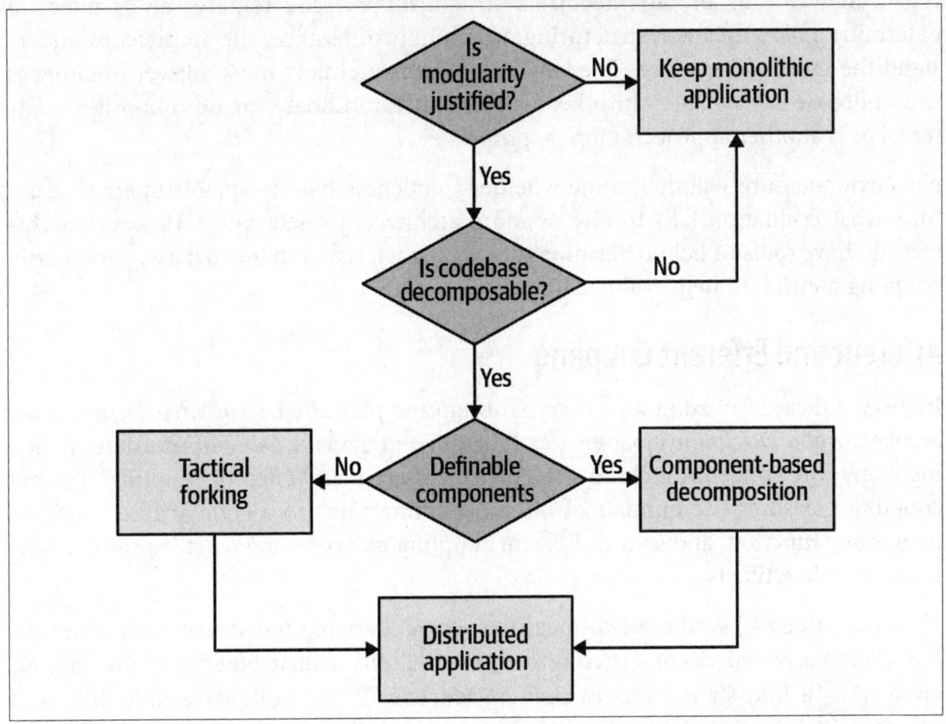

Figure 4-1. The decision tree for selecting a decomposition approach

We describe both of these approaches in this chapter, and then devote an entire chapter (Chapter 5) to describing each of the component-based decomposition patterns in detail.

Is the Codebase Decomposable?

What happens when a codebase lacks internal structure? Can it even be decomposed? Such software has a colloquial name—the Big Ball of Mud Anti-Pattern (*https://oreil.ly/7WkHf*), coined by Brian Foote in a same-named essay (*http://www.laputan.org/mud*) in 1999. For example, a complex web application with event handlers wired directly to database calls and no modularity can be considered a Big Ball of Mud architecture. Generally, architects don't spend much time creating patterns for these kinds of systems; software architecture concerns internal structure, and these systems lack that defining feature.

Unfortunately, without careful governance, many software systems degrade into big balls of mud, leaving it to subsequent architects (or perhaps a despised former self) to repair. Step one in any architecture restructuring exercise requires an architect to determine a *plan* for the restructuring, which in turn requires the architect to understand the internal structure. The key question the architect must answer becomes is this codebase salvageable? In other words, is it a candidate for decomposition patterns, or is another approach more appropriate?

No single measure will determine whether a codebase has reasonable internal structure—that evaluation falls to one or more architects to determine. However, architects do have tools to help determine macro characteristics of a codebase, particularly coupling metrics, to help evaluate internal structure.

Afferent and Efferent Coupling

In 1979, Edward Yourdon and Larry Constantine published *Structured Design: Fundamentals of a Discipline of Computer Program and Systems Design* (Yourdon), defining many core concepts, including the metrics afferent and efferent coupling. *Afferent* coupling measures the number of *incoming* connections to a code artifact (component, class, function, and so on). *Efferent* coupling measures the *outgoing* connections to other code artifacts.

Note the value of just these two measures when changing the structure of a system. For example, when deconstructing a monolith into a distributed architecture, an architect will find shared classes such as `Address`. When building a monolith, it is common and encouraged for developers to reuse core concepts such as `Address`, but when pulling the monolith apart, now the architect must determine how many other parts of the system use this shared asset.

Virtually every platform has tools that allow architects to analyze the coupling characteristics of code in order to assist in restructuring, migrating, or understanding a codebase. Many tools exist for various platforms that provide a matrix view of class and/or component relationships, as illustrated in Figure 4-2.

In this example, the Eclipse plug-in provides a visualization of the output of JDepend, which includes coupling analysis per package, along with some aggregate metrics highlighted in the next section.

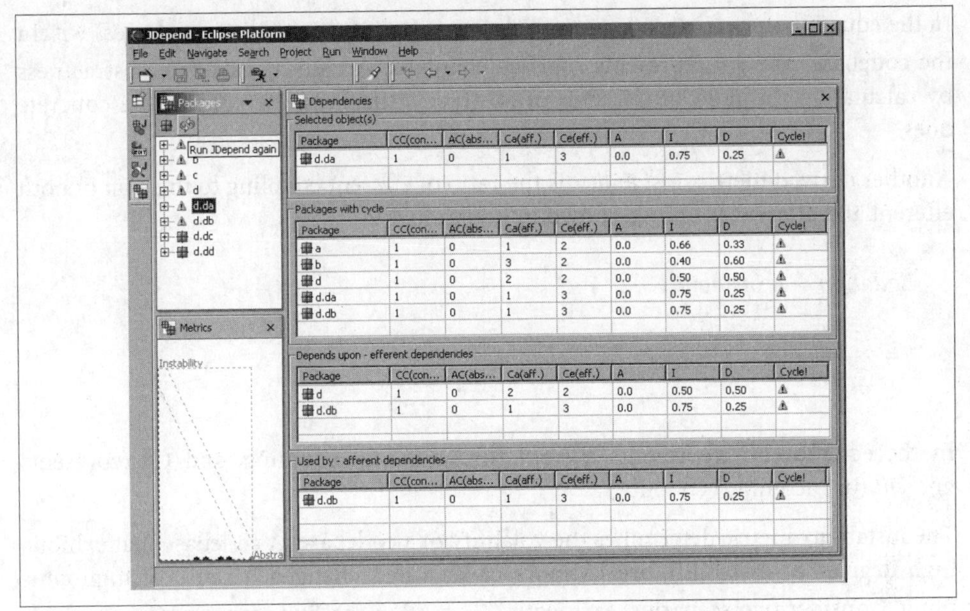

Figure 4-2. JDepend in Eclipse analysis view of coupling relationships

Abstractness and Instability

Robert Martin, a well-known figure in the software architecture world, created some derived metrics for a C++ book in the late 1990s that are applicable to any object-oriented language. These metrics—abstractness and instability—measure the balance of the internal characteristics of a codebase.

Abstractness is the ratio of abstract artifacts (abstract classes, interfaces, and so on) to concrete artifacts (implementation classes). It represents a measure of *abstract* versus *implementation*. Abstract elements are features of a codebase that allow developers to understand the overall function better. For example, a codebase consisting of a single `main()` method and 10,000 lines of code would score nearly zero on this metric and be quite hard to understand.

The formula for abstractness appears in Equation 4-1.

Equation 4-1. Abstractness

$$A = \frac{\sum m^a}{\sum m^c + \sum m^a}$$

In the equation, m^a represents *abstract* elements (interfaces or abstract classes) within the codebase, and m^c represents *concrete* elements. Architects calculate abstractness by calculating the ratio of the sum of abstract artifacts to the sum of the concrete ones.

Another derived metric, *instability*, is the ratio of efferent coupling to the sum of both efferent and afferent coupling, shown in Equation 4-2.

Equation 4-2. Instability

$$I = \frac{C^e}{C^e + C^a}$$

In the equation, C^e represents *efferent* (or outgoing) coupling, and C^a represents *afferent* (or incoming) coupling.

The instability metric determines the volatility of a codebase. A codebase that exhibits high degrees of instability breaks more easily when changed because of high coupling. Consider two scenarios, each with C^a of 2. For the first scenario, $C^e = 0$, yielding an instability score of zero. In the other scenario, $C^e = 3$, yielding an instability score of 3/5. Thus, the measure of instability for a component reflects how many potential changes might be forced by changes to related components. A component with an instability value near one is highly unstable, a value close to zero may be either stable or rigid: it is stable if the module or component contains mostly abstract elements, and rigid if it comprises mostly concrete elements. However, the trade-off for high stability is lack of reuse—if every component is self contained, duplication is likely.

A component with an I value close to 1, we can agree, is highly instable. However, a component with a value of I close to 0 may be either stable or rigid. However, if it contains mostly concrete elements, then it is rigid.

Thus, in general, it is important to look at the value of I and A together rather than in isolation. Hence the reason to consider the main sequence presented on the next page.

Distance from the Main Sequence

One of the few holistic metrics architects have for architectural structure is *distance from the main sequence*, a derived metric based on instability and abstractness, shown in Equation 4-3.

Equation 4-3. Distance from the main sequence

$$D = |A + I - 1|$$

In the equation, A = *abstractness* and I = *instability*.

The distance-from-the-main-sequence metric imagines an ideal relationship between abstractness and instability; components that fall near this idealized line exhibit a healthy mixture of these two competing concerns. For example, graphing a particular component allows developers to calculate the distance-from-the-main-sequence metric, illustrated in Figure 4-3.

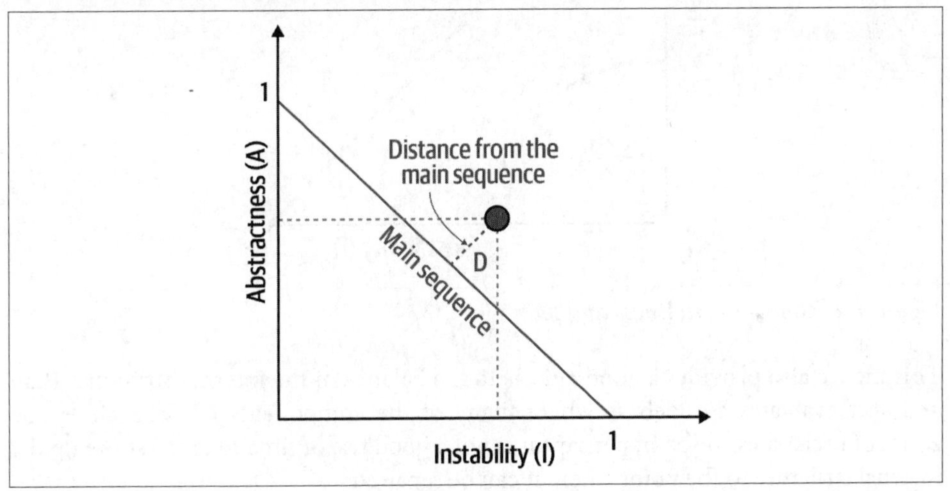

Figure 4-3. Normalized distance from the main sequence for a particular component

Developers graph the candidate component, then measure the distance from the idealized line. The closer to the line, the better balanced the component. Components that fall too far into the upper-right corner enter into what architects call the *zone of uselessness*: code that is too abstract becomes difficult to use. Conversely, code that falls into the lower-left corner enter the *zone of pain*: code with too much implementation and not enough abstraction becomes brittle and hard to maintain, illustrated in Figure 4-4.

Tools exist in many platforms to provide these measures, which assist architects when analyzing codebases because of unfamiliarity, migration, or technical debt assessment.

What does the distance-from-the-main-sequence metric tell architects looking to restructure applications? Just as in construction projects, moving a large structure that has a poor foundation presents risks. Similarly, if an architect aspires to restructure an application, improving the internal structure will make it easier to move the entity.

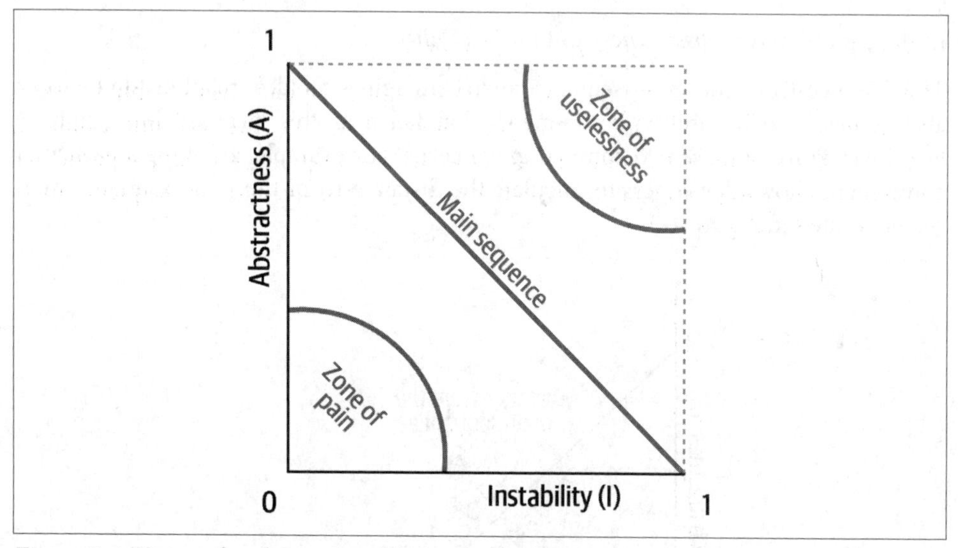

Figure 4-4. Zones of uselessness and pain

This metric also provides a good clue as to the balance of the internal structure. If an architect evaluates a codebase where many of the components fall into either the zones of uselessness or pain, perhaps it is not a good use of time to try to shore up the internal structure to the point where it can be repaired.

Following the flowchart in Figure 4-1, once an architect decides that the codebase is decomposable, the next step is to determine what approach to take to decompose the application. The following sections describe the two approaches for decomposing an application: *component-based decomposition* and *tactical forking*.

Component-Based Decomposition

It has been our experience that most of the difficulty and complexity involved with migrating monolithic applications to highly distributed architecture like microservices comes from poorly defined architectural components. Here we define a *component* as a building block of the application that has a well-defined role and responsibility in the system and a well-defined set of operations. Components in most applications are manifested through namespaces or directory structures and are implemented through component files (or source files). For example, in Figure 4-5, the directory structure *penultimate/ss/ticket/assign* would represent a component called Ticket Assign with the namespace `penultimate.ss.ticket.assign`.

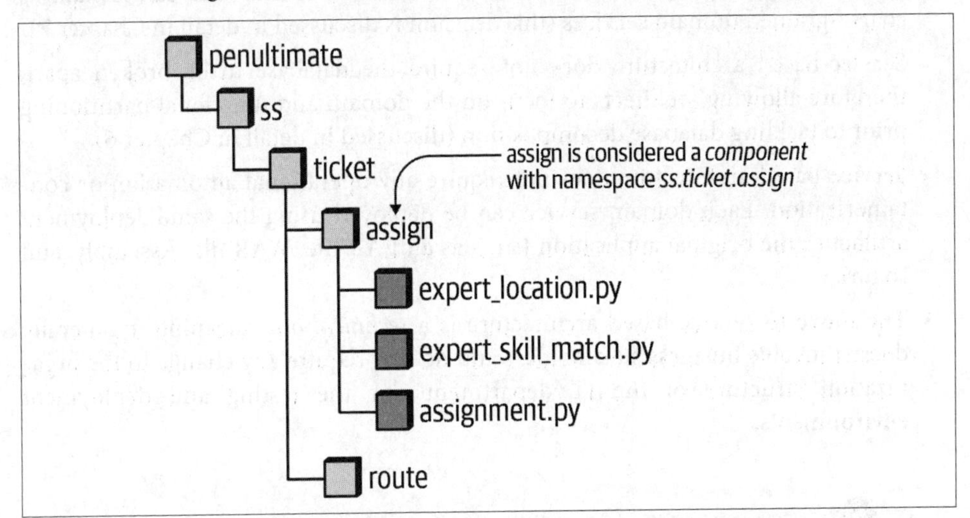

Figure 4-5. The directory structure of a codebase becomes the namespace of the component

When breaking monolithic applications into distributed architectures, build services from *components*, not individual classes.

Throughout many collective years of migrating monolithic applications to distributed architectures (such as microservices), we've developed a set of component-based decomposition patterns described in Chapter 5 that help prepare a monolithic application for migration. These patterns involve the refactoring of source code to arrive at a set of well-defined components that can eventually become services, easing the effort needed to migrate applications to distributed architectures.

These component-based decomposition patterns essentially enable the migration of a monolithic architecture to a service-based architecture, which is defined in Chapter 2 and described in more detail in *Fundamentals of Software Architecture*. Service-based architecture is a hybrid of the microservices architecture style where an application is broken into *domain services*, which are coarse-grained, separately deployed services containing all of the business logic for a particular domain.

Moving to a service-based architecture is suitable as a final target or as a stepping-stone to microservices:

- As a stepping-stone, it allows an architect to determine which domains require further levels of granularity into microservices and which ones can remain as coarse-grained domain services (this decision is discussed in detail in Chapter 7).
- Service-based architecture does not require the database to be broken apart, therefore allowing architects to focus on the domain and functional partitioning prior to tackling database decomposition (discussed in detail in Chapter 6).
- Service-based architecture does not require any operational automation or containerization. Each domain service can be deployed using the same deployment artifact as the original application (such as an EAR file, WAR file, Assembly, and so on).
- The move to service-based architecture is a *technical one*, meaning it generally doesn't involve business stakeholders and doesn't require any change to the organization structure of the IT department nor the testing and deployment environments.

When migrating monolithic applications to microservices, consider moving to a service-based architecture first as a stepping-stone to microservices.

But what if the codebase is an unstructured big ball of mud and doesn't contain very many observable components? That's where tactical forking comes in.

Tactical Forking

The *tactical forking* pattern was named by Fausto De La Torre (*https://faustodelatog.wordpress.com*) as a pragmatic approach to restructuring architectures that are basically big balls of mud.

Generally, when architects think about restructuring a codebase, they think of extracting pieces, as illustrated in Figure 4-6.

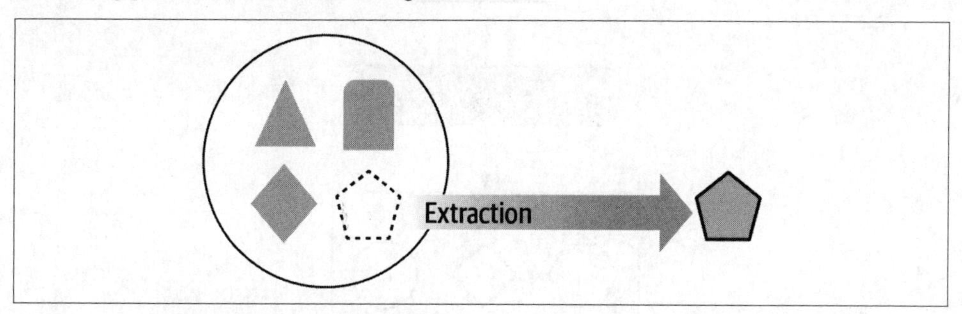

Figure 4-6. Extracting a part of a system

However, another way to think of isolating one part of a system involves *deleting* the parts no longer needed, as illustrated in Figure 4-7.

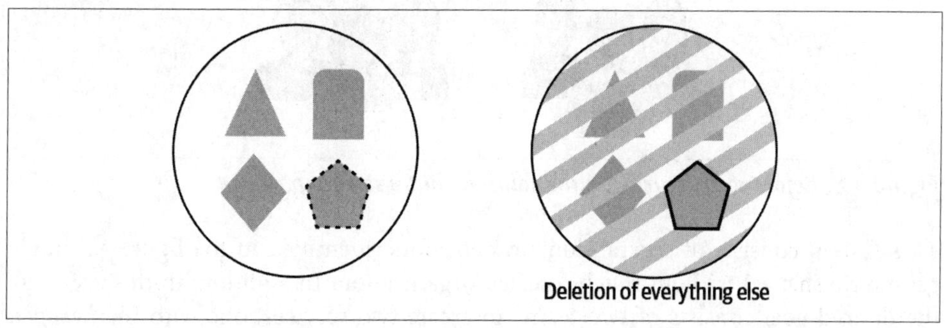

Figure 4-7. Deleting what's not wanted is another way to isolate parts of a system

In Figure 4-6, developers have to constantly deal with the exuberant strands of coupling that define this architecture; as they extract pieces, they discover that more and more of the monolith must come along because of dependencies. In Figure 4-7, developers delete what code isn't needed, but the dependencies remain, avoiding the constant unraveling effect of extraction.

The difference between *extraction* and *deletion* inspires the tactical forking pattern. For this decomposition approach, the system starts as a single monolithic application, as shown in Figure 4-8.

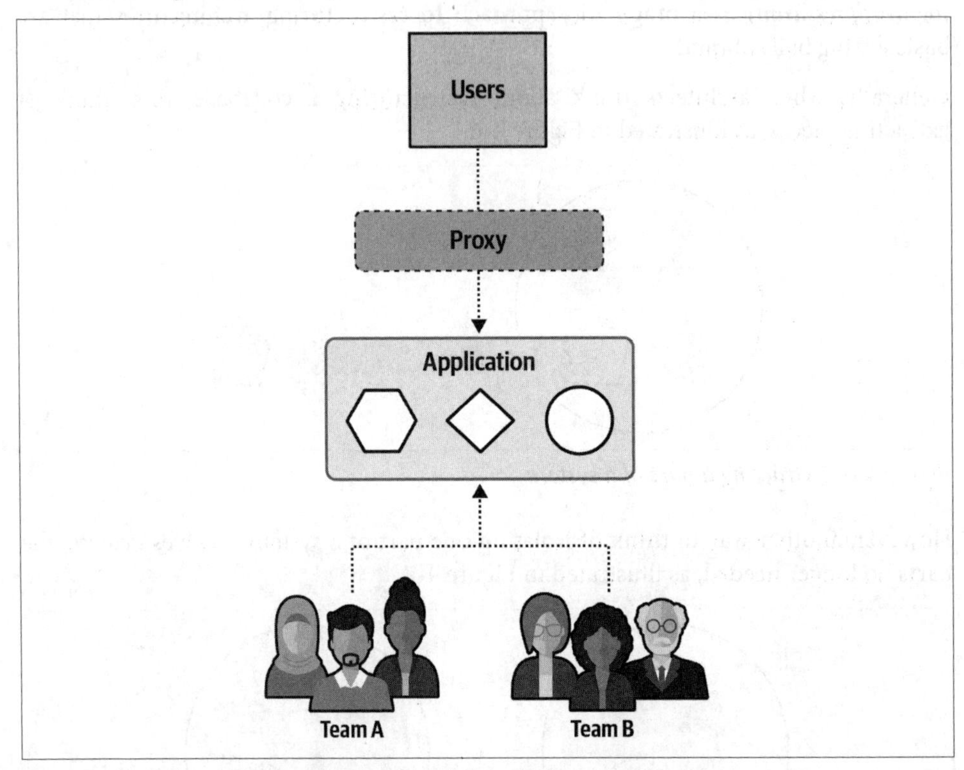

Figure 4-8. Before restructuring, a monolith includes several parts

This system consists of several domain behaviors (identified in the figure as simple geometric shapes) without much internal organization. In addition, in this scenario, the desired goal consists of two teams to create two services, one with the *hexagon* and *square* domain, and another with the *circle* domain, from the existing monolith.

The first step in tactical forking involves cloning the entire monolith, and giving each team a copy of the entire codebase, as illustrated in Figure 4-9.

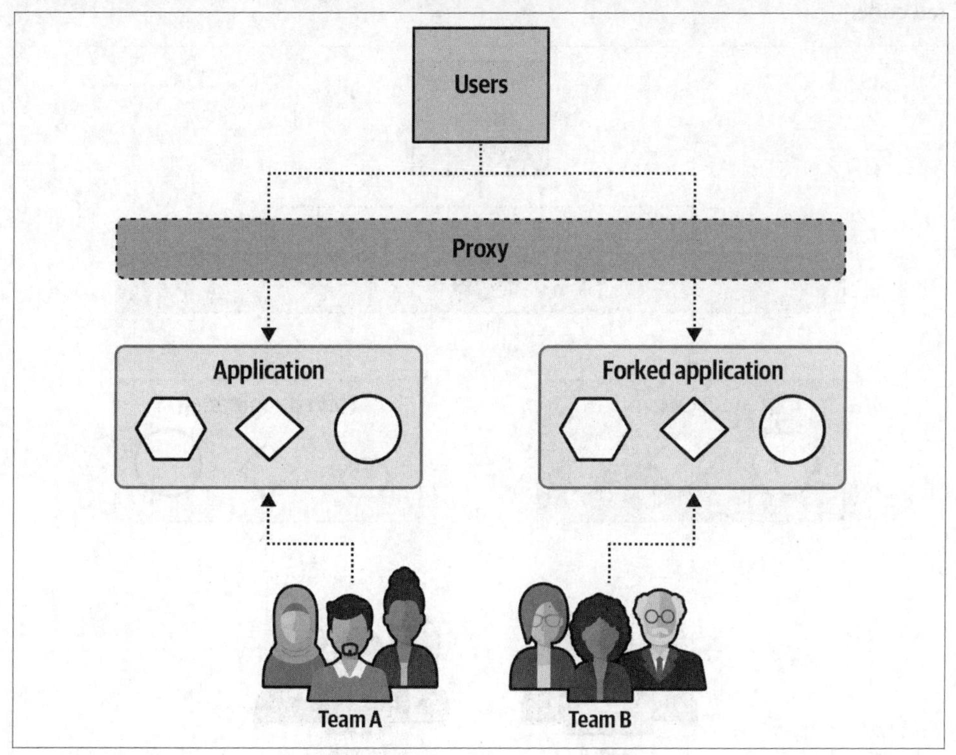

Figure 4-9. Step one clones the monolith

Each team receives a copy of the entire codebase, and they start *deleting* (as illustrated previously in Figure 4-7) the code they don't need rather than extract the desirable code. Developers often find this easier in a tightly coupled codebase because they don't have to worry about extracting the large number of dependencies that high coupling creates. Rather, in the *deletion* strategy, once functionality has been isolated, delete any code that doesn't break anything.

As the pattern continues to progress, teams begin to isolate the target portions, as shown in Figure 4-10. Then each team continues the gradual elimination of unwanted code.

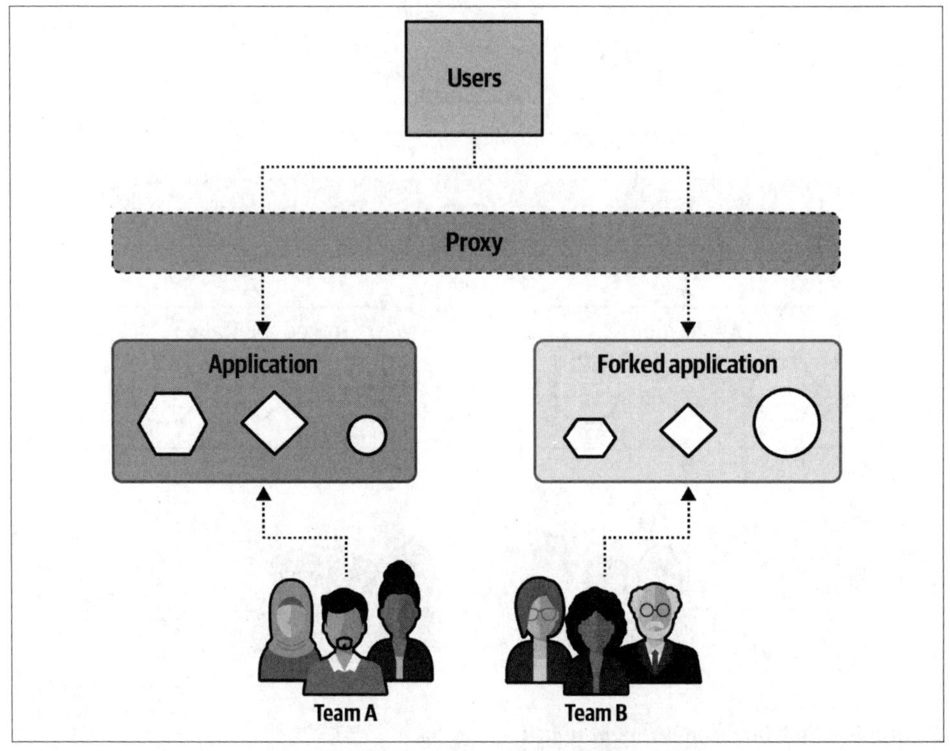

Figure 4-10. Teams constantly refactor to remove unwanted code

At the completion of the tactical forking pattern, teams have split the original monolithic application into two parts, preserving the coarse-grained structure of the behavior in each part, as illustrated in Figure 4-11.

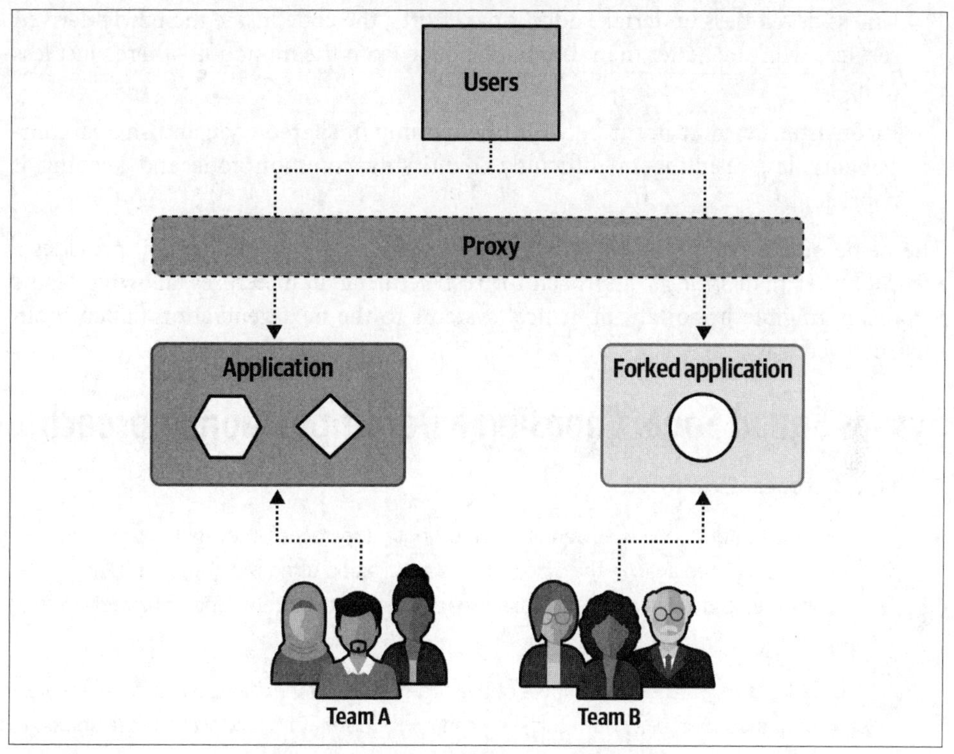

Figure 4-11. The end state of tactical forking features two services

Now the restructuring is complete, leaving two coarse-grained services as the result.

Trade-Offs

Tactical forking is a viable alternative to a more formal decomposition approach, most suited to codebases that have little or no internal structure. Like all practices in architecture, it has its share of trade-offs:

Benefits

- Teams can start working right away with virtually no up-front analysis.
- Developers find it easier to delete code rather than extract it. Extracting code from a chaotic codebase presents difficulties because of high coupling, whereas code not needed can be verified by compilation or simple testing.

Shortcomings

- The resulting services will likely still contain a large amount of mostly latent code left over from the monolith.
- Unless developers undertake additional efforts, the code inside the newly derived services won't be better than the chaotic code from the monolith—there's just less of it.
- Inconsistencies may occur between the naming of shared code and shared component files, resulting in difficultly identifying common code and keeping it consistent.

The name of this pattern is apt (as all good pattern names should be)—it provides a *tactical* rather than *strategic* approach for restructuring architectures, allowing teams to quickly migrate important or critical systems to the next generation (albeit in an unstructured way).

Sysops Squad Saga: Choosing a Decomposition Approach

`Friday, October 29, 10:01`

Now that Addison and Austen understood both approaches, they met in the main conference room to analyze the Sysops Squad application using the abstractness and instability metrics to determine which approach would be the most appropriate given their situation.

"Look at this," said Addison. "Most of the code lies along the main sequence. There are a few outliers of course, but I think we can conclude that it's feasible to break apart this application. So the next step is to determine which approach to use."

"I really like the tactical forking approach," said Austen. "It reminds me of famous sculptors, when asked how they were able to carve such beautiful works out of solid marble, who replied that they were merely removing the marble that wasn't supposed to be there. I feel like the Sysops Squad application could be my sculpture!"

"Hold on there, Michelangelo," said Addison. "First sports, and now sculpting? You need to make up your mind about what you like to spend your nonworking time on. The thing I don't like about the tactical forking approach is all the duplicate code and shared functionality within each service. Most of our problems have to do with maintainability, testability, and overall reliability. Can you imagine having to apply the same change to several different services at the same time? That would be a nightmare!"

"But how much shared functionality is there, really?" asked Austen.

"I'm not sure," said Addison, "but I do know there's quite a bit of shared code for the infrastructure stuff like logging and security, and I know a lot of the database calls are shared from the persistence layer of the application."

Austen paused and thought about Addison's argument for a bit. "Maybe you're right. Since we have good component boundaries already defined, I'm OK with doing the slower component-based decomposition approach and giving up my sculpting career. But I'm not giving up sports!"

Addison and Austen came to an agreement that the component decomposition approach would be the appropriate one for the Sysops Squad application. Addison wrote an ADR for this decision, outlining the trade-offs and justification for the component-based decomposition approach.

> *ADR: Migration Using the Component-Based Decomposition Approach*
>
> *Context*
>
> We will be breaking apart the monolithic Sysops Squad application into separately deployed services. The two approaches we considered for the migration to a distributed architecture were tactical forking and component-based decomposition.
>
> *Decision*
>
> We will use the component-based decomposition approach to migrate the existing monolithic Sysops Squad application to a distributed architecture.
>
> The application has well-defined component boundaries, lending itself to the component-based decomposition approach.
>
> This approach reduces the chance of having to maintain duplicate code within each service.
>
> With the tactical forking approach, we would have to define the service boundaries up front to know how many forked applications to create. With the component-based decomposition approach, the service definitions will naturally emerge through component grouping.
>
> Given the nature of the problems we are facing with the current application with regard to reliability, availability, scalability, and workflow, using the component-based decomposition approach provides a safer and more controlled incremental migration than the tactical forking approach does.
>
> *Consequences*
>
> The migration effort will likely take longer with the component-based decomposition approach than with tactical forking. However, we feel the justifications in the previous section outweigh this trade-off.
>
> This approach allows the developers on the team to work collaboratively to identify shared functionality, component boundaries, and domain boundaries. Tactical forking would require us to break apart the team into smaller, separate teams for each forked application and increase the amount of coordination needed between the smaller teams.

CHAPTER 5

Component-Based Decomposition Patterns

Monday, November 1, 11:53

Addison and Austen chose to use the component-based decomposition approach, but were unsure about the details of each decomposition pattern. They tried to research this approach, but did not find much on the internet about it. Once again, they met with Logan in the conference room for advice on what these patterns are all about and how to use them.

"Listen, Logan," said Addison, "I want to start out by saying we both really appreciate the amount of time you have been spending with us to get this migration process started. I know you're super busy on your own firefights."

"No problem," said Logan. "Us firefighters have to stick together. I've been in your shoes before, so I know what it's like flying blind on these sort of things. Besides, this is a highly visible migration effort, and it's important you both get this thing right the first time. Because there won't be a second time."

"Thanks, Logan," said Austen. "I've got a game in about two hours, so we'll try to make this short. You talked earlier about component-based decomposition, and we chose that approach, but we aren't able to find much about it on the internet."

"I'm not surprised," said Logan. "Not much has been written about them yet, but I know a book is coming out describing these patterns in detail sometime later this year. I first learned about these decomposition patterns at a conference about four years ago in a session with an experienced software architect. I was really impressed with the iterative and methodical approach to safely move from a monolithic architecture to a distributed one like service-based architecture and microservices. Since then I've been using these patterns with quite a bit of success."

"Can you show us how these patterns work?" asked Addison.

"Sure," said Logan. "Let's take it one pattern at a time."

Component-based decomposition (introduced in Chapter 4) is a highly effective technique for breaking apart a monolithic application when the codebase is somewhat structured and grouped by namespaces (or directories). This chapter introduces a set of patterns, known as *component-based decomposition patterns*, that describe the refactoring of monolithic source code to arrive at a set of well-defined components that can eventually become services. These decomposition patterns significantly ease the effort of migrating monolithic applications to distributed architectures.

Figure 5-1 shows the road map for the component-based decomposition patterns described in this chapter and how they are used together to break apart a monolithic application. Initially, these patterns are used together in sequence when moving a monolithic application to a distributed one, and then individually as maintenance is applied to the monolithic application during migration. These decomposition patterns are summarized as follows:

"Identify and Size Components Pattern" on page 84
: Typically the first pattern applied when breaking apart a monolithic application. This pattern is used to identify, manage, and properly size components.

"Gather Common Domain Components Pattern" on page 94
: Used to consolidate common business domain logic that might be duplicated across the application, reducing the number of potentially duplicate services in the resulting distributed architecture.

"Flatten Components Pattern" on page 101
: Used to collapse or expand domains, subdomains, and components, thus ensuring that source code files reside only within well-defined components.

"Determine Component Dependencies Pattern" on page 111
: Used to identify component dependencies, refine those dependencies, and determine the feasibility and overall level of effort for a migration from a monolithic architecture to a distributed one.

"Create Component Domains Pattern" on page 120
: Used to group components into logical domains within the application and to refactor component namespaces and/or directories to align with a particular domain.

"Create Domain Services Pattern" on page 126
: Used to physically break apart a monolithic architecture by moving logical domains within the monolithic application to separately deployed domain services.

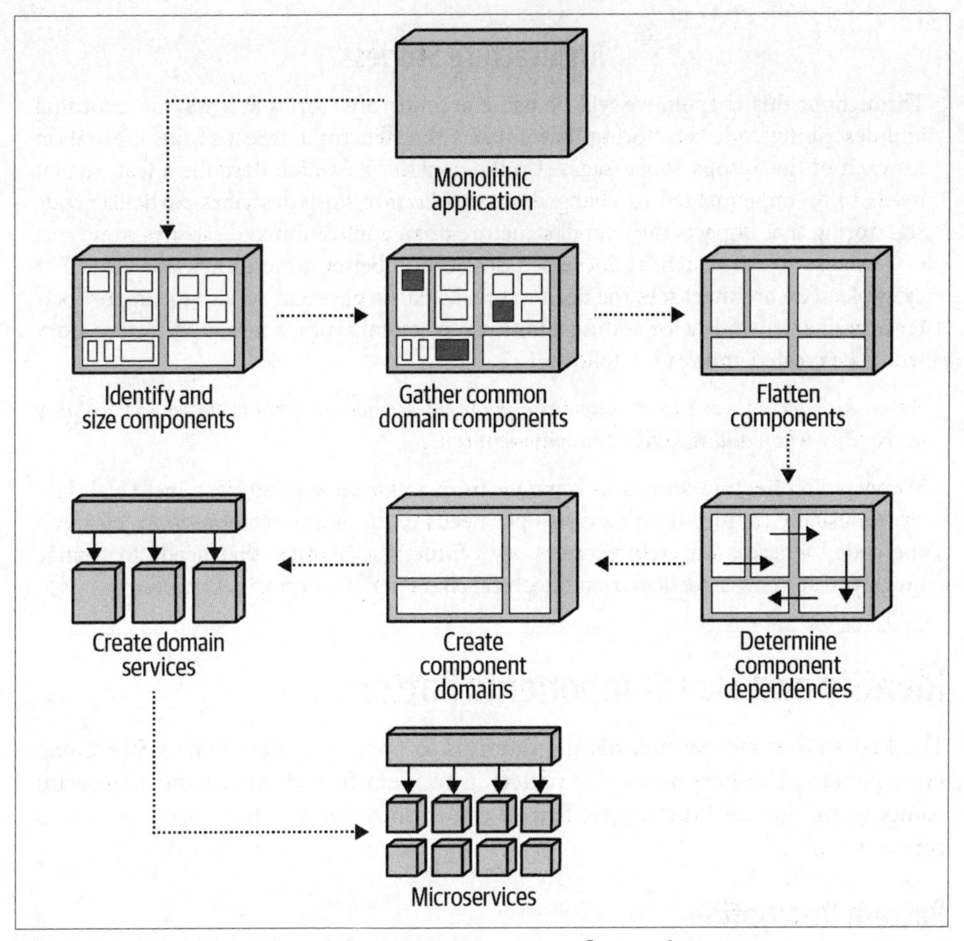

Figure 5-1. Component-based decomposition pattern flow and usage

Each pattern described in this chapter is divided into three sections. The first section, "Pattern Description," describes how the pattern works, why the pattern is important, and what the outcome is of applying the pattern. Knowing that most systems are moving targets during a migration, the second section, "Fitness Functions for Governance," describes the automated governance that can be used after applying the pattern to continually analyze and verify the correctness of the codebase during ongoing maintenance. The third section uses the real-world Sysops Squad application (see "Introducing the Sysops Squad Saga" on page 15) to illustrate the use of the pattern and illustrate the transformations of the application after the pattern has been applied.

> **Architecture Stories**
>
> Throughout this chapter, we will be using architecture stories as a way of recording and describing code refactoring that impacts the structural aspect of the application for each of the Sysops Squad sagas. Unlike *user stories*, which describe a feature that needs to be implemented or changed, an *architecture story* describes particular code refactoring that impacts the overall structure of an application and satisfies some sort of business driver (such as increased scalability, better time-to-market, etc.). For example, if an architect sees the need to break apart a payment service to support better overall extensibility for adding additional payment types, a new architecture story would be created and read as follows:
>
> *As an architect, I need to decouple the payment service to support better extensibility and agility when adding additional payment types.*
>
> We view architecture stories as separate from *technical debt* stories. Technical debt stories usually capture things a developer needs to do in a later iteration to "clean up the code," whereas an architecture story captures something that needs to change quickly to support a particular architectural characteristic or business need.

Identify and Size Components Pattern

The first step in any monolithic migration is to apply the *Identify and Size Components* pattern. The purpose of this pattern is to identify and catalog the architectural components (logical building blocks) of the application and then properly size the components.

Pattern Description

Because services are built from components, it is critical to not only identify the components within an application, but to properly size them as well. This pattern is used to identify components that are either too big (doing too much) or too small (not doing enough). Components that are too large relative to other components are generally more coupled to other components, are harder to break into separate services, and lead to a less modular architecture.

Unfortunately, it is difficult to determine the size of a component. The number of source files, classes, and total lines of code are not good metrics because every programmer designs classes, methods, and functions differently. One metric we've found useful for component sizing is calculating the total number of statements within a given component (the sum of statements within all source files contained within a namespace or directory). A *statement* is a single complete action performed in the source code, usually terminated by a special character (such as a semicolon in languages such as Java, C, C++, C#, Go, and JavaScript; or a newline in languages such as

F#, Python, and Ruby). While not a perfect metric, at least it's a good indicator of how much the component is doing and how complex the component is.

Having a relatively consistent component size within an application is important. Generally speaking, the size of components in an application should fall between one to two standard deviations from the average (or mean) component size. In addition, the percentage of code represented by each component should be somewhat evenly distributed between application components and not vary significantly.

While many static code analysis tools (*https://oreil.ly/XyIgr*) can show the number of statements within a source file, many of them don't accumulate total statement by *component*. Because of this, the architect usually must perform manual or automated post-processing to accumulate total statements by component and then calculate the percentage of code that component represents.

Regardless of the tools or algorithms used, the important information and metrics to gather and calculate for this pattern are shown in Table 5-1 and are defined in the following list.

Table 5-1. Component inventory and component size analysis example

Component name	Component namespace	Percent	Statements	Files
Billing Payment	`ss.billing.payment`	5	4,312	23
Billing History	`ss.billing.history`	4	3,209	17
Customer Notification	`ss.customer.notification`	2	1,433	7

Component name

A descriptive name and identifier of the component that is consistent throughout application diagrams and documentation. The component name should be clear enough to be as self-describing as possible. For example, the component Billing History shown in Table 5-1 is clearly a component that contains source code files used to manage a customer's billing history. If the distinct role and responsibility of the component isn't immediately identifiable, consider changing the component (and potentially the corresponding namespace) to a more descriptive one. For example, a component named Ticket Manager leaves too many unanswered questions about its role and responsibility in the system, and should be renamed to better describe its role.

Component namespace

The physical (or logical) identification of the component representing where the source code files implementing that component are grouped and stored. This identifier is usually denoted through a namespace, package structure (Java), or directory structure. When a directory structure is used to denote the component, we usually convert the file separator to a dot (.) and create a corresponding logical namespace. For example, the component namespace for source code files in the *ss/customer/notification* directory structure would have the namespace value `ss.customer.notification`. Some languages require that the namespace match the directory structure (such as Java with a *package*), whereas other languages (such as C# with a *namespace*) do not enforce this constraint. Whatever namespace identifier is used, make sure the type of identifier is consistent across all of the components in the application.

Percent
: The relative size of the component based on its percentage of the overall source code containing that component. The percent metric is helpful in identifying components that appear too large or too small in the overall application. This metric is calculated by taking the total number of statements within the source code files representing that component and dividing that number by the total number of statements in the entire codebase of the application. For example, the percent value of 5 for the `ss.billing.payment` component in Table 5-1 means that this component constitutes 5% of the overall codebase.

Statements
: The sum of the total number of source code statements in all source files contained within that component. This metric is useful for determining not only the relative size of the components within an application, but also for determining the overall complexity of the component. For example, a seemingly simple single-purpose component named Customer Wishlist might have a total of 12,000 statements, indicating that the processing of wish list items is perhaps more complex than it looks. This metric is also necessary for calculating the percent metric previously described.

Files
: The total number of source code files (such as classes, interfaces, types, and so on) that are contained within the component. While this metric has little to do with the size of a component, it does provide additional information about the component from a class structure standpoint. For example, a component with 18,409 statements and only 2 files is a good candidate for refactoring into smaller, more contextual classes.

When resizing a large component, we recommend using a functional decomposition approach or a domain-driven approach to identify subdomains that might exist within the large component. For example, assume the Sysops Squad application has a Trouble Ticket component containing 22% of the codebase that is responsible for ticket creation, assignment, routing, and completion. In this case, it might make sense to break the single Trouble Ticket component into four separate components (Ticket Creation, Ticket Assignment, Ticket Routing, and Ticket Completion), reducing the percentage of code each component represents, therefore creating a more modular application. If no clear subdomains exist within a large component, then leave the component as is.

Fitness Functions for Governance

Once this decomposition pattern has been applied and components have been identified and sized correctly, it's important to apply some sort of automated governance to identify new components and to ensure components don't get too large during normal application maintenance and create unwanted or unintended dependencies. Automated holistic fitness functions can be triggered during deployment to alert the architect if specified constraints are exceeded (such as the percent metric discussed previously or use of standard deviations to identify outliers).

Fitness functions can be implemented through custom-written code or through the use of open source or COTS tools as part of a CI/CD pipeline. Some of the automated fitness functions that can be used to help govern this decomposition pattern are as follows.

Fitness function: Maintain component inventory

This automated holistic fitness function, usually triggered on deployment through a CI/CD pipeline, helps keep the inventory of components current. It's used to alert an architect of components that might have been added or removed by the development team. Identifying new or removed components is not only critical for this pattern, but for the other decomposition patterns as well. Example 5-1 shows the pseudocode and algorithm for one possible implementation of this fitness function.

Example 5-1. Pseudocode for maintaining component inventory

```
# Get prior component namespaces that are stored in a datastore
LIST prior_list = read_from_datastore()

# Walk the directory structure, creating namespaces for each complete path
LIST current_list = identify_components(root_directory)

# Send an alert if new or removed components are identified
LIST added_list = find_added(current_list, prior_list)
LIST removed_list = find_removed(current_list, prior_list)
IF added_list NOT EMPTY {
  add_to_datastore(added_list)
  send_alert(added_list)
}
IF removed_list NOT EMPTY {
  remove_from_datastore(removed_list)
  send_alert(removed_list)
}
```

Fitness function: No component shall exceed <some percent> of the overall codebase

This automated holistic fitness function, usually triggered on deployment through a CI/CD pipeline, identifies components that exceed a given threshold in terms of the percentage of overall source code represented by that component, and alerts the architect if any component exceeds that threshold. As mentioned earlier in this chapter, the threshold percentage value will vary depending on the size of the application, but should be set so as to identify significant outliers. For example, for a relatively small application with only 10 components, setting the percentage threshold to something like 30% would sufficiently identify a component that is too large, whereas for a large application with 50 components, a threshold of 10% would be more appropriate. Example 5-2 shows the pseudocode and algorithm for one possible implementation of this fitness function.

Example 5-2. Pseudocode for maintaining component size based on percent of code

```
# Walk the directory structure, creating namespaces for each complete path
LIST component_list = identify_components(root_directory)

# Walk through all of the source code to accumulate total statements
total_statements = accumulate_statements(root_directory)

# Walk through the source code for each component, accumulating statements
# and calculating the percentage of code each component represents. Send
# an alert if greater than 10%
FOREACH component IN component_list {
  component_statements = accumulate_statements(component)
  percent = component_statements / total_statements
  IF percent > .10 {
    send_alert(component, percent)
  }
}
```

Fitness function: No component shall exceed <some number of standard deviations> from the mean component size

This automated holistic fitness function, usually triggered on deployment through a CI/CD pipeline, identifies components that exceed a given threshold in terms of the number of standard deviations from the mean of all component sizes (based on the total number of statements in the component), and alerts the architect if any component exceeds that threshold.

Standard deviation is a useful means of determining outliers in terms of component size. Standard deviation is calculated as follows:

$$s = \sqrt{\frac{1}{N-1}\sum_{i=1}^{N}(x_i - \bar{x})^2}$$

where N is the number of observed values, x_i is the observed values, and $\bar{x}$ is the mean of the observed values. The mean of observed values ($\bar{x}$) is calculated as follows:

$$\bar{x} = \frac{1}{N}\sum_{i=1}^{N} x_i$$

The standard deviation can then be used along with the difference from the mean to determine the number of standard deviations the component size is from the mean. Example 5-3 shows the pseudocode for this fitness function, using three standard deviations from the mean as a threshold.

Example 5-3. Pseudocode for maintaining component size based on number of standard deviations

```
# Walk the directory structure, creating namespaces for each complete path
LIST component_list = identify_components(root_directory)

# Walk through all of the source code to accumulate total statements and number
# of statements per component
SET total_statements TO 0
MAP component_size_map
FOREACH component IN component_list {
  num_statements = accumulate_statements(component)
  ADD num_statements TO total_statements
  ADD component,num_statements TO component_size_map
}

# Calculate the standard deviation
SET square_diff_sum TO 0
num_components = get_num_entries(component_list)
mean = total_statements / num_components
FOREACH component,size IN component_size_map {
  diff = size - mean
  ADD square(diff) TO square_diff_sum
}
std_dev = square_root(square_diff_sum / (num_components - 1))

# For each component calculate the number of standard deviations from the
# mean. Send an alert if greater than 3
FOREACH component,size IN component_size_map {
  diff_from_mean = absolute_value(size - mean);
  num_std_devs = diff_from_mean / std_dev
  IF num_std_devs > 3 {
    send_alert(component, num_std_devs)
  }
}
```

Sysops Squad Saga: Sizing Components

Tuesday, November 2, 09:12

After the discussion with Logan (the lead architect) about component-based decomposition patterns, Addison decided to apply the Identify and Size Components pattern to identify all of the components in the Sysops Squad ticketing application and calculate the size of each component based on the total number of statements in each component.

Addison gathered all the necessary component information and put this information into Table 5-2, calculating the percentage of code for each component based on the total number of statements in the entire application (in this case, 82,931 statements).

Table 5-2. Component size analysis for the Sysops Squad application

Component name	Component namespace	Percent	Statements	Files
Login	`ss.login`	2	1865	3
Billing Payment	`ss.billing.payment`	5	4,312	23
Billing History	`ss.billing.history`	4	3,209	17
Customer Notification	`ss.customer.notification`	2	1,433	7
Customer Profile	`ss.customer.profile`	5	4,012	16
Expert Profile	`ss.expert.profile`	6	5,099	32
KB Maint	`ss.kb.maintenance`	2	1,701	14
KB Search	`ss.kb.search`	3	2,871	4
Reporting	`ss.reporting`	**33**	**27,765**	**162**
Ticket	`ss.ticket`	8	7,009	45
Ticket Assign	`ss.ticket.assign`	9	7,845	14
Ticket Notify	`ss.ticket.notify`	2	1,765	3
Ticket Route	`ss.ticket.route`	2	1,468	4
Support Contract	`ss.supportcontract`	5	4,104	24
Survey	`ss.survey`	3	2,204	5
Survey Notify	`ss.survey.notify`	2	1,299	3
Survey Templates	`ss.survey.templates`	2	1,672	7
User Maintenance	`ss.users`	4	3,298	12

Addison noticed that most of the components listed in Table 5-2 are about the same size, with the exception of the Reporting component (`ss.reporting`) which consisted of 33% of the codebase. Since the Reporting component was significantly larger than the other components (illustrated in Figure 5-2), Addison chose to break this component apart to reduce its overall size.

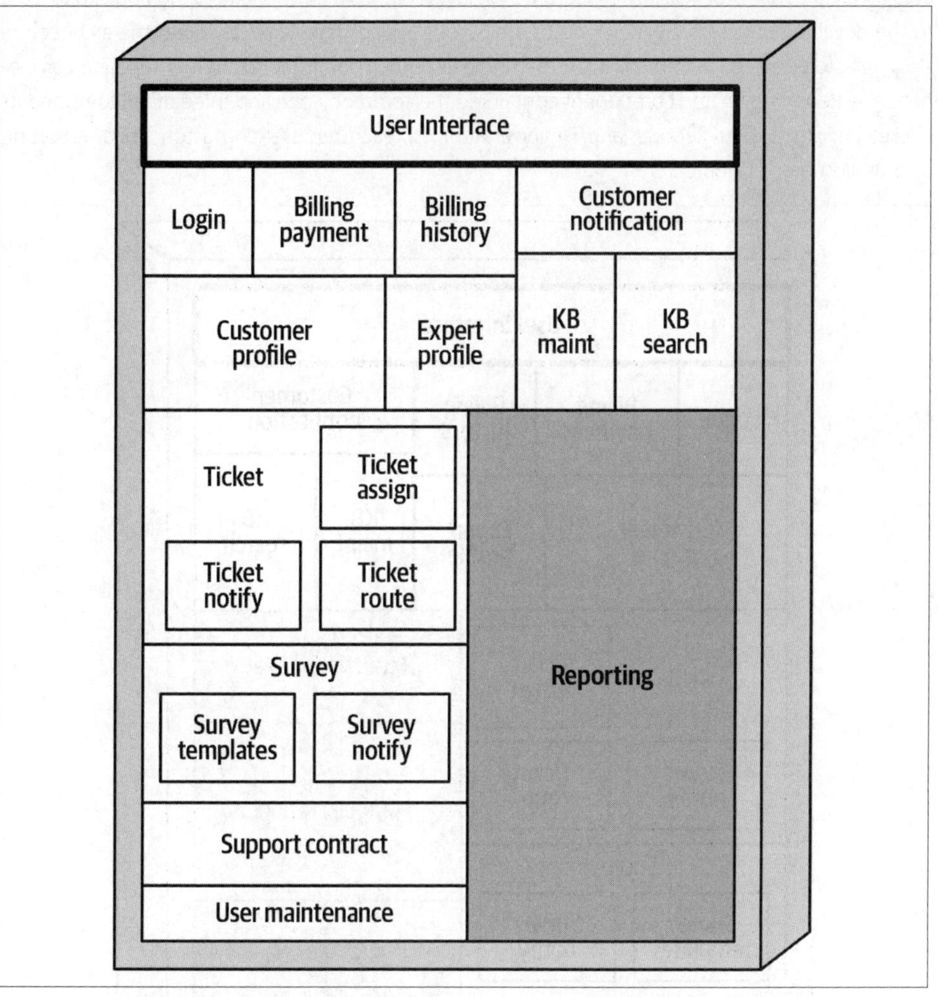

Figure 5-2. The Reporting component is too big and should be broken apart

After doing some analysis, Addison found that the reporting component contained source code that implemented three categories of reports:

- Ticketing reports (ticket demographics reports, tickets per day/week/month reports, ticket resolution time reports, and so on)
- Expert reports (expert utilization reports, expert distribution reports, and so on)

- Financial reports (repair cost reports, expert cost reports, profit reports, and so on)

Addison also identified common (shared) code that all reporting categories used, such as common utilities, calculators, shared data queries, report distribution, and shared data formatters. Addison created an architecture story (see "Architecture Stories" on page 84) for this refactoring and explained it to the development team. Sydney, one of the Sysops Squad developers assigned the architecture story, refactored the code to break apart the single Reporting component into four separate components—a Reporting Shared component containing the common code and three other components (Ticket Reports, Expert Reports, and Financial Reports), each representing a functional reporting area, as illustrated in Figure 5-3.

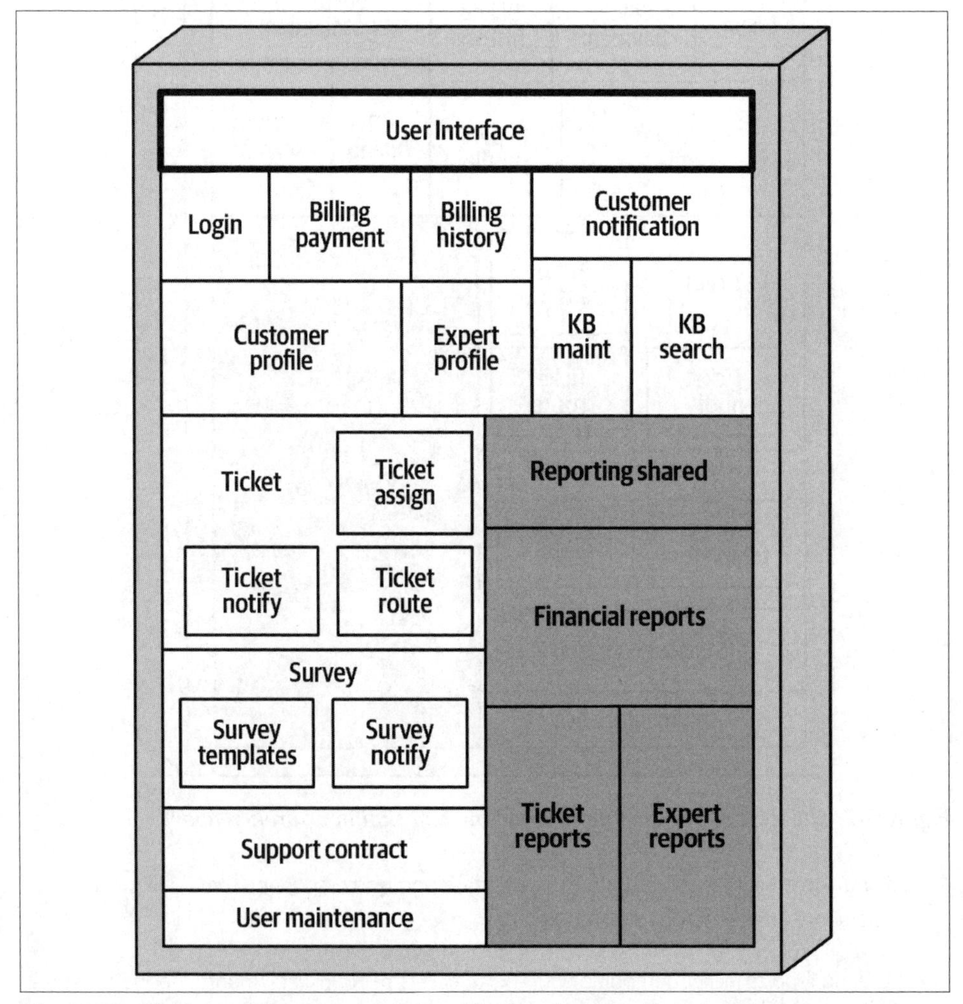

Figure 5-3. The large Reporting component broken into smaller reporting components

After Sydney committed the changes, Addison reanalyzed the code and verified that all of the components were now fairly equally distributed in size. Addison recorded the results of applying this decomposition pattern in Table 5-3.

Table 5-3. Component size after applying the Identify and Size Components pattern

Component name	Component namespace	Percent	Statements	Files
Login	ss.login	2	1865	3
Billing Payment	ss.billing.payment	5	4,312	23
Billing History	ss.billing.history	4	3,209	17
Customer Notification	ss.customer.notification	2	1,433	7
Customer Profile	ss.customer.profile	5	4,012	16
Expert Profile	ss.expert.profile	6	5,099	32
KB Maint	ss.kb.maintenance	2	1,701	14
KB Search	ss.kb.search	3	2,871	4
Reporting Shared	ss.reporting.shared	**7**	**5,309**	**20**
Ticket Reports	ss.reporting.tickets	**8**	**6,955**	**58**
Expert Reports	ss.reporting.experts	**9**	**7,734**	**48**
Financial Reports	ss.reporting.financial	**9**	**7,767**	**36**
Ticket	ss.ticket	8	7,009	45
Ticket Assign	ss.ticket.assign	9	7,845	14
Ticket Notify	ss.ticket.notify	2	1,765	3
Ticket Route	ss.ticket.route	2	1,468	4
Support Contract	ss.supportcontract	5	4,104	24
Survey	ss.survey	3	2,204	5
Survey Notify	ss.survey.notify	2	1,299	3
Survey Templates	ss.survey.templates	2	1,672	7
User Maintenance	ss.users	4	3,298	12

Notice in the preceding Sysops Squad Saga that Reporting no longer exists as a component in Table 5-3 or Figure 5-3. Although the *namespace* still exists (`ss.reporting`), it is no longer considered a component, but rather a subdomain. The refactored components listed in Table 5-3 will be used when applying the next decomposition pattern, Gather Common Domain Components.

Gather Common Domain Components Pattern

When moving from a monolithic architecture to a distributed one, it is often beneficial to identify and consolidate common domain functionality to make common services easier to identify and create. The *Gather Common Domain Components* pattern is used to identify and collect common domain logic and centralize it into a single component.

Pattern Description

Shared *domain* functionality is distinguished from shared *infrastructure* functionality in that domain functionality is part of the business processing logic of an application (such as notification, data formatting, and data validation) and is common to only some processes, whereas infrastructure functionality is operational in nature (such as logging, metrics gathering, and security) and is common to all processes.

Consolidating common domain functionality helps eliminate duplicate services when breaking apart a monolithic system. Often there are only very subtle differences among common domain functionality that is duplicated throughout the application, and these differences can be easily resolved within a single common service (or shared library).

Finding common domain functionality is mostly a manual process, but some automation can be used to assist in this effort (see "Fitness Functions for Governance" on page 95). One hint that common domain processing exists in the application is the use of shared classes across components or a common inheritance structure used by multiple components. Take, for example, a class file named *SMTPConnection* in a large codebase that is used by five classes, all contained within different namespaces (components). This scenario is a good indication that common email notification functionality is spread throughout the application and might be a good candidate for consolidation.

Another way of identifying common domain functionality is through the name of a logical component or its corresponding namespace. Consider the following components (represented as namespaces) in a large codebase:

- Ticket Auditing (`penultimate.ss.ticket.audit`)
- Billing Auditing (`penultimate.ss.billing.audit`)
- Survey Auditing (`penultimate.ss.survey.audit`)

Notice how each of these components (Ticket Auditing, Billing Auditing, and Survey Auditing) all have the same thing in common—writing the action performed and the user requesting the action to an audit table. While the context may be different, the final outcome is the same—inserting a row in an audit table. This common domain functionality can be consolidated into a new component called `penultimate.ss.shared.audit`, resulting in less duplication of code and also fewer services in the resulting distributed architecture.

Not all common domain functionality necessarily becomes a shared service. Alternatively, common code could be gathered into a *shared library* that is bound to the code during compile time. The pros and cons of using a shared service rather than a shared library are discussed in detail in Chapter 8.

Fitness Functions for Governance

Automating the governance of shared domain functionality is rather difficult because of the subjectiveness of identifying shared functionality and classifying it as domain functionality versus infrastructure functionality. For the most part, the fitness functions used to govern this pattern are therefore somewhat manual. That said, there are some ways to automate the governance to assist in the manual interpretation of common domain functionality. The following fitness functions can assist in finding common domain functionality.

Fitness function: Find common names in leaf nodes of component namespace

This automated holistic fitness function can be triggered on deployment through a CI/CD pipeline to locate common names within the namespace of a component. When a common ending namespace node name is found between two or more components, the architect is alerted and can analyze the functionality to determine if it is common domain logic. So that the same alert isn't continuously sent as a "false positive," an exclusion file can be used to store those namespaces that have common ending node names but are not deemed common domain logic (such as multiple namespaces ending in `.calculate` or `.validate`). Example 5-4 shows the pseudocode for this fitness function.

Example 5-4. Pseudocode for finding common namespace leaf node names

```
# Walk the directory structure, creating namespaces for each complete path
LIST component_list = identify_components(root_directory)

# Locate possible duplicate component node names that are not in the exclusion
# list stored in a datastore
LIST excluded_leaf_node_list = read_datastore()
LIST leaf_node_list
LIST common_component_list
```

```
FOREACH component IN component_list {
  leaf_name = get_last_node(component)
  IF leaf_name IN leaf_node_list AND
     leaf_name NOT IN excluded_leaf_node_list {
    ADD component TO common_component_list
  } ELSE {
    ADD leaf_name TO leaf_node_list
  }
}

# Send an alert if any possible common components were found
IF common_component_list NOT EMPTY {
  send_alert(common_component_list)
}
```

Fitness function: Find common code across components

This automated holistic fitness function can be triggered on deployment through a CI/CD pipeline to locate common classes used between namespaces. While not always accurate, it does help in alerting an architect of possible duplicate domain functionality. Like the previous fitness function, an exclusion file is used to reduce the number of "false positives" for known common code that is not considered duplicate domain logic. Example 5-5 shows the pseudocode for thisfitness function.

Example 5-5. Pseudocode for finding common source files between components

```
# Walk the directory structure, creating namespaces for each complete path and a list
# of source file names for each component
LIST component_list = identify_components(root_directory)
LIST source_file_list = get_source_files(root_directory)
MAP component_source_file_map
FOREACH component IN component_list {
  LIST component_source_file_list = get_source_files(component)
  ADD component, component_source_file_list TO component_source_file_map
}

# Locate possible common source file usage across components that are not in
# the exclusion list stored in a datastore
LIST excluded_source_file_list = read_datastore()
LIST common_source_file_list
FOREACH source_file IN source_file_list {
  SET count TO 0
  FOREACH component,component_source_file_list IN component_source_file_map {
    IF source_file IN component_source_file_list {
      ADD 1 TO count
    }
  }
  IF count > 1 AND source_file NOT IN excluded_source_file_list {
    ADD source_file TO common_source_file_list
  }
```

```
}

# Send an alert if any source files are used in multiple components
IF common_source_file_list NOT EMPTY {
        send_alert(common_source_file_list)
}
```

Sysops Squad Saga: Gathering Common Components

Friday, November 5, 10:34

Having identified and sized the components in the Sysops Squad application, Addison applied the Gather Common Domain Components pattern to see if any common functionality existed between components. From the list of components in Table 5-3, Addison noticed there were three components all related to notifying a Sysops Squad customer, and listed these in Table 5-4.

Table 5-4. Sysops Squad components with common domain functionality

Component	Namespace	Responsibility
Customer Notification	`ss.customer.notification`	General notification
Ticket Notify	`ss.ticket.notify`	Notify that expert is en route
Survey Notify	`ss.survey.notify`	Send survey email

While each of these notification components had a different context for notifying a customer, Addison realized they all have one thing in common—they all sent information to a customer. Figure 5-4 illustrates these common notification components within the Sysops Squad application.

Noticing that the source code contained in these components was also very similar, Addison consulted with Austen (the other Sysops Squad architect). Austen liked the idea of a single notification component, but was concerned about impacting the overall level of coupling between components. Addison agreed that this might be an issue and investigated this trade-off further.

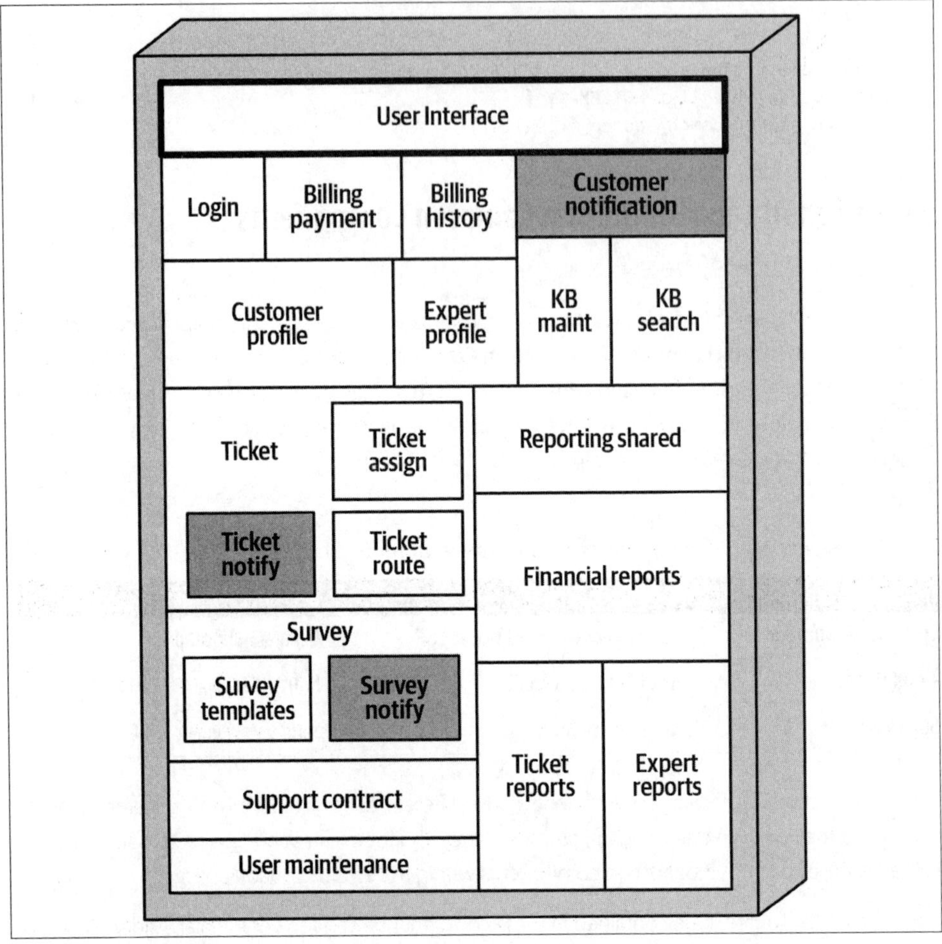

Figure 5-4. Notification functionality is duplicated throughout the application

Addison analyzed the incoming (afferent) coupling level for the existing Sysops Squad notification components and came up with the resulting coupling metrics listed in Table 5-5, with "CA" representing the number of other components requiring that component (afferent coupling).

Table 5-5. Sysops Squad coupling analysis before component consolidation

Component	CA	Used by
Customer Notification	2	Billing Payment, Support Contract
Ticket Notify	2	Ticket, Ticket Route
Survey Notify	1	Survey

Addison then found that if the customer notification functionality was consolidated into a single component, the coupling level for the resulting single component increased to an incoming coupling level of 5, as shown in Table 5-6.

Table 5-6. Sysops Squad coupling analysis after component consolidation

Component	CA	Used by
Notification	5	Billing Payment, Support Contract, Ticket, Ticket Route, Survey

Addison brought these findings to Austen, and they discussed the results. What they found is that, while the new consolidated component had a fairly high level of incoming coupling, it didn't affect the overall afferent (incoming) coupling level for notifying a customer. In other words, the three separate components had a total incoming coupling level of 5, but so did the single consolidated component.

Addison and Austen both realized how important it was to analyze the coupling level after consolidating common domain functionality. In some cases, combining common domain functionality into a single consolidated component increased the incoming coupling level of that component, thus resulting in too many dependencies on a single shared component within the application. However, in this case both Addison and Austen were comfortable with the coupling analysis, and agreed to consolidate the notification functionality to reduce the duplication of both code and functionality.

Addison wrote an architecture story to combine all of the notification functionality into a single namespace representing a common Notification component. Sydney, assigned to the architecture story, refactored the source code, creating a single component for customer notification, as illustrated in Figure 5-5.

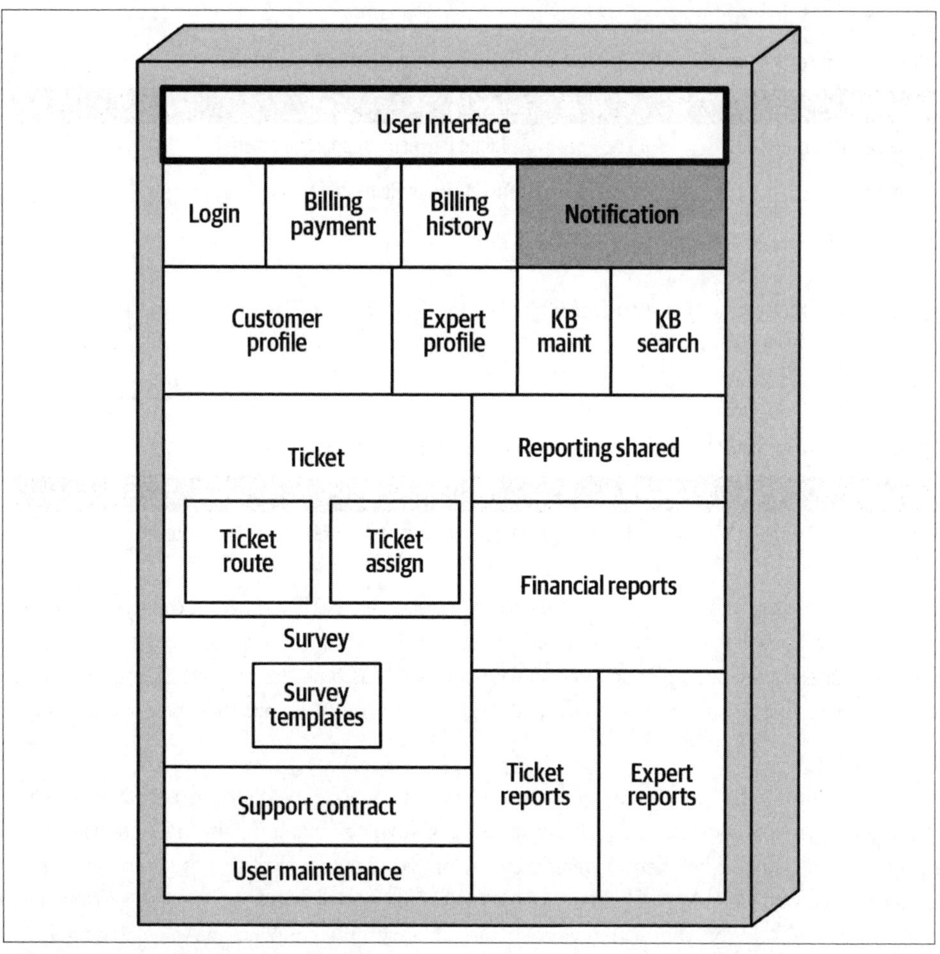

Figure 5-5. Notification functionality is consolidated into a new single component called Notification

Table 5-7 shows the resulting components after Sydney implemented the architecture story Addison created. Notice that the Customer Notification component (`ss.customer.notification`), Ticket Notify component (`ss.ticket.notify`), and Survey Notify components (`ss.survey.notify`) were removed, and the source code moved to the new consolidated Notification component (`ss.notification`).

Table 5-7. Sysops Squad components after applying the Gather Common Domain Components pattern

Component	Namespace	Responsibility
Login	ss.login	User and customer login
Billing Payment	ss.billing.payment	Customer monthly billing
Billing History	ss.billing.history	Payment history
Customer Profile	ss.customer.profile	Maintain customer profile
Expert Profile	ss.expert.profile	Maintain expert profile
KB Maint	ss.kb.maintenance	Maintain & view knowledge base
KB Search	ss.kb.search	Search knowledge base
Notification	ss.notification	**All customer notification**
Reporting Shared	ss.reporting.shared	Shared functionality
Ticket Reports	ss.reporting.tickets	Create ticketing reports
Expert Reports	ss.reporting.experts	Create expert reports
Financial Reports	ss.reporting.financial	Create financial reports
Ticket	ss.ticket	Ticket creation & maintenance
Ticket Assign	ss.ticket.assign	Assign expert to ticket
Ticket Route	ss.ticket.route	Send ticket to expert
Support Contract	ss.supportcontract	Support contract maintenance
Survey	ss.survey	Send and receive surveys
Survey Templates	ss.survey.templates	Maintain survey templates
User Maintenance	ss.users	Maintain internal users

Flatten Components Pattern

As mentioned previously, components—the building blocks of an application—are usually identified through namespaces, package structures, or directory structures and are implemented through class files (or source code files) contained within these structures. However, when components are built on top of other components, which are in turn built on top of other components, they start to lose their identity and stop becoming components as per our definition. The *Flatten Components* pattern is used to ensure that components are not built on top of one another, but rather flattened and represented as leaf nodes in a directory structure or namespace.

Pattern Description

When the namespace representing a particular component gets extended (in other words, another node is added to the namespace or directory structure), the prior namespace or directory no longer represents a component, but rather a subdomain. To illustrate this point, consider the customer survey functionality within the Sysops Squad application represented by two components: Survey (`ss.survey`) and Survey Templates (`ss.survey.templates`). Notice in Table 5-8 how the `ss.survey` namespace, which contains five class files used to manage and collect the surveys, is extended with the `ss.survey.templates` namespace to include seven classes representing each survey type send out to customers.

Table 5-8. The Survey component contains orphaned classes and should be flattened

Component name	Component namespace	Files
→ Survey	`ss.survey`	5
Survey Templates	`ss.survey.templates`	7

While this structure might seem to make sense from a developer's standpoint in order to keep the template code separate from survey processing, it does create some problems because Survey Templates, as a component, would be considered part of the Survey component. One might be tempted to consider Survey Templates as a *subcomponent* of Survey, but then issues arise when trying to form services from these components—should both components reside in a single service called Survey, or should the Survey Templates be a separate service from the Survey service?

We've resolved this dilemma by defining a component as the last node (or leaf node) of the namespace or directory structure. With this definition, `ss.survey.templates` is a component, whereas `ss.survey` would be considered a *subdomain*, not a component. We further define namespaces such as `ss.survey` as *root namespaces* because they are extended with other namespace nodes (in this case, `.templates`).

Notice how the `ss.survey` root namespace in Table 5-8 contains five class files. We call these class files *orphaned classes* because they do not belong to any definable component. Recall that a component is identified by a leaf node namespace containing source code. Because the `ss.survey` namespace was extended to include `.templates`, `ss.survey` is no longer considered a component and therefore should not contain any class files.

The following terms and corresponding definitions are important for understanding and applying the Flatten Components decomposition pattern:

Component
: A collection of classes grouped within a *leaf node namespace* that performs some sort of specific functionality in the application (such as payment processing or customer survey functionality).

Root namespace
: A namespace node that has been extended by another namespace node. For example, given the namespaces `ss.survey` and `ss.survey.templates`, `ss.survey` would be considered a root namespace because it is extended by `.templates`. Root namespaces are also sometimes referred to as *subdomains*.

Orphaned classes
: Classes contained within a root namespace, and hence have no definable component associated with them.

These definitions are illustrated in Figure 5-6, where the box with a C represents source code contained within that namespace. This diagram (and all others like it) are purposely drawn from the bottom up to emphasize the notion of *hills* in the application, as well as emphasize the notion of namespaces *building upon* each other.

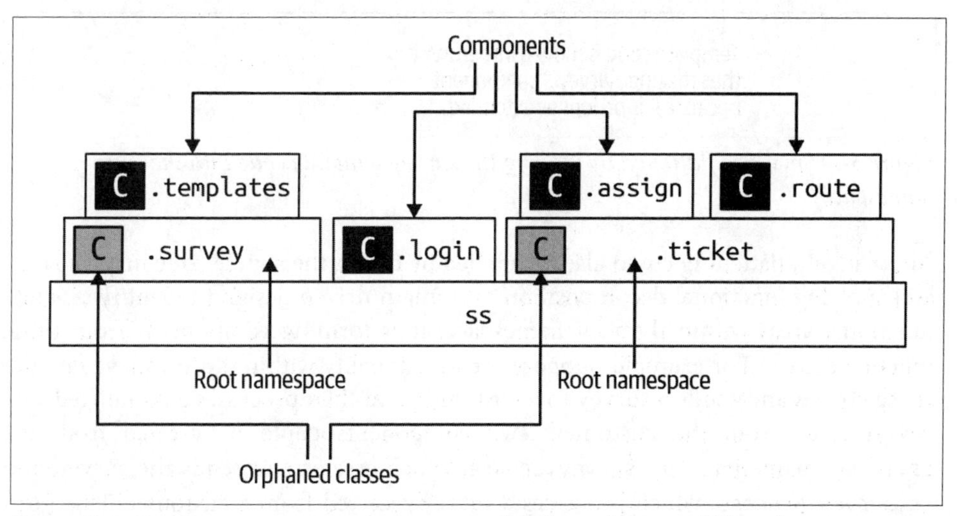

Figure 5-6. Components, root namespaces, and orphaned classes (C box denotes source code)

Notice that since both `ss.survey` and `ss.ticket` are extended through other namespace nodes, those namespaces are considered root namespaces, and the classes contained in those root namespaces are hence orphaned classes (belonging to no defined component). Thus, the only components denoted in Figure 5-6 are `ss.survey.templates`, `ss.login`, `ss.ticket.assign`, and `ss.ticket.route`.

The Flatten Components decomposition pattern is used to move orphaned classes to create well-defined components that exist only as leaf nodes of a directory or namespace, creating well-defined subdomains (root namespaces) in the process. We refer to the *flattening* of components as the breaking down (or building up) of namespaces within an application to remove orphaned classes. For example, one way of flattening the `ss.survey` root namespace in Figure 5-6 and remove orphaned classes is to move the source code contained in the `ss.survey.templates` namespace down to the `ss.survey` namespace, thereby making `ss.survey` a single component (`.survey` is now the leaf node of that namespace). This flattening option is illustrated in Figure 5-7.

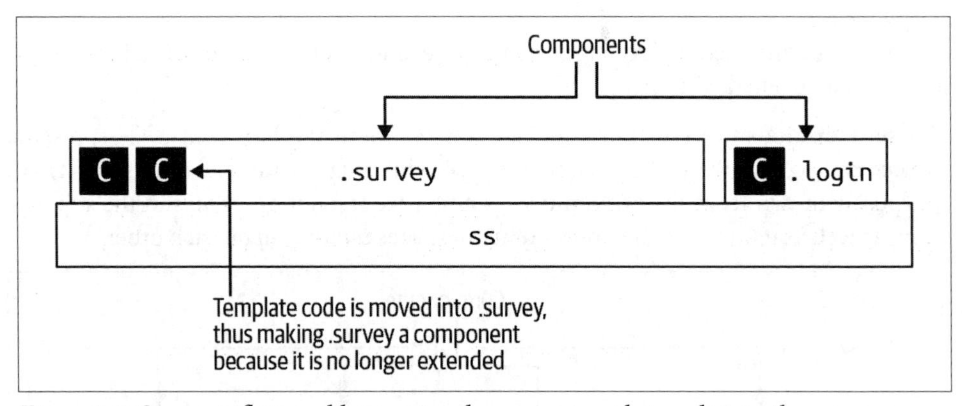

Figure 5-7. Survey is flattened by moving the survey template code into the .survey namespace

Alternatively, flattening could also be applied by taking the source code in `ss.survey` and applying functional decomposition or domain-driven design to identify separate functional areas within the root namespace, thus forming components from those functional areas. For example, suppose the functionality within the `ss.survey` namespace creates and sends a survey to a customer, and then processes a completed survey received from the customer. Two components could be created from the `ss.survey` namespace: `ss.survey.create`, which creates and sends the survey, and `ss.survey.process`, which processes a survey received from a customer. This form of flattening is illustrated in Figure 5-8.

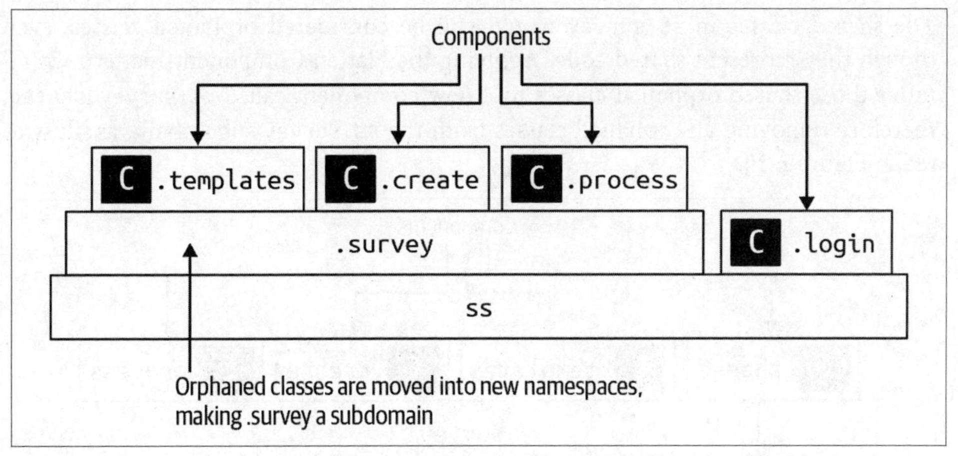

Figure 5-8. Survey is flattened by moving the orphaned classes to new leaf nodes (components)

Regardless of the direction of flattening, make sure source code files reside only in leaf node namespaces or directories so that source code can always be identified within a specific component.

Another common scenario where orphaned source code might reside in a root namespace is when code is shared by other components within that namespace. Consider the example in Figure 5-9 where customer survey functionality resides in three components (`ss.survey.templates`, `ss.survey.create`, and `ss.survey.process`), but common code (such as interfaces, abstract classes, common utilities) resides in the root namespace `ss.survey`.

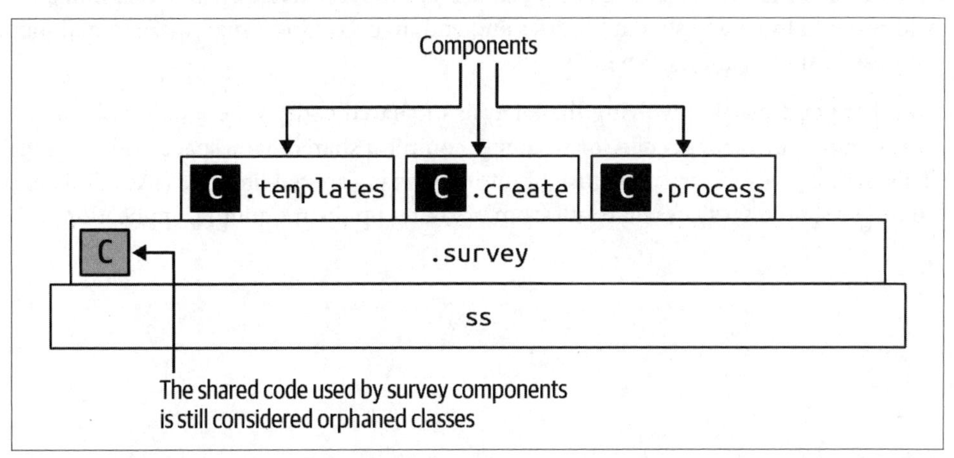

Figure 5-9. Shared code in .survey is considered orphaned classes and should be moved

The shared classes in `ss.survey` would still be considered orphaned classes, even though they represent shared code. Applying the Flatten Components pattern would move those shared orphaned classes to a new component called `ss.survey.shared`, therefore removing all orphaned classes from the `ss.survey` subdomain, as illustrated in Figure 5-10.

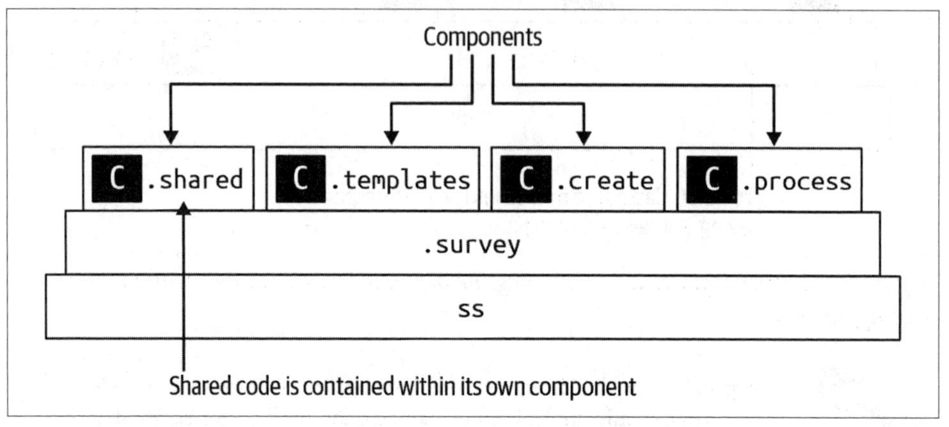

Figure 5-10. Shared survey code is moved into its own component

Our advice when moving shared code to a separate component (leaf node namespace) is to pick a word that is not used in any existing codebase in the domain, such as `.sharedcode`, `.commoncode`, or some such unique name. This allows the architect to generate metrics based on the number of shared components in the codebase, as well as the *percentage* of source code that is shared in the application. This is a good indicator as to the feasibility of breaking up the monolithic application. For example, if the sum of all the statements in all namespaces ending with `.sharedcode` constitutes 45% of the overall source code, chances are moving to a distributed architecture will result in too many shared libraries and end up becoming a nightmare to maintain because of shared library dependencies.

Another good metric involving the analysis of shared code is the *number* of components ending in `.sharedcode` (or whatever common shared namespace node is used). This metric gives the architect insight into how many shared libraries (JAR, DLL, and so on) or shared services will result from breaking up the monolithic application.

Fitness Functions for Governance

Applying the Flatten Components decomposition pattern involves a fair amount of subjectivity. For example, should code from leaf nodes be consolidated into the root namespace, or should code in a root namespace be moved into leaf nodes? That said, the following fitness function can assist in automating the governance of keeping components flat (only in leaf nodes).

Fitness function: No source code should reside in a root namespace

This automated holistic fitness function can be triggered on deployment through a CI/CD pipeline to locate orphaned classes—classes that reside in a root namespace. Use of this fitness function helps keep components flat when undergoing a monolithic migration, especially when performing ongoing maintenance to the monolithic application during the migration effort. Example 5-6 shows the pseudocode that alerts an architect when orphaned classes appear anywhere in the codebase.

Example 5-6. Pseudocode for finding code in root namespaces

```
# Walk the directory structure, creating namespaces for each complete path
LIST component_list = identify_components(root_directory)

# Send an alert if a non-leaf node in any component contains source files
FOREACH component IN component_list {
  LIST component_node_list = get_nodes(component)
  FOREACH node IN component_node_list {
    IF contains_code(node) AND NOT last_node(component_node_list) {
      send_alert(component)
    }
  }
}
```

Sysops Squad Saga: Flattening Components

`Wednesday, November 10, 11:10`

After applying the "Gather Common Domain Components Pattern" on page 94, Addison analyzed the results in Table 5-7 and observed that the Survey and Ticket components contained orphaned classes. Addison highlighted these components in Table 5-9 and in Figure 5-11.

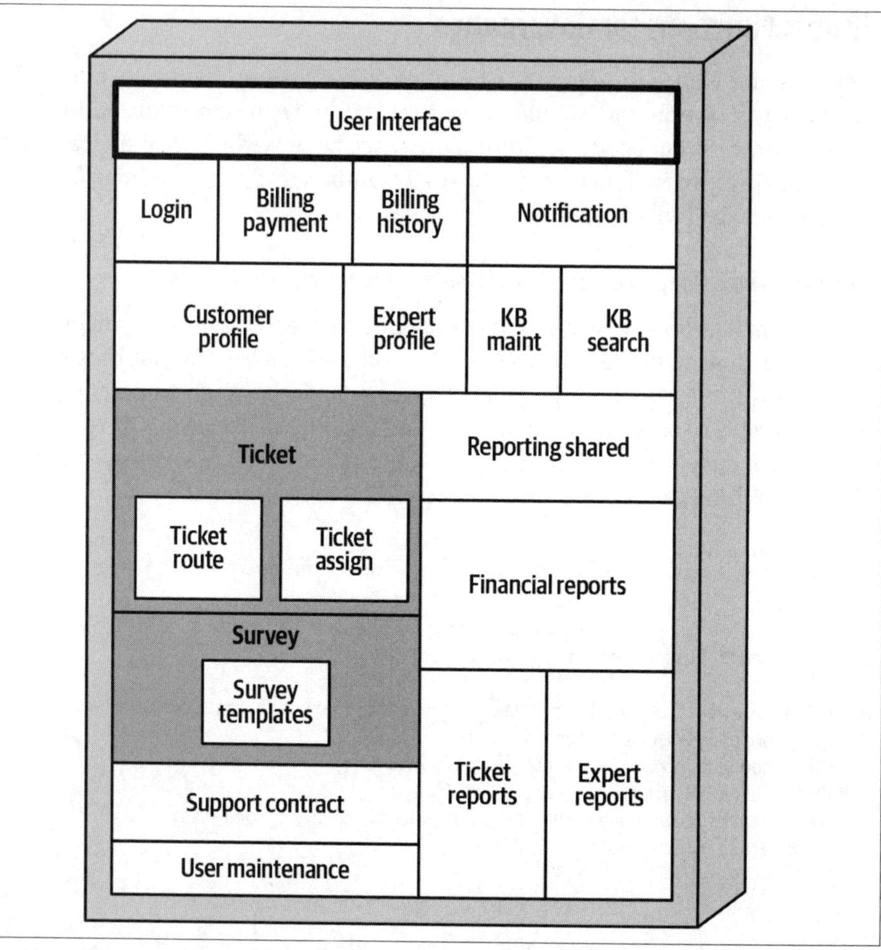

Figure 5-11. The Survey and Ticket components contain orphaned classes and should be flattened

Table 5-9. Sysops Squad Ticket and Survey components should be flattened

Component name	Component namespace	Statements	Files
Ticket	`ss.ticket`	**7,009**	**45**
Ticket Assign	`ss.ticket.assign`	7,845	14
Ticket Route	`ss.ticket.route`	1,468	4
Survey	`ss.survey`	**2,204**	**5**
Survey Templates	`ss.survey.templates`	1,672	7

Addison decided to address the ticketing components first. Knowing that flattening components meant getting rid of source code in nonleaf nodes, Addison had two choices: consolidate the code contained in the ticket assignment and ticket routing components into the `ss.ticket` component, or break up the 45 classes in the `ss.ticket` component into separate components, thus making `ss.ticket` a subdomain. Addison discussed these options with Sydney (one of the Sysops Squad developers), and based on the complexity and frequent changes in the ticket assignment functionality, decided to keep those components separate and move the orphaned code from the `ss.ticket` root namespace into other namespaces, thus forming new components.

With help from Sydney, Addison found that the 45 orphaned classes contained in the `ss.ticket` namespace implemented the following ticketing functionality:

- Ticket creation and maintenance (creating a ticket, updating a ticket, canceling a ticket, etc.)
- Ticket completion logic
- Shared code common to most of the ticketing functionality

Since ticket assignment and ticket routing functionality were already in their own components (`ss.ticket.assign` and `ss.ticket.route`, respectively), Addison created an architecture story to move the source code contained in the `ss.ticket` namespace to three new components, as shown in Table 5-10.

Table 5-10. The prior Sysops Squad Ticket component broken into three new components

Component	Namespace	Responsibility
Ticket Shared	`ss.ticket.shared`	Common code and utilities
Ticket Maintenance	`ss.ticket.maintenance`	Add and maintain tickets
Ticket Completion	`ss.ticket.completion`	Complete ticket and initiate survey
Ticket Assign	`ss.ticket.assign`	Assign expert to ticket
Ticket Route	`ss.ticket.route`	Send ticket to expert

Addison then considered the survey functionality. Working with Sydney, Addison found that the survey functionality rarely changed and was not overly complicated. Sydney talked with Skyler, the Sysops Squad developer who originally created the `ss.survey.templates` namespace, and found there was no compelling reason to separate the survey templates into their own namespace ("It just seemed like a good idea at the time," said Skyler). With this information, Addison created an architecture story to move the seven class files from `ss.survey.templates` into the `ss.survey` namespace and removed the `ss.survey.template` component, as shown in Table 5-11.

Table 5-11. The prior Sysops Squad Survey components flattened into a single component

Component	Namespace	Responsibility
Survey	`ss.survey`	Send and seceive surveys

After applying the Flatten Components pattern (illustrated in Figure 5-12), Addison observed that there were no "hills" (component upon component) or orphaned classes and that all of the components were contained only in the leaf nodes of the corresponding namespace.

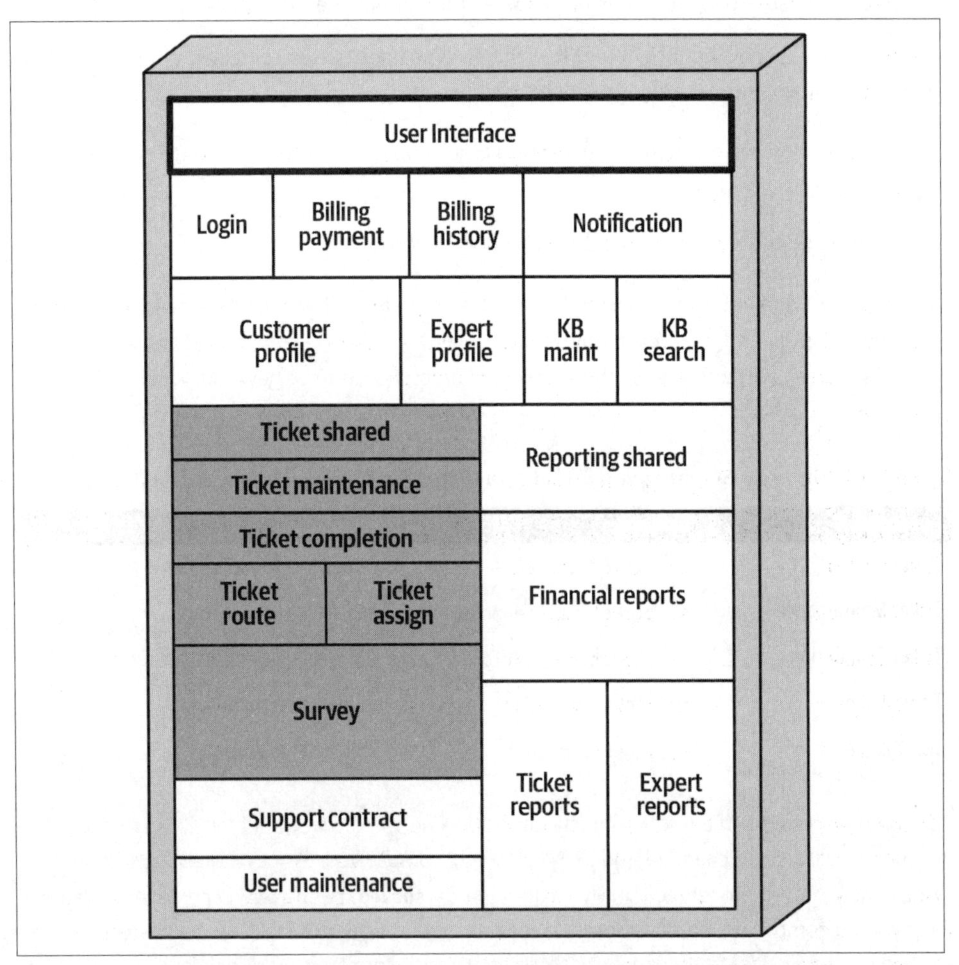

Figure 5-12. The Survey component was flattened into a single component, whereas the Ticket component was raised up and flattened, creating a Ticket subdomain

Addison recorded the results of the refactoring efforts thus far in applying these decomposition patterns and listed them in Table 5-12.

Table 5-12. Sysops Squad components after applying the Flatten Components pattern

Component	Namespace
Login	ss.login
Billing Payment	ss.billing.payment
Billing History	ss.billing.history
Customer Profile	ss.customer.profile
Expert Profile	ss.expert.profile
KB Maint	ss.kb.maintenance
KB Search	ss.kb.search
Notification	ss.notification
Reporting Shared	ss.reporting.shared
Ticket Reports	ss.reporting.tickets
Expert Reports	ss.reporting.experts
Financial Reports	ss.reporting.financial
Ticket Shared	ss.ticket.shared
Ticket Maintenance	ss.ticket.maintenance
Ticket Completion	ss.ticket.completion
Ticket Assign	ss.ticket.assign
Ticket Route	ss.ticket.route
Support Contract	ss.supportcontract
Survey	ss.survey
User Maintenance	ss.users

Determine Component Dependencies Pattern

Three of the most common questions asked when considering a migration from a monolithic application to a distributed architecture are as follows:

1. Is it feasible to break apart the existing monolithic application?
2. What is the rough overall level of effort for this migration?
3. Is this going to require a rewrite of the code or a refactoring of the code?

One of your authors was engaged several years ago in a large migration effort to move a complex monolithic application to microservices. On the first day of the project, the CIO wanted to know only one thing—was this migration effort a golfball, basketball, or an airliner? Your author was curious about the sizing comparisons, but the CIO insisted that the answer to this simple question shouldn't be that difficult given that kind of coarse-grained sizing. As it turned out, applying the *Determine Component Dependencies* pattern quickly and easily answered this question for the CIO—the effort was unfortunately an airliner, but only a small Embraer 190 migration rather than a large Boeing 787 Dreamliner migration.

Pattern Description

The purpose of the *Determine Component Dependencies* pattern is to analyze the incoming and outgoing dependencies (coupling) between *components* to determine what the resulting service dependency graph might look like after breaking up the monolithic application. While there are many factors in determining the right level of granularity for a service (see Chapter 7), each component in the monolithic application is potentially a service candidate (depending on the target distributed architecture style). For this reason, it is critical to understand the interactions and dependencies between components.

It's important to note that this pattern is about component dependencies, not individual class dependencies within a component. A *component dependency* is formed when a class from one component (namespace) interacts with a class from another component (namespace). For example, suppose the `CustomerSurvey` class in the `ss.survey` component invokes a method in the `CustomerNotification` class in the `ss.notification` component to send out the customer survey, as illustrated in the pseudocode in Example 5-7.

Example 5-7. Pseudocode showing a dependency between the Survey and Notification components

```
namespace ss.survey
class CustomerSurvey {
   function createSurvey {
      ...
   }

   function sendSurvey {
      ...
      ss.notification.CustomerNotification.send(customer_id, survey)
   }
}
```

Notice the dependency between the Survey and Notification components, because the `CustomerNotification` class used by the `CustomerSurvey` class resides outside the `ss.survey` namespace. Specifically, the Survey component would have an efferent (or outgoing) dependency on the Notification component, and the Notification component would have an afferent (or incoming) dependency on the Survey component.

Note that the classes *within* a particular component may be a highly coupled mess of numerous dependencies, but that doesn't matter when applying this pattern—what matters is only those dependencies *between* components.

Several tools (*https://oreil.ly/XyIgr*) are available that can assist in applying this pattern and visualizing component dependencies. In addition, many modern IDEs have plug-ins that will produce dependency diagrams of the components, or namespaces, within a particular codebase. These visualizations can be useful in answering the three key questions posed at the start of this section.

For example, consider the dependency diagram shown in Figure 5-13, where the boxes represent components (not classes), and the lines represent coupling points between the components. Notice there is only a single dependency between the components in this diagram, making this application a good candidate for breaking apart since the components are functionally independent from one another.

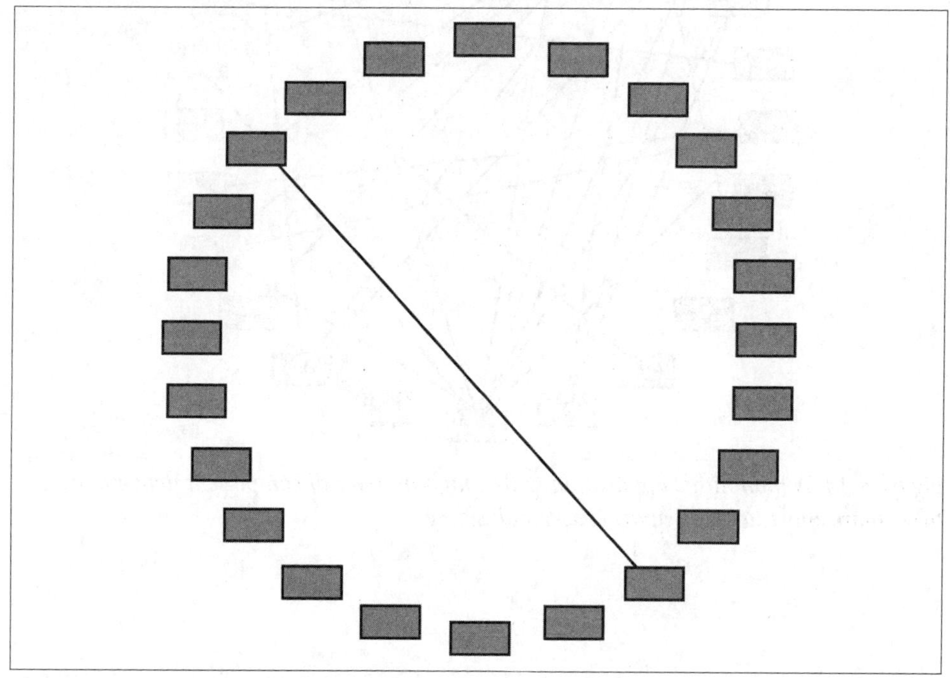

Figure 5-13. A monolithic application with minimal component dependencies takes less effort to break apart (golf ball sizing)

With a dependency diagram like Figure 5-13, the answers to the three key questions are as follows:

1. Is it feasible to break apart the existing monolithic application? *Yes*
2. What is the rough overall level of effort for this migration? *A golf ball (relatively straightforward)*
3. Is this going to be a rewrite of the code or a refactoring of the code? *Refactoring (moving existing code into separately deployed services)*

Now look at the dependency diagram shown in Figure 5-14. Unfortunately, this diagram is typical of the dependencies between components in most business applications. Notice in particular how the lefthand side of this diagram has the highest level of coupling, whereas the righthand side looks much more feasible to break apart.

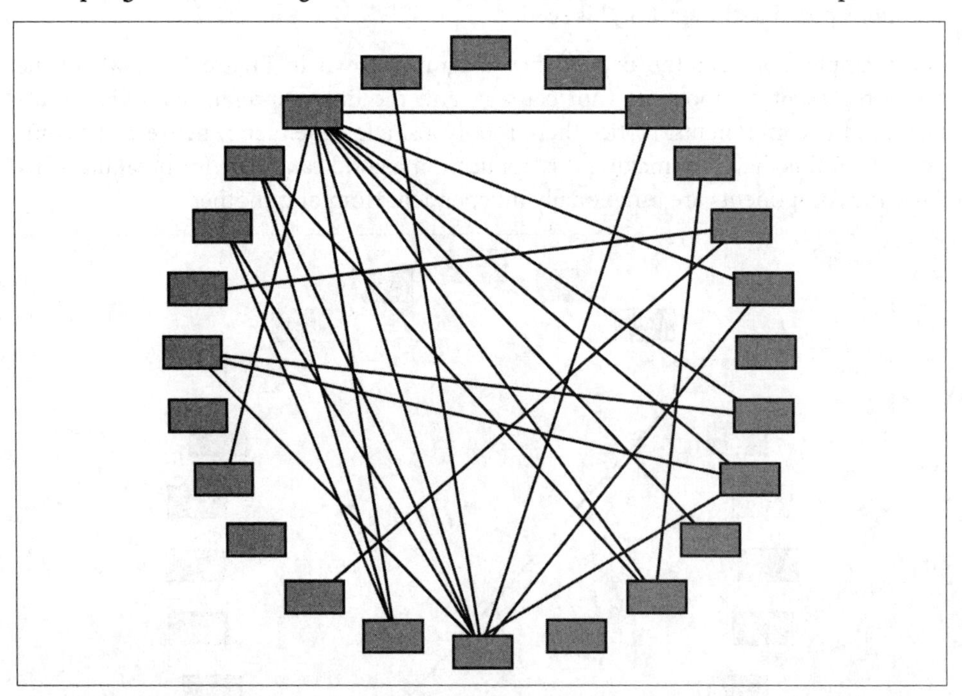

Figure 5-14. A monolithic application with a high number of component dependencies takes more effort to break apart (basketball sizing)

With this level of tight coupling between components, the answers to the three key questions are not very encouraging:

1. Is it feasible to break apart the existing monolithic application? *Maybe...*
2. What is the rough overall level of effort for this migration? *A basketball (much harder)*
3. Is this going to be a rewrite of the code or a refactoring of the code? *Likely a combination of some refactoring and some rewriting of the existing code*

Finally, consider the dependency diagram illustrated in Figure 5-15. In this case, the architect should turn around and run in the opposite direction as fast as they can!

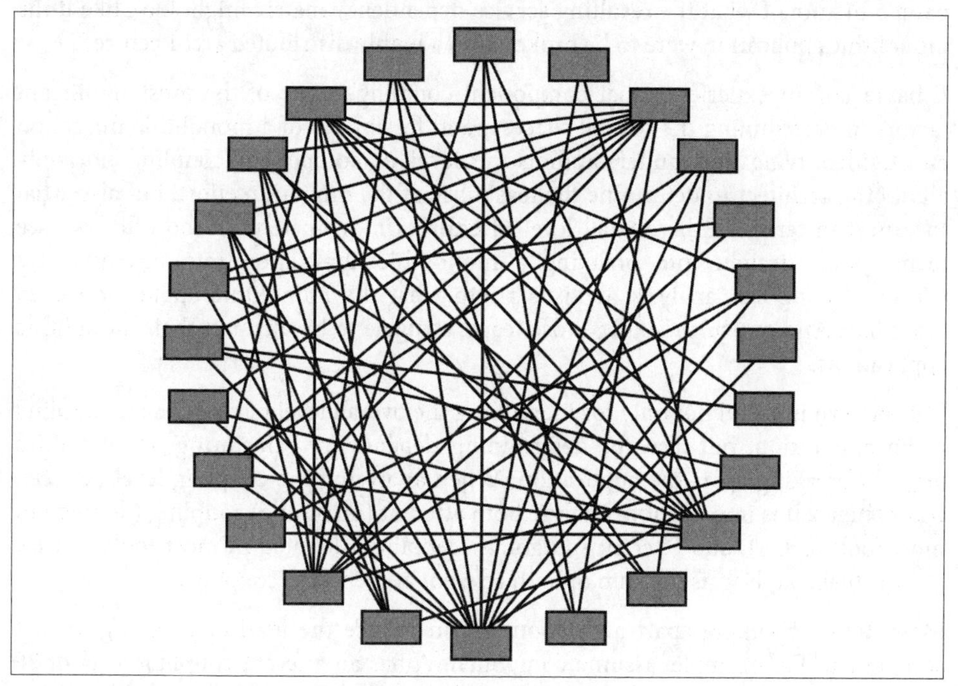

Figure 5-15. A monolithic application with too many component dependencies is not feasible to break apart (airliner sizing)

The answers to the three key questions for applications with this sort of component dependency matrix are not surprising:

1. Is it feasible to break apart the existing monolithic application? *No*
2. What is the rough overall level of effort for this migration? *An airliner*
3. Is this going to be a rewrite of the code or a refactoring of the code? *Total rewrite of the application*

We cannot stress enough the importance of these kinds of visual diagrams when breaking apart a monolithic application. In essence these diagrams form a *radar* from which to determine where the enemy (high component coupling) is located, and also paint a picture of what the resulting service dependency matrix might look like if the monolithic application were to be broken into a highly distributed architecture.

It has been our experience that component coupling is one of the most significant factors in determining the overall success (and feasibility) of a monolithic migration effort. Identifying and understanding the level of component coupling not only allows the architect to determine the feasibility of the migration effort, but also what to expect in terms of the overall level of effort. Unfortunately, all too often we see teams jump straight into breaking a monolithic application into microservices without having any analysis or visuals into what the monolithic application even looks like. And not surprisingly, those teams struggle to break apart their monolithic applications.

This pattern is useful not only for identifying the overall level of component coupling in an application, but also for determining dependency refactoring opportunities prior to breaking apart the application. When analyzing the coupling level between components, it is important to analyze both afferent (incoming) coupling (denoted in most tools as *CA*), and efferent (outgoing) coupling (denoted in most tools as *CE*). *CT*, or total coupling, is the sum of both afferent and efferent coupling.

Many times, breaking apart a component can reduce the level of coupling of that component. For example, assume component A has an afferent coupling level of 20 (meaning, 20 other components are dependent on the functionality of the component). This does not necessarily mean that all 20 of the other components require *all* of the functionality from component A. Maybe 14 of the other components require only a small part of the functionality contained in component A. Breaking component A into two different components (component A1 containing the smaller, coupled functionality, and component A2 containing the majority of the functionality) reduces the afferent coupling in component A2 to 6, with component A1 having an afferent coupling level of 14.

Fitness Functions for Governance

Two ways to automate the governance for component dependencies are to make sure no component has "too many" dependencies, and to restrict certain components from being coupled to other components. The fitness functions described next are some ways of governing these type of dependencies.

Fitness function: No component shall have more than <some number> of total dependencies

This automated holistic fitness function can be triggered on deployment through a CI/CD pipeline to make sure that the coupling level of any given component doesn't exceed a certain threshold. It is up to the architect to determine that this maximum threshold should be based on the overall level of coupling within the application and the number of components. An alert generated from this fitness function allows the architect to discuss any sort of increase in coupling with the development team, possibly promoting action to break apart components to reduce coupling. This fitness function could also be modified to generate an alert for a threshold limit of incoming only, outgoing only, or both (as separate fitness functions). Example 5-8 shows the pseudocode for sending an alert if the total coupling (incoming and outgoing) exceeds a combined level of 15, which for most applications would be considered relatively high.

Example 5-8. Pseudocode for limiting the total number of dependencies of any given component

```
# Walk the directory structure, gathering components and the source code files
# contained within those components
LIST component_list = identify_components(root_directory)
MAP component_source_file_map
FOREACH component IN component_list {
  LIST component_source_file_list = get_source_files(component)
  ADD component, component_source_file_list TO component_source_file_map
}

# Determine how many references exist for each source file and send an alert if
# the total dependency count is greater than 15
FOREACH component,component_source_file_list IN component_source_file_map {
  FOREACH source_file IN component_source_file_list {
    incoming count = used_by_other_components(source_file, component_source_file_map) {
    outgoing_count = uses_other_components(source_file) {
    total_count = incoming count + outgoing count
  }
  IF total_count > 15 {
    send_alert(component, total_count)
  }
}
```

Fitness function: <some component> should not have a dependency on <another component>

This automated holistic fitness function can be triggered on deployment through a CI/CD pipeline to restrict certain components from having a dependency on other ones. In most cases, there will be one fitness function for each dependency restriction so that, if there were 10 different component restrictions, there would be 10 different fitness functions, one for each component in question. Example 5-9 shows an example using ArchUnit (*https://www.archunit.org*) for ensuring that the Ticket Maintenance component (`ss.ticket.maintenance`) does not have a dependency on the Expert Profile component (`ss.expert.profile`).

Example 5-9. ArchUnit code for governing dependency restrictions between components

```
public void ticket_maintenance_cannot_access_expert_profile() {
   noClasses().that()
   .resideInAPackage("..ss.ticket.maintenance..")
   .should().accessClassesThat()
   .resideInAPackage("..ss.expert.profile..")
   .check(myClasses);
}
```

Sysops Squad Saga: Identifying Component Dependencies

`Monday, November 15, 09:45`

After reading about the Determine Component Dependencies pattern, Addison wondered what the Sysops Squad application dependency matrix looked like and whether it was feasible to even break the application apart. Addison used an IDE plug-in to generate a component dependency diagram of the current Sysops Squad application. Initially, Addison felt a bit discouraged because Figure 5-16 showed a lot of dependencies between the Sysops Squad application components.

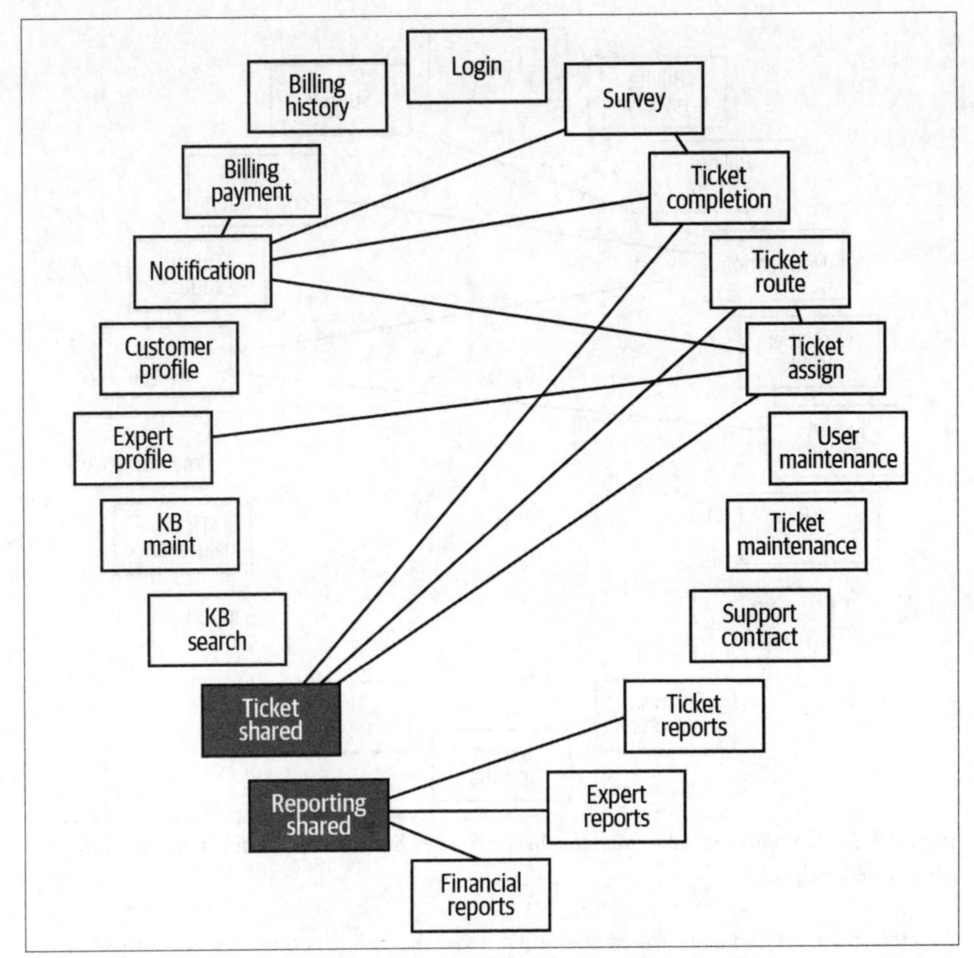

Figure 5-16. Component dependencies in the Sysops Squad application

However, after further analysis, Addison saw that the Notification component had the most dependencies, which was not surprising given that it's a shared component. However, Addison also saw lots of dependencies within the Ticketing and Reporting components. Both of these domain areas have a specific component for shared code (interfaces, helper classes, entity classes, and so on). Realizing that both the ticketing and reporting shared code contains mostly compile-based class references and would likely be implemented as shared libraries rather than services, Addison filtered out these components to get a better view of the dependencies between the core functionality of the application, which is illustrated in Figure 5-17.

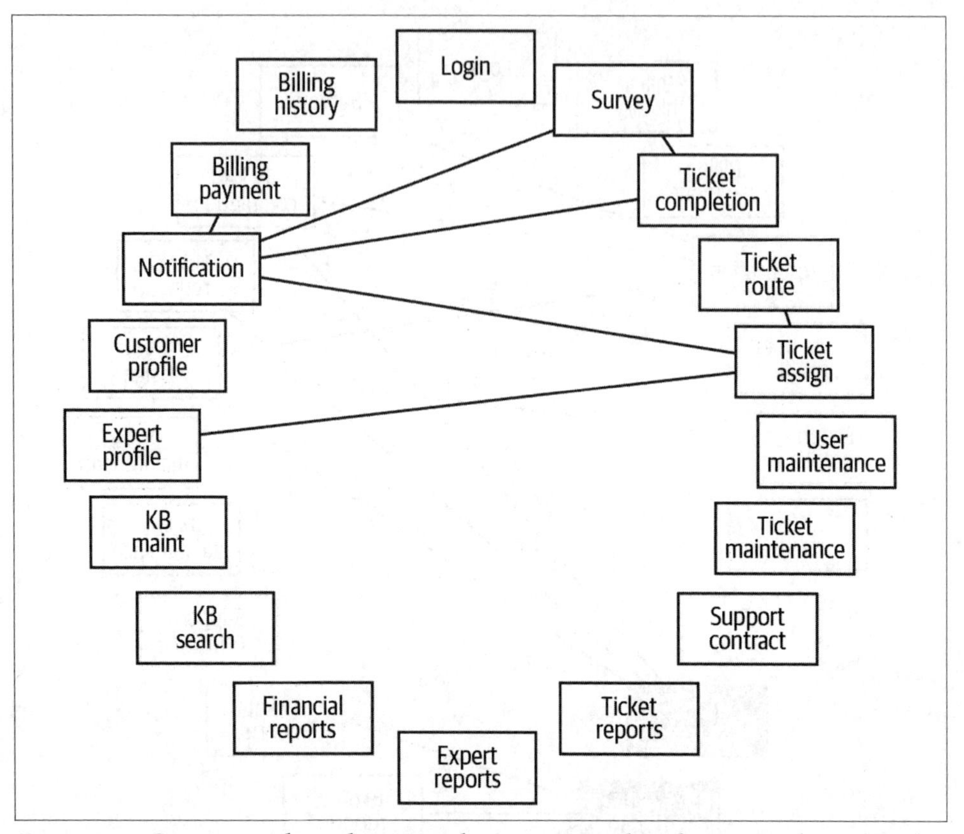

Figure 5-17. Component dependencies in the Sysops Squad application without shared library dependencies

With the shared components filtered out, Addison saw that the dependencies were fairly minimal. Addison showed these results to Austen, and they both agreed that most of the components were relatively self-contained and it appeared that the Sysops Squad application was a good candidate for breaking apart into a distributed architecture.

Create Component Domains Pattern

While each component identified within a monolithic application can be considered a possible candidate for a separate service, in most cases the relationship between a service and components is a one-to-many relationship—that is, a single service may contain one or more components. The purpose of the *Create Component Domains* pattern is to logically group components together so that more coarse-grained domain services can be created when breaking up an application.

Pattern Description

Identifying component domains—the grouping of components that perform some sort of related functionality—is a critical part of breaking apart any monolithic application. Recall the advice from Chapter 4:

> *When breaking apart monolithic applications, consider first moving to service-based architecture as a stepping-stone to other distributed architectures.*

Creating component domains is an effective way of determining what will eventually become domain services in a service-based architecture.

Component domains are physically manifested in an application through component namespaces (or directories). Because namespace nodes are hierarchical in nature, they become an excellent way of representing the domains and subdomains of functionality. This technique is illustrated in Figure 5-18, where the second node in the namespace (`.customer`) refers to the *domain*, the third node represents a *subdomain* under the customer domain (`.billing`), and the leaf node (`.payment`) refers to the *component*. The `.MonthlyBilling` at the end of this namespace refers to a class file contained within the Payment component.

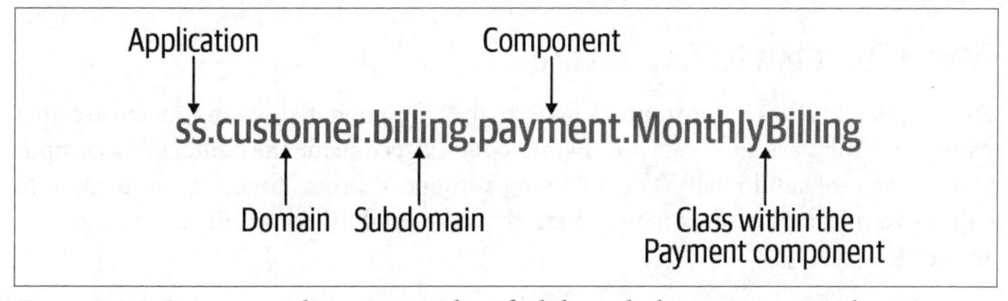

Figure 5-18. Component domains are identified through the namespace nodes

Since many older monolithic applications were implemented prior to the widespread use of domain-driven design (*https://oreil.ly/AaKR2*), in many cases refactoring of the namespaces is needed to structurally identify domains within the application. For example, consider the components listed in Table 5-13 that make up the Customer domain within the Sysops Squad application.

Table 5-13. Components related to the Customer domain before refactoring

Component	Namespace
Billing Payment	`ss.billing.payment`
Billing History	`ss.billing.history`
Customer Profile	`ss.customer.profile`
Support Contract	`ss.supportcontract`

Notice how each component is related to customer functionality, but the corresponding namespaces don't reflect that association. To properly identify the Customer domain (manifested through the namespace `ss.customer`), the namespaces for the Billing Payment, Billing History, and Support Contract components would have to be modified to add the `.customer` node at the beginning of the namespace, as shown in Table 5-14.

Table 5-14. Components related to the Customer domain after refactoring

Component	Namespace
Billing Payment	`ss.customer.billing.payment`
Billing History	`ss.customer.billing.history`
Customer Profile	`ss.customer.profile`
Support Contract	`ss.customer.supportcontract`

Notice in the prior table that all of the customer-related functionality (billing, profile maintenance, and support contract maintenance) is now grouped under `.customer`, aligning each component with that particular domain.

Fitness Functions for Governance

Once refactored, it's important to govern the component domains to ensure that namespace rules are enforced and that no code exists outside the context of a component domain or subdomain. The following automated fitness function can be used to help govern component domains once they are established within the monolithic application.

Fitness function: All namespaces under <root namespace node> should be restricted to <list of domains>

This automated holistic fitness function can be triggered on deployment through a CI/CD pipeline to restrict the domains contained within an application. This fitness function helps prevent additional domains from being inadvertently created by development teams and alerts the architect if any new namespaces (or directories) are created outside the approved list of domains. Example 5-10 shows an example using ArchUnit (*https://www.archunit.org*) for ensuring that only the ticket, customer, and admin domains exist within an application.

Example 5-10. ArchUnit code for governing domains within an application

```
public void restrict_domains() {
   classes()
      .should().resideInAPackage("..ss.ticket..")
      .orShould().resideInAPackage("..ss.customer..")
      .orShould().resideInAPackage("..ss.admin..")
      .check(myClasses);
}
```

Sysops Squad Saga: Creating Component Domains

`Thursday, November 18, 13:15`

Addison and Austen consulted with Parker, the Sysops Squad product owner, and together they identified five main domains within the application: a Ticketing domain (`ss.ticket`) containing all ticket-related functionality, including ticket processing, customer surveys, and knowledge base (KB) functionality; a Reporting domain (`ss.reporting`) containing all reporting functionality; a Customer domain (`ss.customer`) containing customer profile, billing, and support contracts; an Admin domain (`ss.admin`) containing maintenance of users and Sysops Squad experts; and finally, a Shared domain (`ss.shared`) containing login and notification functionality used by the other domains.

Addison created a domain diagram (see Figure 5-19) showing the various domains and corresponding groups of components within each domain, and was satisfied with this grouping as no component was left out, and there was good cohesion between the components within each domain.

The exercise Addison did in diagramming and grouping the components was an important one as it validated the identified domain candidates and also demonstrated the need for collaboration with business stakeholders (such as the product owner or business application sponsor). Had the components not lined up properly or Addison was left with components that didn't belong anywhere, more collaboration with Parker (the product owner) would have been necessary.

Satisfied that all of the components fit nicely into these domains, Addison then looked at the various component namespaces in Table 5-12 after applying the "Flatten Components Pattern" on page 101 and identified the component domain refactoring that needed to take place.

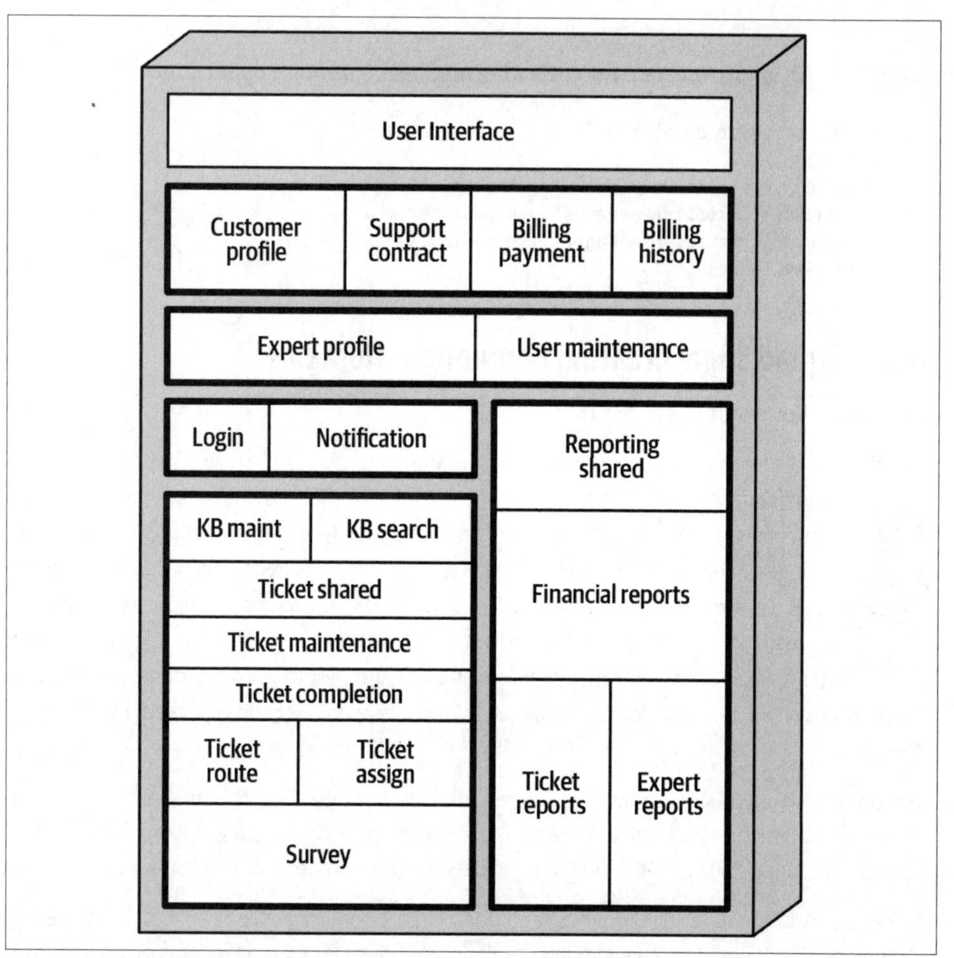

Figure 5-19. The five domains identified (with darkened borders) within the Sysops Squad application

Addison started with the Ticket domain and saw that while the core ticket functionality started with the namespace `ss.ticket`, the survey and knowledge base components did not. Therefore, Addison wrote an architecture story to refactor the components listed in Table 5-15 to align with the ticketing domain.

Table 5-15. Sysops Squad component refactoring for the Ticket domain

Component	Domain	Current namespace	Target namespace
KB Maint	Ticket	`ss.kb.maintenance`	`ss.ticket.kb.maintenance`
KB Search	Ticket	`ss.kb.search`	`ss.ticket.kb.search`
Ticket Shared	Ticket	`ss.ticket.shared`	Same (no change)
Ticket Maintenance	Ticket	`ss.ticket.maintenance`	Same (no change)
Ticket Completion	Ticket	`ss.ticket.completion`	Same (no change)
Ticket Assign	Ticket	`ss.ticket.assign`	Same (no change)
Ticket Route	Ticket	`ss.ticket.route`	Same (no change)
Survey	Ticket	`ss.survey`	`ss.ticket.survey`

Next Addison considered the customer-related components, and found that the billing and survey components needed to be refactored to include them under the Customer domain, creating a Billing subdomain in the process. Addison wrote an architecture story for the refactoring of the Customer domain functionality, shown in Table 5-16.

Table 5-16. Sysops Squad component refactoring for the Customer domain

Component	Domain	Current namespace	Target namespace
Billing Payment	Customer	`ss.billing.payment`	`ss.customer.billing.payment`
Billing History	Customer	`ss.billing.history`	`ss.customer.billing.history`
Customer Profile	Customer	`ss.customer.profile`	Same (no change)
Support Contract	Customer	`ss.supportcontract`	`ss.customer.supportcontract`

By applying the "Identify and Size Components Pattern" on page 84, Addison found that the reporting domain was already aligned, and no further action was needed with the reporting components listed in Table 5-17.

Table 5-17. Sysops Squad Reporting components are already aligned with the Reporting domain

Component	Domain	Current namespace	Target namespace
Reporting Shared	Reporting	`ss.reporting.shared`	Same (no change)
Ticket Reports	Reporting	`ss.reporting.tickets`	Same (no change)
Expert Reports	Reporting	`ss.reporting.experts`	Same (no change)
Financial Reports	Reporting	`ss.reporting.financial`	Same (no change)

Addison saw that both the Admin and Shared domains needed alignment as well, and decided to create a single architecture story for this refactoring effort and listed these components in Table 5-18. Addison also decided to rename the `ss.expert.profile` namespace to `ss.experts` to avoid an unnecessary Expert subdomain under the Admin domain.

Table 5-18. Sysops Squad component refactoring for the Admin and Shared domains

Component	Domain	Current namespace	Target namespace
Login	Shared	`ss.login`	`aa.shared.login`
Notification	Shared	`ss.notification`	`ss.shared.notification`
Expert Profile	Admin	`ss.expert.profile`	`ss.admin.experts`
User Maintenance	Admin	`ss.users`	`ss.admin.users`

With this pattern complete, Addison realized they were now prepared to structurally break apart the monolithic application and move to the first stage of a distributed architecture by applying the Create Domain Services pattern (described in the next section).

Create Domain Services Pattern

Once components have been properly sized, flattened, and grouped into domains, those domains can then be moved to separately deployed *domain services*, creating what is known as a service-based architecture (see Appendix A). Domain services are coarse-grained, separately deployed units of software containing all of the functionality for a particular domain (such as Ticketing, Customer, Reporting, and so on).

Pattern Description

The previous "Create Component Domains Pattern" on page 120 forms well-defined component domains within a monolithic application and manifests those domains through the component namespaces (or directory structures). This pattern takes those well-defined component domains and extracts those component groups into separately deployed services, known as a *domain services*, thus creating a service-based architecture.

In its simplest form, service-based architecture consists of a user interface that remotely accesses coarse-grained domain services, all sharing a single monolithic database. Although there are many topologies within service-based architecture (such as breaking up the user interface, breaking up the database, adding an API gateway, and so on), the basic topology shown in Figure 5-20 is a good starting point for migrating a monolithic application.

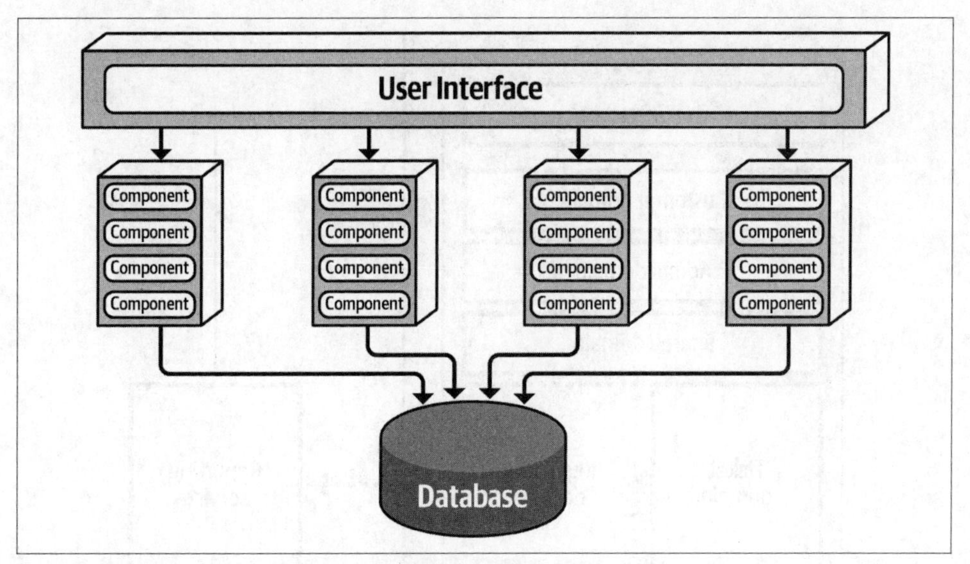

Figure 5-20. The basic topology for a service-based architecture

In addition to the benefits mentioned in "Component-Based Decomposition" on page 71, moving to service-based architecture first allows the architect and development team to learn more about each domain service to determine whether it should be broken into smaller services within a microservices architecture or left as a larger domain service. Too many teams make the mistake of starting out too fine-grained, and as a result must embrace all of the trappings of microservices (such as data decomposition, distributed workflows, distributed transactions, operational automation, containerization, and so on) without the need for all of those fine-grained microservices.

Figure 5-21 illustrates how the *Create Domain Services* pattern works. Notice in the diagram how the Reporting component domain defined in the "Create Component Domains Pattern" on page 120 is extracted from of the monolithic application, forming its own separately deployed Reporting service.

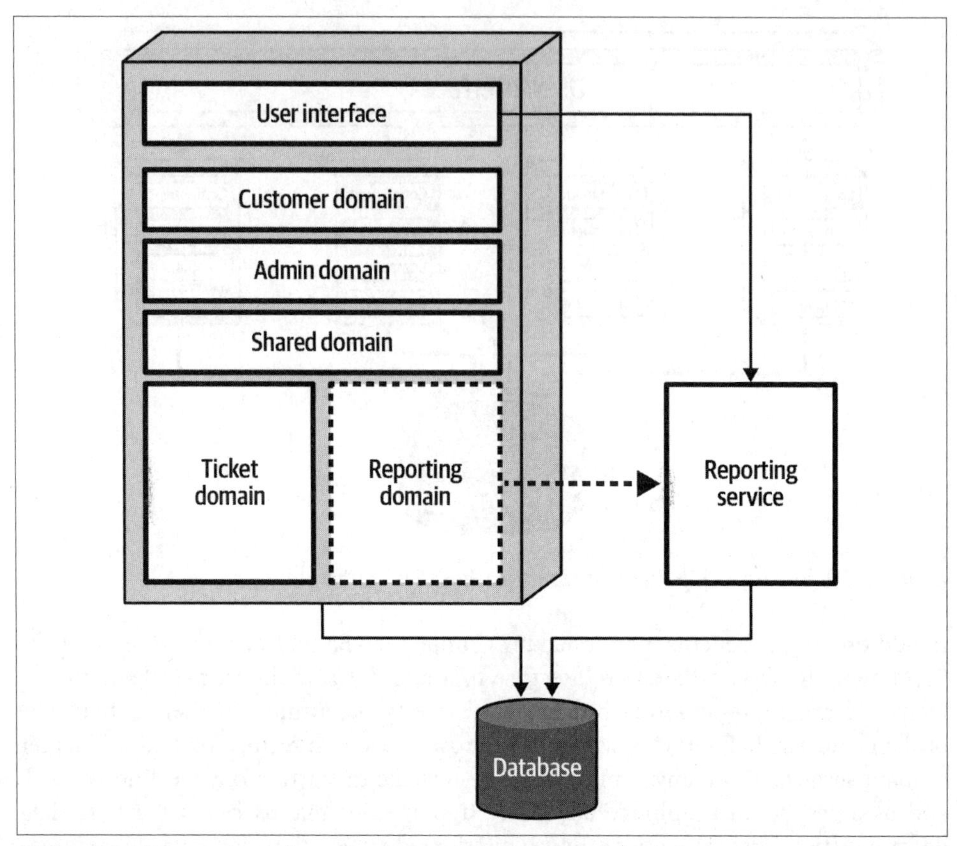

Figure 5-21. Component domains are moved to external domain services

A word of advice, however: don't apply this pattern until *all* of the component domains have been identified and refactored. This helps reduce the amount of modification needed to each domain service when moving components (and hence source code) around. For example, suppose all of the ticketing and knowledge base functionality in the Sysops Squad application was grouped and refactored into a Ticket domain, and a new Ticket service created from that domain. Now suppose that the customer survey component (identified through the `ss.customer.survey` namespace) was deemed part of the Ticket domain. Since the Ticket domain had already been migrated, the Ticket service would now have to be modified to include the Survey component. Better to align and refactor all of the components into component domains first, *then* start migrating those component domains to domain services.

Fitness Functions for Governance

It is important to keep the components within each domain service aligned with the domain, particularly if the domain service will be broken into smaller microservices. This type of governance helps keep domain services from becoming their own unstructured monolithic service. The following fitness function ensures that the namespace (and hence components) are kept consistent within a domain service.

Fitness function: All components in <some domain service> should start with the same namespace

This automated holistic fitness function can be triggered on deployment through a CI/CD pipeline to make sure the namespaces for components within a domain service remain consistent. For example, all components within the Ticket domain service should start with `ss.ticket`. Example 5-11 uses ArchUnit for ensuring this constraint. Each domain service would have its own corresponding fitness function based on its particular domain.

Example 5-11. ArchUnit code for governing components within the Ticket domain service

```
public void restrict_domain_within_ticket_service() {
   classes().should().resideInAPackage("..ss.ticket..")
   .check(myClasses);
}
```

Sysops Squad Saga: Creating Domain Services

`Tuesday, November 23, 09:04`

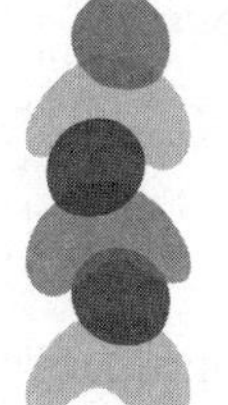

Addison and Austen worked closely with the Sysops Squad development team to develop a migration plan to stage the migration from component domains to domain services. They realized this effort not only required the code within each component domain to be extracted from the monolith and moved to a new project workspace, but also for the user interface to now remotely access the functionality within that domain.

Working from the component domains identified previously in Figure 5-19, the team migrated each component, one at a time, eventually arriving at a service-based architecture, as shown in Figure 5-22. Notice how each domain area identified in the previous pattern now becomes a separately deployed service.

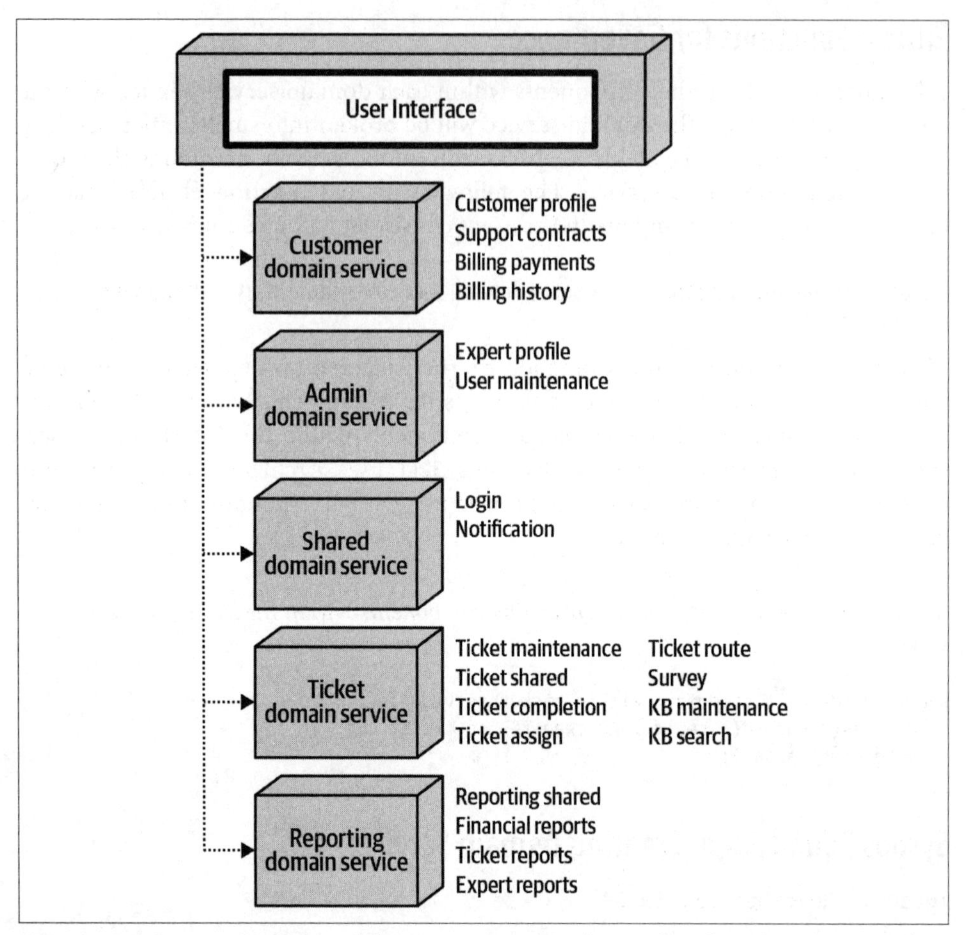

Figure 5-22. Separately deployed domain services result in a distributed Sysops Squad application

Summary

It has been our experience that "seat-of-the-pants" migration efforts rarely produce positive results. Applying these component-based decomposition patterns provides a structured, controlled, and incremental approach for breaking apart monolithic architectures. Once these patterns are applied, teams can now work to decompose monolithic data (see Chapter 6) and begin breaking apart domain services into more fine-grained microservices (see Chapter 7) as needed.

CHAPTER 6

Pulling Apart Operational Data

Thursday, October 7, 08:55

Now that the Sysops Squad application was successfully broken into separately deployed domain services, Addison and Austen both realized that it was time to start thinking about breaking apart the monolithic Sysops Squad database. Addison agreed to start this effort, while Austen began to work on enhancing the CI/CD deployment pipeline. Addison met with Dana, the Sysops Squad data architect, and also Devon, one of the DBAs supporting the Penultimate Electronics databases.

"I'd like your opinions on how we might go about breaking up the Sysops Squad database," said Addison.

"Wait a minute," said Dana. "Who said anything about breaking apart the database?"

"Addison and I agreed last week that we needed to break up the Sysops Squad database," said Devon. "As you know, the Sysops Squad application has been going through a major overhaul, and breaking apart the data is part of that overhaul."

"I think the monolithic database is just fine," said Dana. "I see no reason why it should be broken apart. Unless you can convince me otherwise, I'm not going to budge on this issue. Besides, do you know how hard it would be to break apart that database?"

"Of course it will be difficult," said Devon, "but I know of a five-step process leveraging what are known as data domains that would work really well on this database. That way, we can even start investigating using different kinds of databases for certain parts of the application, like the knowledge base and even the customer survey functionality."

"Let's not get ahead of ourselves," said Dana. "And let's also not forget that I am the one who is responsible for all of these databases."

Addison quickly realized things were spiraling out of control, and quickly put some key negotiation and facilitation skills to use. "OK," said Addison, "we should have included you in our initial discussions, and for that I apologize. I should have known better. What can we do to bring you on board and help us decompose the Sysops Squad database?"

"That's easy," said Dana. "Convince me that the Sysops Squad database really does need to be broken apart. Provide me with a solid justification. If you can do that, then we'll talk about Devon's five-step process. Otherwise, it stays as it is."

Breaking apart a database is hard—much harder, in fact, than breaking apart application functionality. Because data is generally the most important asset in the company, there is greater risk of business and application disruption when breaking apart or restructuring data. Also, data tends to be highly coupled to application functionality, making it harder to identify well-defined seams within a large data model.

In the same way a monolithic application is broken into separate deployment units, there are times when it is desirable (or even necessary) to break up a monolithic database as well. Some architecture styles, such as microservices, *require* data to be broken apart to form well-defined bounded contexts (where each service owns its own data), whereas other distributed architectures, such as service-based architecture, allow services to share a single database.

Interestingly enough, some of the same techniques used to break apart application functionality can be applied to breaking apart data as well. For example, components translate to data domains, class files translate to database tables, and coupling points between classes translate to database artifacts such as foreign keys, views, triggers, or even stored procedures.

In this chapter, we explore some of the drivers for decomposing data and show techniques for how to effectively break apart monolithic data into separate data domains, schemas, and even separate databases in an iterative and controlled fashion. Knowing that the database world is not all relational, we also discuss various types of databases (relational, graph, document, key-value, columnar, NewSQL, and cloud native) and outline the various trade-offs associated with each of these database types.

Data Decomposition Drivers

Breaking apart a monolithic database can be a daunting task, and as such it's important to understand if (and when) a database should be decomposed, as illustrated in Figure 6-1. Architects can justify a data decomposition effort by understanding and analyzing *data disintegrators* (drivers that justify breaking apart data) and *data integrators* (drivers that justify keeping data together). Striving for a balance between these two driving forces and analyzing the trade-offs of each is the key to getting data granularity right.

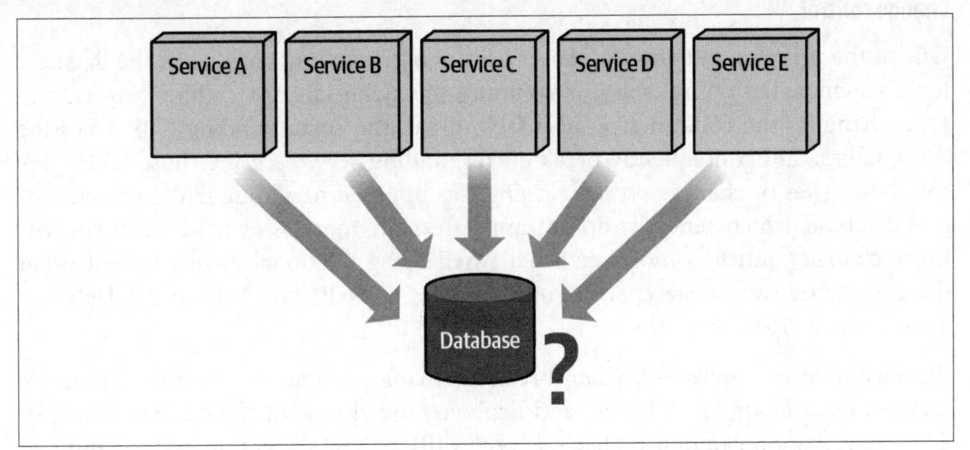

Figure 6-1. Under what circumstances should a monolithic database be decomposed?

In this section, we will explore the data disintegrators and data integrators used to help make the right choice when considering breaking apart monolithic data.

Data Disintegrators

Data disintegration drivers provide answers and justifications for the question "when should I consider breaking apart my data?" The six main disintegration drivers for breaking apart data include the following:

Change control
: How many services are impacted by a database table change?

Connection management
: Can my database handle the connections needed from multiple distributed services?

Scalability
: Can the database scale to meet the demands of the services accessing it?

Fault tolerance
: How many services are impacted by a database crash or maintenance downtime?

Architectural quanta
: Is a single shared database forcing me into an undesirable single architecture quantum?

Database type optimization
: Can I optimize my data by using multiple database types?

Each of these disintegration drivers is discussed in detail in the following sections.

Change control

One of the primary data disintegration drivers is controlling changes in the database table schemas. Dropping tables or columns, changing table or column names, and even changing the column type in a table break the corresponding SQL accessing those tables, and consequently break corresponding services using those tables. We call these types of changes *breaking changes* as opposed to adding tables or columns in a database, which generally do not impact existing queries or writes. Not surprisingly, change control is most impacted when using relational databases, but other database types can create change control issues as well (see "Selecting a Database Type" on page 161).

As illustrated in Figure 6-2, when breaking changes occur to a database, multiple services must be updated, tested, and deployed together with the database changes. This coordination can quickly become both difficult and error prone as the number of separately deployed services sharing the same database increases. Imagine trying to coordinate 42 separately deployed services for a single breaking database change!

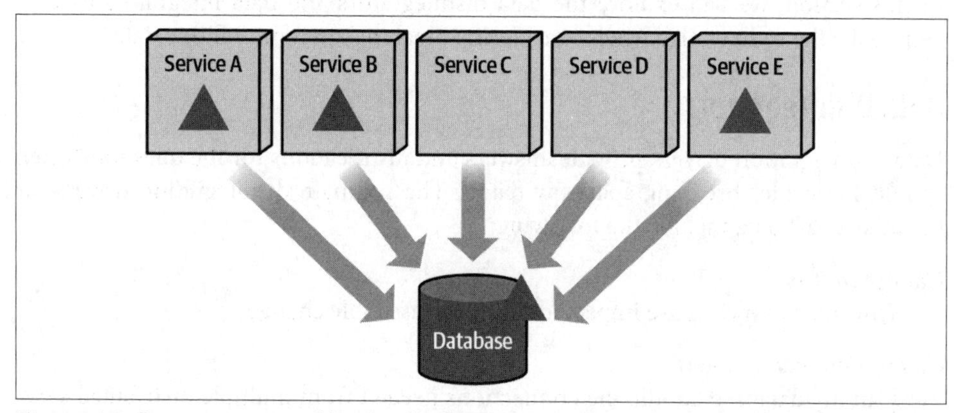

Figure 6-2. Services impacted by the database change must be deployed together with the database

Coordinating changes to multiple distributed services for a shared database change is only half the story. The real danger of changing a shared database in any distributed architecture is forgetting about services that access the table just changed. As illustrated in Figure 6-3, those services become nonoperational *in production* until they can be changed, tested, and redeployed.

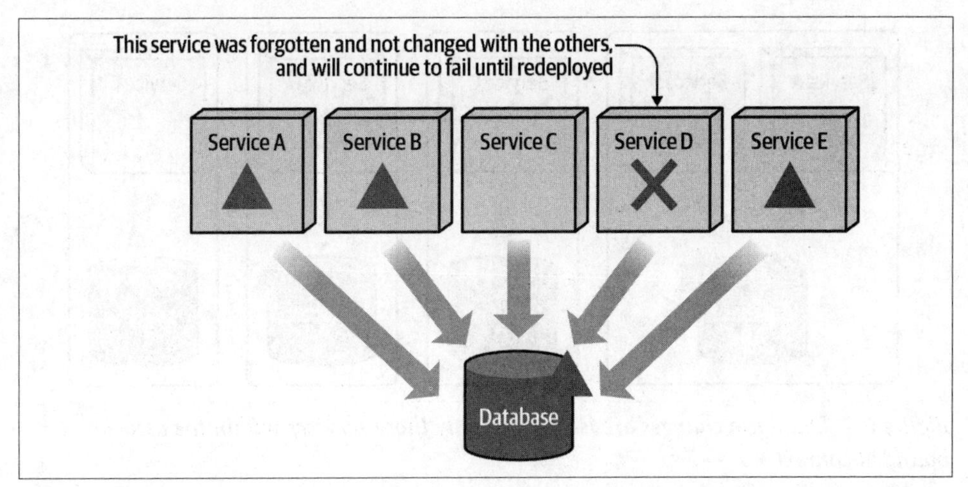

Figure 6-3. Services impacted by a database change but forgotten will continue to fail until redeployed

In most applications, the danger of forgotten services is mitigated by diligent impact analysis and agressive regression testing. However, consider a microservices ecosystem with 400 services, all sharing the same monolithic highly available clustered relational database. Imagine running around to all the development teams in many domain areas, trying to find out which services use the table being changed. Also imagine having to then coordinate, test, and deploy all of these services *together* as a single unit, along with the database. Thinking about this scenario starts to become a mind-numbing exercise, usually leading to some degree of insanity.

Breaking apart a database into well-defined bounded contexts (*https://oreil.ly/Q8mI7*) significantly helps control breaking database changes. The bounded context concept comes from the seminal book *Domain-Driven Design* by Eric Evans (Addison-Wesley) and describes the source code, business logic, data structures, and data all bound together—encapsulated—within a specific context. As illustrated in Figure 6-4, well-formed bounded contexts around services and their corresponding data helps control change, because change is isolated to just those services within that bounded context.

Most typically, bounded contexts are formed around services and the data the services owns. By "own" we mean a service that writes to the database (as opposed to having read-only access to the data). We discuss distributed data ownership in more detail in Chapter 9.

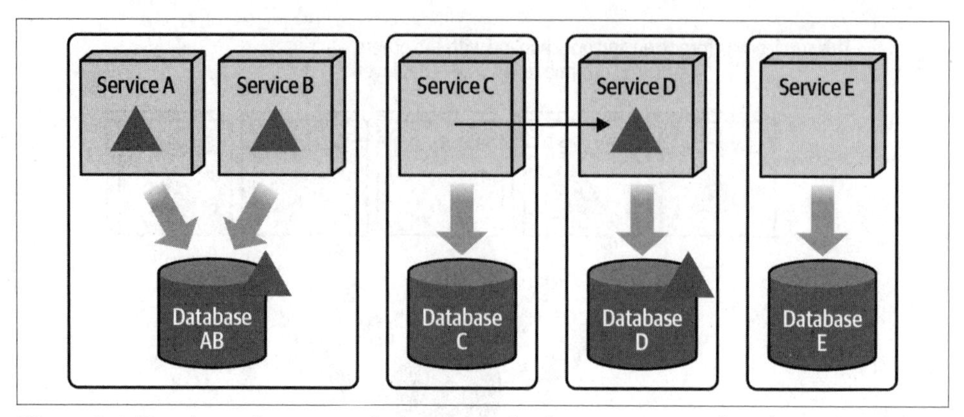

Figure 6-4. Database changes are isolated to only those services within the associated bounded context

Notice in Figure 6-4 that Service C needs access to some of the data in Database D that is contained in a bounded context with Service D. Since Database D is in a different bounded context, Service C cannot directly access the data. This would not only violate the bounded context rule, but also create a mess with regard to change control. Therefore, Service C must *ask* Service D for the data. There are many ways of accessing data a service doesn't own while still maintaining a bounded context. These techniques are discussed in detail in Chapter 10.

One important aspect of a bounded context related to the scenario between Service C needing data and Service D owning that data within its bounded context is that of *database abstraction*. Notice in Figure 6-5 that Service D is sending data that was requested by Service C through some sort of *contract* (such as JSON, XML, or maybe even an object).

The advantage of the bounded context is that the data sent to Service C can be a different contract than the schema for Database D. This means that a breaking change to some table in Database D impacts only Service D and not necessarily the contract of the data sent to Service C. In other words, Service C is abstracted from the actual schema structure of Database D.

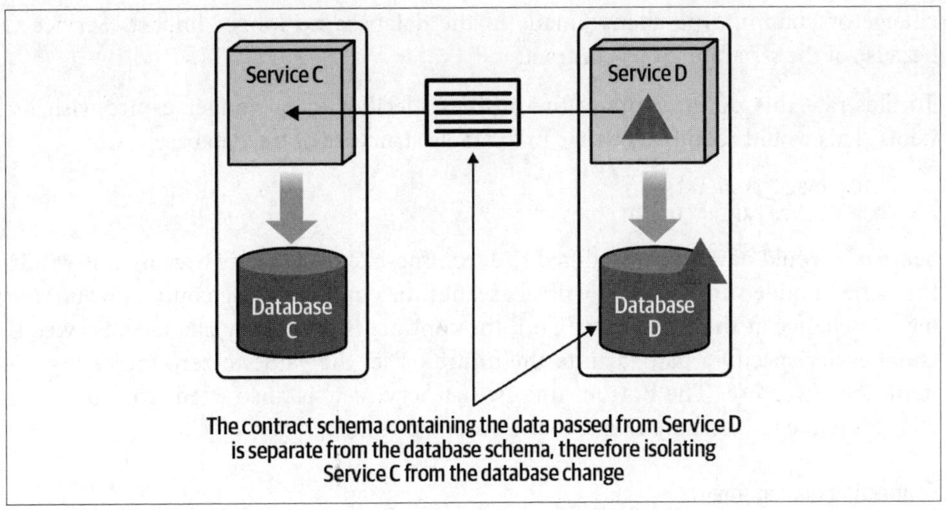

Figure 6-5. The contract from a service call abstracts the caller from the underlying database schema

To illustrate the power of this bounded context abstraction within a distributed architecture, assume Database D has a Wishlist table with the following structure:

```
CREATE TABLE Wishlist
(
CUSTOMER_ID VARCHAR(10),
ITEM_ID VARCHAR(20),
QUANTITY INT,
EXPIRATION_DT DATE
);
```

The corresponding JSON contract that Service D sends to Service C requesting wish list items is as follows:

```
{
  "$schema": "http://json-schema.org/draft-04/schema#",
  "properties": {
    "cust_id": {"type": "string"},
    "item_id": {"type": "string"},
    "qty": {"type": "number"},
    "exp_dt": {"type": "number"}
  },
}
```

Notice how the expiration data field (`exp_dt`) in the JSON schema is named differently than the database column name and is specified as a number (a long value representing the epoch time—the number of milliseconds since midnight on 1 January 1970), whereas in the database it is represented as a `DATE` field. Any column name change or column type change made in the database no longer impacts Service C because of the separate JSON contract.

To illustrate this point, suppose the business decides to no longer expire wish list items. This would require a change in the table structure of the database:

```
ALTER TABLE Wishlist
DROP COLUMN EXPIRATION_DT;
```

Service D would have to be modified to accommodate this change because it is within the same bounded context as the database, but the corresponding contract would not have to change at the same time. Until the contract is eventually changed, Service D could either specify a date far into the future or set the value to zero indicating the item doesn't expire. The bottom line is that Service C is abstracted from breaking changes made to Database D due to the bounded context.

Connection management

Establishing a connection to a database is an expensive operation. A database connection pool is often used not only to increase performance, but also to limit the number of concurrent connections an application is allowed to use. In monolithic applications, the database connection pool is usually owned by the application (or application server). However, in distributed architectures, each service—or more specifically, each service instance—typically has its own connection pool. As illustrated in Figure 6-6, when multiple services share the same database, the number of connections can quickly become saturated, particularly as the number of services or service instances increase.

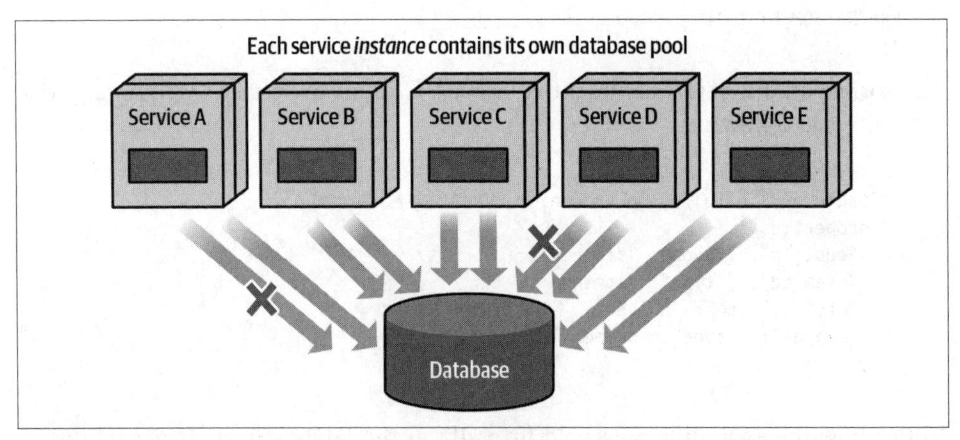

Figure 6-6. Database connections can quickly get saturated with multiple service instances

Reaching (or exceeding) the maximum number of available database connections is yet another driver to consider when deciding whether to break apart a database. Frequent connection waits (the amount of time it takes waiting for a connection to become available) is usually the first sign that the maximum number of database connections has been reached. Since connection waits can also manifest themselves as request time-outs or tripped circuit breakers, looking for connection waits is usually the first thing we recommend if these conditions frequently occur when using a shared database.

To illustrate the issues associated with database connections and distributed architecture, consider the following example: a monolithic application with 200 database connections is broken into a distributed architecture consisting of 50 services, each with 10 database connections in its connection pool.

Original monolithic application	200 connections
Distributed services	50
Connections per service	10
Minimum service instances	2
Total service connections	1,000

Notice how the number of database connections within the same application context grew from 200 to 1,000, and the services haven't even started scaling yet! Assuming half of the services scale to an average of 5 instances each, the number of database connections quickly grows to 1,700.

Without some sort of connection strategy or governance plan, services will try to use as many connections as possible, frequently starving other services from much needed connections. For this reason, it's important to govern how database connections are used in a distributed architecture. One effective approach is to assign each service a *connection quota* to govern the distribution of available database connections across services. A connection quota specifies the maximum number of database connections a service is allowed to use or make available in its connection pool.

By specifying a connection quota, services are not allowed to create more database connections than are allocated to it. If a service reaches the maximum number of database connections in its quota, it must wait for one of the connections it's using to become available. This method can be implemented using two approaches: evenly distributing the same connection quota to every service, or assigning a different connection quota to each service based on its needs.

The even distribution approach is typically used when first deploying services, and it is not known yet how many connections each service will need during normal and peak operations. While simple, this approach is not overly efficient because some services may need more connections than others, while some connections held by other services may go unused.

While more complex, the variable distribution approach is much more efficient for managing database connections to a shared database. With this approach, each service is assigned a different connection quota based on its functionality and scalability requirements. The advantage of this approach is that it optimizes the use of available database connections across distributed services, making sure those services that require more database connections have them available for use. However, the disadvantage is that it requires knowledge about the nature of the functionality and the scalability requirements of each service.

We usually recommend starting out with the even distribution approach and creating fitness functions to measure the concurrent connection usage for each service. We also recommend keeping the connection quota values in an external configuration server (or service) so that the values can be easily adjusted either manually or programmatically through simple machine learning algorithms. This technique not only helps mitigate connection saturation risk, but also properly balances available database connections between distributed services to ensure that no idle connections are wasted.

Table 6-1 shows an example of starting out using the even distribution approach for a database that can support a maximum of 100 concurrent connections. Notice that Service A has only ever needed a maximum of 5 connections, Service C only 15 connections, and Service E only 14 connections, whereas Service B and Service D have reached their max connection quota and have experienced connection waits.

Table 6-1. Connection quota allocations evenly distributed

	Service	Quota	Max used	Waits
	A	20	5	No
→	B	20	20	Yes
	C	20	15	No
→	D	20	20	Yes
	E	20	14	No

Since Service A is well below its connection quota, this is a good place to start reallocating connections to other services. Moving five database connections to Service B and five database connections to Service D yields the results shown in Table 6-2.

Table 6-2. Connection quota allocations with varying distributions

	Service	Quota	Max used	Waits
	A	10	5	No
→	B	25	25	Yes
	C	20	15	No
	D	25	25	No
	E	20	14	No

This is better, but Service B is still experiencing connection waits, indicating that it requires more connections than it has in its connection quota. Readjusting the quotas even further by taking two connections each from Service A and Service E yields much better results, as shown in Table 6-3.

Table 6-3. Further connection quota tuning results in no connection waits

Service	Quota	Max used	Waits
A	8	5	No
B	29	27	No
C	20	15	No
D	25	25	No
E	18	14	No

This analysis, which can be derived from continuous fitness functions that gather streamed metrics data from each service, can also be used to determine how close the maximum number of connections used is to the maximum number of connections available, and also how much buffer exists for each service in terms of its quota and maximum connections used.

Scalability

One of the many advantages of a distributed architecture is scalability—the ability for services to handle increases in request volume while maintaining a consistent response time. Most cloud-based and on-prem infrastructure-related products do a good job at ensuring that services, containers, HTTP servers, and virtual machines scale to satisfy increases in demand. But what about the database?

As illustrated in Figure 6-7, service scalability can put a tremendous strain on the database, not only in terms of database connections (as discussed in the prior section), but also on throughput and database capacity. In order for a distributed system to scale, *all* parts of the system need to scale—including the database.

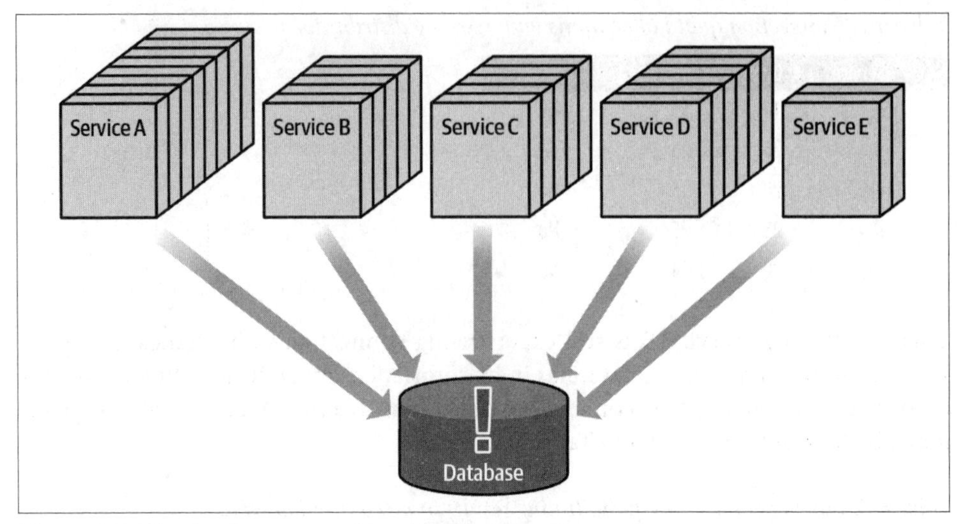

Figure 6-7. The database must also scale when services scale

Scalability is another data disintegration driver to consider when thinking about breaking apart a database. Database connections, capacity, throughput, and performance are all factors in determining whether a shared database can meet the demands of multiple services within a distributed architecture.

Consider the refined variable database connection quotas in Table 6-3 in the prior section. When services scale by adding multiple instances, the picture changes dramatically, as shown in Table 6-4, where the total number of database connections is 100.

Table 6-4. When services scale, more connection are used than are available

Service	Quota	Max used	Instances	Total used
A	8	5	2	10
B	29	27	3	81
C	20	15	3	45
D	25	25	2	50
E	18	14	4	56
TOTAL	100	86	14	242

Notice that even though the connection quota is distributed to match the 100 database connections available, once services start to scale, the quota is no longer valid because the total number of connections used increases to 242, which is 142 more connections than are available in the database. This will likely result in connection waits, which in turn will result in overall performance degradation and request time-outs.

Breaking data into separate data domains or even a database-per-service, as illustrated in Figure 6-8, requires fewer connections to each database, hence providing better database scalability and performance as the services scale.

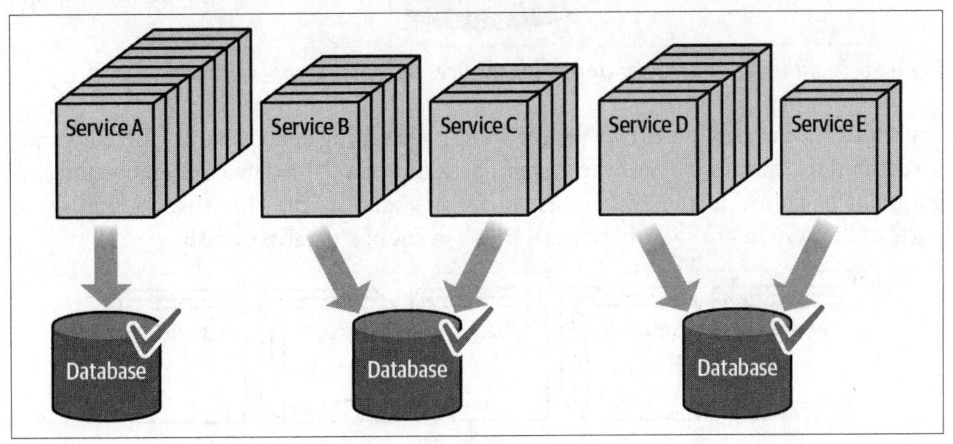

Figure 6-8. Breaking apart the database provides better database scalability

In addition to database connections, another factor to consider with respect to scalability is the load placed on the database. By breaking apart a database, less load is placed on each database, thereby also improving overall performance and scalability.

Fault tolerance

When multiple services share the same database, the overall system becomes less fault tolerant because the database becomes a single point of failure (SPOF). Here, we are defining fault tolerance as the ability of some parts of the system to continue uninterrupted when a service or database fails. Notice in Figure 6-9 that when sharing a single database, overall fault tolerance is low because if the database goes down, *all* services become nonoperational.

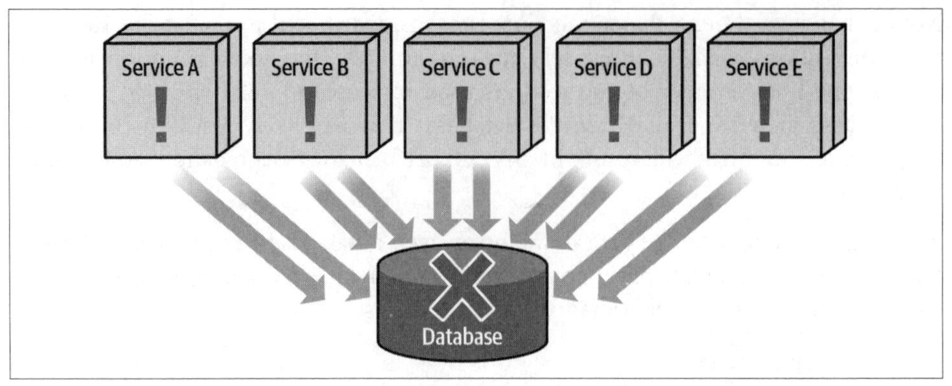

Figure 6-9. If the database goes down, all services become nonoperational

Fault tolerance is another driver for considering breaking apart data. If fault tolerance is required for certain parts of the system, breaking apart the data can remove the single point of failure in the system, as shown in Figure 6-10. This ensures that some parts of the system are still operational in the event of a database crash.

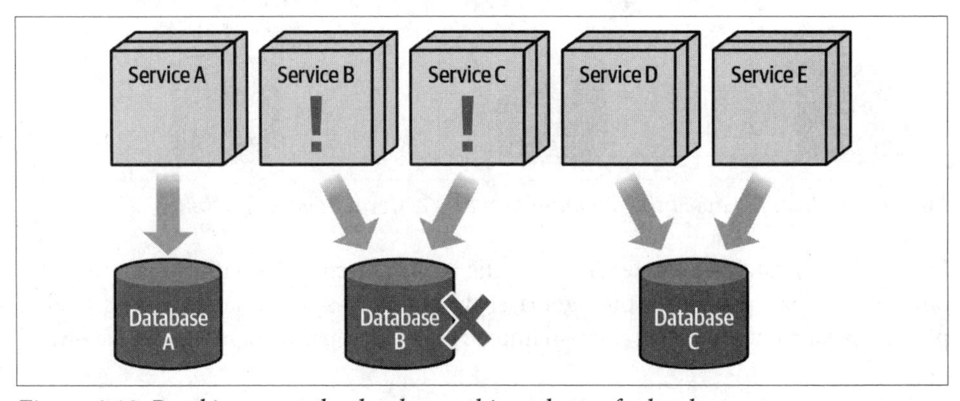

Figure 6-10. Breaking apart the database achieves better fault tolerance

Notice that since the data is now broken apart, if Database B goes down, only Service B and Service C are impacted and become nonoperational, whereas the other services continue to operate uninterrupted.

Architectural quantum

Recall from Chapter 2 that an architectural quantum is defined as an independently deployable artifact with high functional cohesion, high static coupling, and synchronous dynamic coupling. The architecture quantum helps provide guidance in terms of when to break apart a database, making it another data disintegration driver.

Consider the services in Figure 6-11, where Service A and Service B require different architectural characteristics than the other services. Notice in the diagram that although Service A and Service B are grouped together, they do not form a separate quantum from the other services because of a single shared database. Thus, all five services, along with the database, form a single architectural quantum.

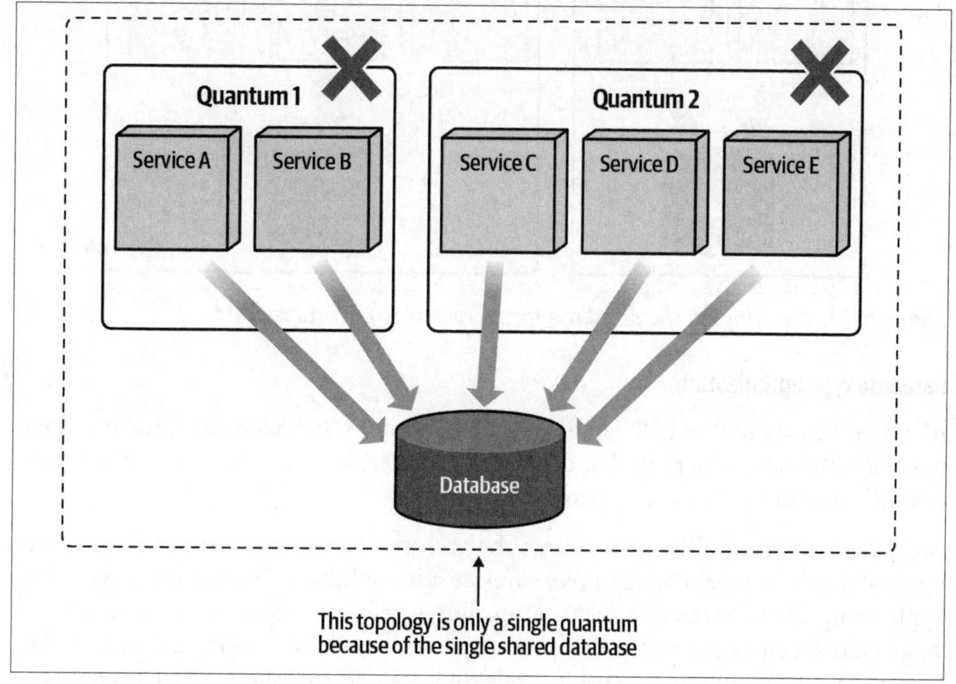

Figure 6-11. The database is part of the architectural quantum

Because the database is included in the *functional cohesion* part of the architecture quantum definition, it is necessary to break apart the data so that each resulting part can be in its own quantum. Notice in Figure 6-12 that since the database is broken apart, Service A and Service B, along with the corresponding data, are now a separate quantum from the one formed with services C, D, and E.

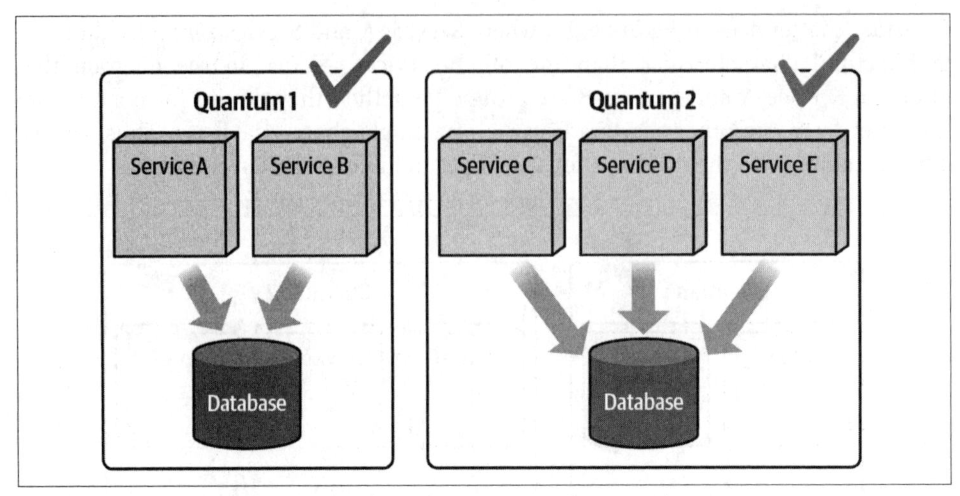

Figure 6-12. Breaking up the database forms two architectural quanta

Database type optimization

It's often the case that not all data is treated the same. When using a monolithic database, *all* data must adhere to that database type, therefore producing potentially suboptimal solutions for certain types of data.

Breaking apart monolithic data allows the architect to move certain data to a more optimal database type. For example, suppose a monolithic relational database stored application-related transactional data, including reference data in the form of key-value pairs (such as country codes, product codes, warehouse codes, and so on). This type of data is difficult to manage in a relational database because the data is not relational in nature, but rather key-value. Hence, a *key-value* database (see "Key-Value Databases" on page 165) would produce a more optimal solution than a relational database.

Data Integrators

Data integrators do the exact opposite of the data disintegrators discussed in the prior section. These drivers provide answers and justifications for the question "when should I consider putting data back together?" Along with data disintegrators, data integrators provide the balance and trade-offs for analyzing when to break apart data and when not to.

The two main integration drivers for pulling data back together are the following:

Data relationships
: Are there foreign keys, triggers, or views that form close relationships between the tables?

Database transactions
: Is a single transactional unit of work necessary to ensure data integrity and consistency?

Each of these integration drivers is discussed in detail in the following sections.

Data relationships

Like components within an architecture, database tables can be coupled as well, particularly with regard to relational databases. Artifacts like foreign keys, triggers, views, and stored procedures tie tables together, making it difficult to pull data apart; see Figure 6-13.

Imagine walking up to your DBA or data architect and telling them that since the database must be broken apart to support tightly formed bounded contexts within a microservices ecosystem, every foreign key and view in the database needs to be removed! That's not a likely (or even feasible) scenario, yet that is precisely what would need to happen to support a database-per-service pattern in microservices.

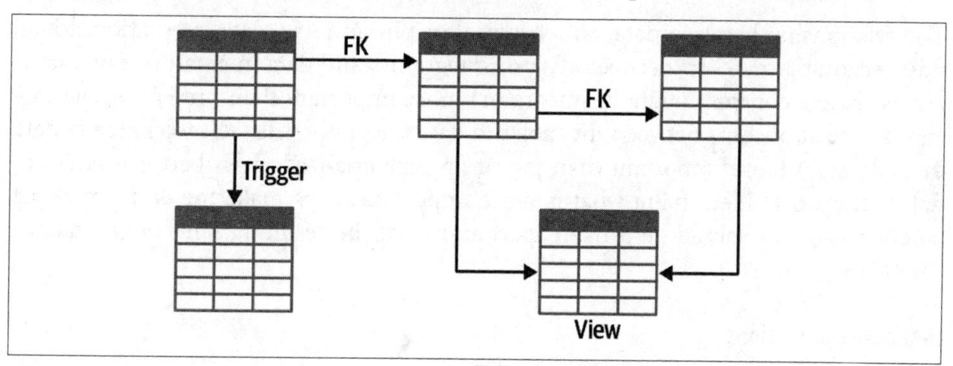

Figure 6-13. Foreign keys (FK), triggers, and views create tightly coupled relationships between data

These artifacts are necessary in most relational databases to support data consistency and data integrity. In addition to these physical artifacts, data may also be logically related, such as a problem ticket table and its corresponding problem ticket status table. However, as illustrated in Figure 6-14, these artifacts must be removed when moving data to another schema or database to form bounded contexts.

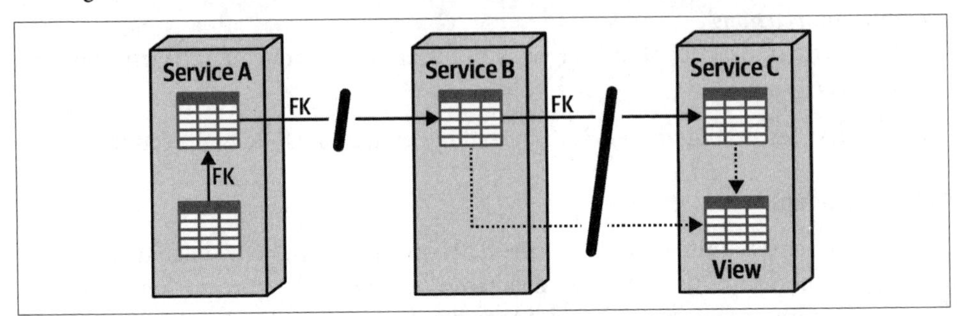

Figure 6-14. Data artifacts must be removed when breaking apart data

Notice that the foreign key (FK) relationship between the tables in Service A can be preserved because the data is in the same bounded context, schema, or database. However, the foreign keys (FK) between the tables in Service B and Service C must be removed (as well as the view that is used in Service C) because those tables are associated with different databases or schemas.

The relationship between data, either logical or physical, is a data integration driver, thus creating a trade-off between data disintegrators and data integrators. For example, is change control (a data disintegrator) more important than preserving the foreign key relationships between the tables (a data integrator)? Is fault tolerance (a data disintegrator) more important than preserving materialized views between tables (a data integrator)? Identifying what is more important helps make the decision about whether the data should be broken apart and what the resulting schema granularity should be.

Database transactions

Another data integrator is that of database transactions, something we discuss in detail in "Distributed Transactions" on page 263. As shown in Figure 6-15, when a single service does multiple database write actions to separate tables in the same database or schema, those updates can be done within an Atomicity, Consistency, Isolation, Durability (ACID) transaction and either committed or rolled back as a single unit of work.

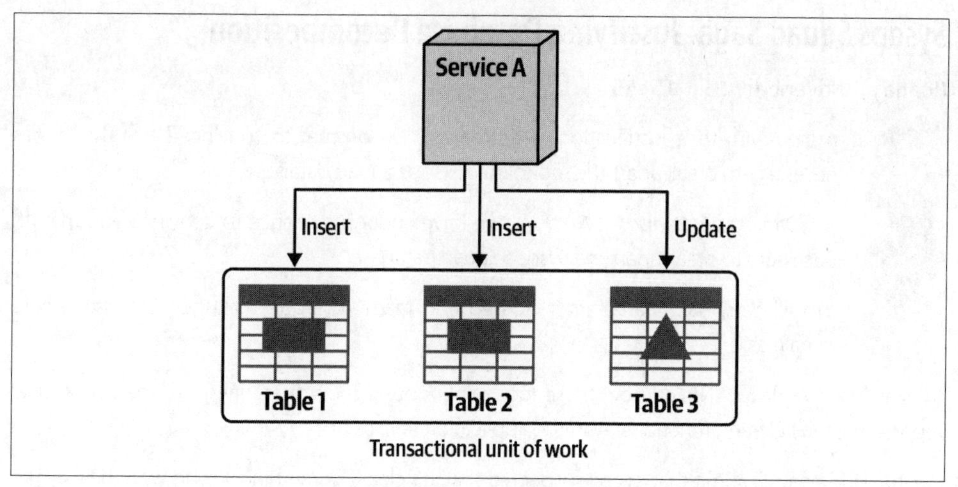

Figure 6-15. A single transactional unit of work exists when the data is together

However, when data is broken apart into either separate schemas or databases, as illustrated in Figure 6-16, a single transactional unit of work no longer exists because of the remote calls between services. This means that an insert or update can be committed in one table, but not in the other tables because of error conditions, resulting in data consistency and integrity issues.

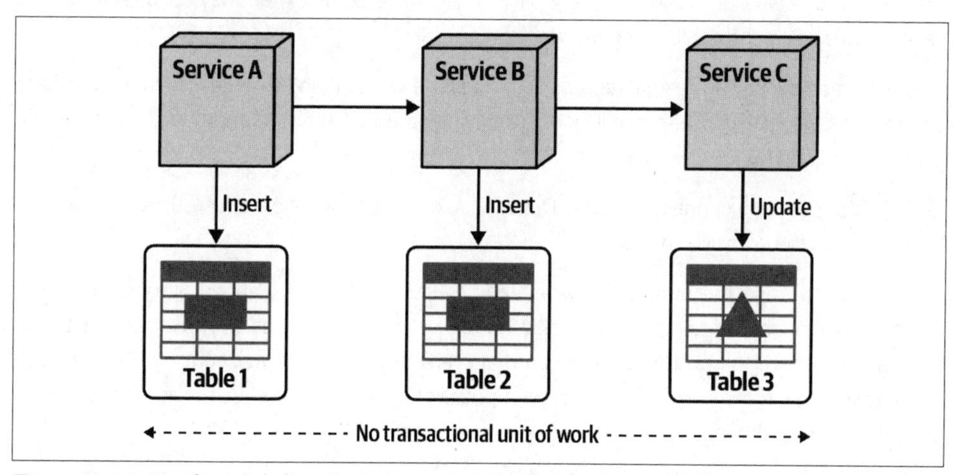

Figure 6-16. Single unit of work transactions don't exist when data is broken apart

While we dive into the details of distributed transaction management and transactional sagas in Chapter 12, the point here is to emphasize that database transactions are yet another data integration driver, and should be taken into account when considering breaking apart a database.

Sysops Squad Saga: Justifying Database Decomposition

Monday, November 15, 15:55

Armed with their justifications, Addison and Devon met to convince Dana that it was necessary to break apart the monolithic Sysops Squad database.

"Hi, Dana," said Addison. "We think we have enough evidence to convince you that it's necessary to break apart the Sysops Squad database."

"I'm all ears," said Dana, arms crossed and ready to argue that the database should remain as is.

"I'll start," said Addison. "Notice how these logs continuously show that whenever the operational reports run, the ticketing functionality in the application freezes up?"

"Yeah," said Dana, "I'll admit that even I suspected that. It's clearly something wrong with the way the ticketing functionality is accessing the database, not reporting."

"Actually," said Addison, "it's a combination of both ticketing *and* reporting. Look here."

Addison showed Dana metrics and logs that demonstrated some of the queries were necessarily wrapped in threads, and that the queries from the ticketing functionality were timing out because of a wait state when the reporting queries were run. Addison also showed how the reporting part of the system used parallel threads to query parts of the more complex reports concurrently, essentially taking up all of the database connections.

"OK, I can see how having a separate reporting database would help the situation from a database connection perspective. But that still doesn't convince me that the nonreporting data should be broken apart," said Dana.

"Speaking of database connections," said Devon, "look at this connection pool estimate as we start breaking apart the domain services."

Devon showed Dana the number of estimated services in the final planned Sysops Squad distributed application, including the projected number of instances for each of the services as the application scales. Dana explained to Devon that the connection pool was contained within each separate service instance, not like in the current phase of the migration where the application server owned the connection pool.

"So you see, Dana," said Devon, "with these projected estimates, we will need an additional 2,000 connections to the database to provide the scalability we need to handle the ticket load, and we simply do not have them with a single database."

Dana took a moment to look over the numbers. "Do you agree with these numbers, Addison?"

"I do," said Addison. "Devon and I came up with them ourselves after a lot of analysis based on the amount of HTTP traffic as well as the projected growth rates supplied by Parker."

"I must admit," said Dana, "this is good stuff you've both prepared. I particularly like that you've already thought about not having services connect to multiple databases or schemas. As you know, in my book that's a no-go."

"Us, too. However, we have one more justification to talk to you about," said Addison. "As you may or may not know, we've been having lots of issues with regard to the system not being available for our customers. While breaking apart the services provides us with some level of fault tolerance, if a monolithic database should go down for either maintenance or a server crash, all services would become nonoperational."

"What Addison is saying," added Devon, "is that by breaking apart the database, we can provide better fault tolerance by creating domain silos for the data. In other words, if the survey database were to go down, ticketing functionality would still be available."

"We call that an architectural quantum," said Addison. "In other words, since the database is part of the static coupling of a system, breaking it apart would make the core ticketing functionality standalone and not synchronously dependent on other parts of the system."

"Listen," said Dana, "you've convinced me that there's good reasons to break apart the Sysops Squad database, but explain to me how you can even think about doing that. Do you realize how many foreign keys and views there are in that database? There's no way you're going to be able remove all of those things."

"We don't necessarily have to remove all of those artifacts. That's where data domains and the five-step process come into play," said Devon. "Here, let me explain..."

Decomposing Monolithic Data

Decomposing a monolithic database is hard, and requires an architect to collaborate closely with the database team to safely and effectively break apart the data. One particularly effective technique for breaking apart data is to leverage what is known as the *five-step process*. As illustrated in Figure 6-17, this evolutionary and iterative process leverages the concept of a data domain as a vehicle for methodically migrating data into separate schemas, and consequently different physical databases.

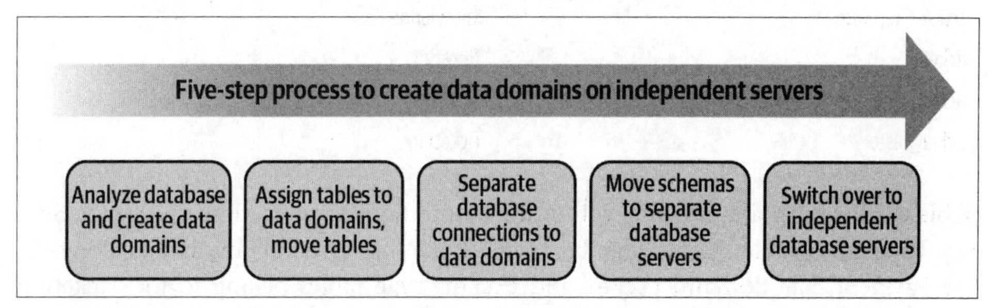

Figure 6-17. Five-step process for decomposing a monolithic database

A *data domain* is a collection of coupled database artifacts—tables, views, foreign keys, and triggers—that are all related to a particular domain and frequently used together within a limited functional scope. To illustrate the concept of a data domain, consider the Sysops Squad tables introduced in Table 1-2 and the corresponding proposed data domain assignments shown in Table 6-5.

Table 6-5. Existing Sysops Squad database tables assigned to data domains

Table	Proposed data domains
customer	Customer
customer_notification	Customer
survey	Survey
question	Survey
survey_administered	Survey
survey_question	Survey
survey_response	Survey
billing	Payment
contract	Payment
payment_method	Payment
payment	Payment
sysops_user	Profile
profile	Profile
expert_profile	Profile
expertise	Profile
location	Profile
article	Knowledge Base
tag	Knowledge Base
keyword	Knowledge Base
article_tag	Knowledge Base
article_keyword	Knowledge Base
ticket	Ticketing
ticket_type	Ticketing
ticket_history	Ticketing

Table 6-5 lists six data domains within the Sysops Squad application: Customer, Survey, Payment, Profile, Knowledge base, and Ticketing. The `billing` table belongs to the Payment data domain, `ticket` and `ticket_type` tables belong to the Ticketing data domain, and so on.

One way to conceptually think about data domains is to think about the database as a soccer ball, where each white hexagon represents a separate data domain. As illustrated in Figure 6-18, each white hexagon of the soccer ball contains a collection of domain-related tables along with all of the coupling artifacts (such as foreign keys, views, stored procedures, and so on).

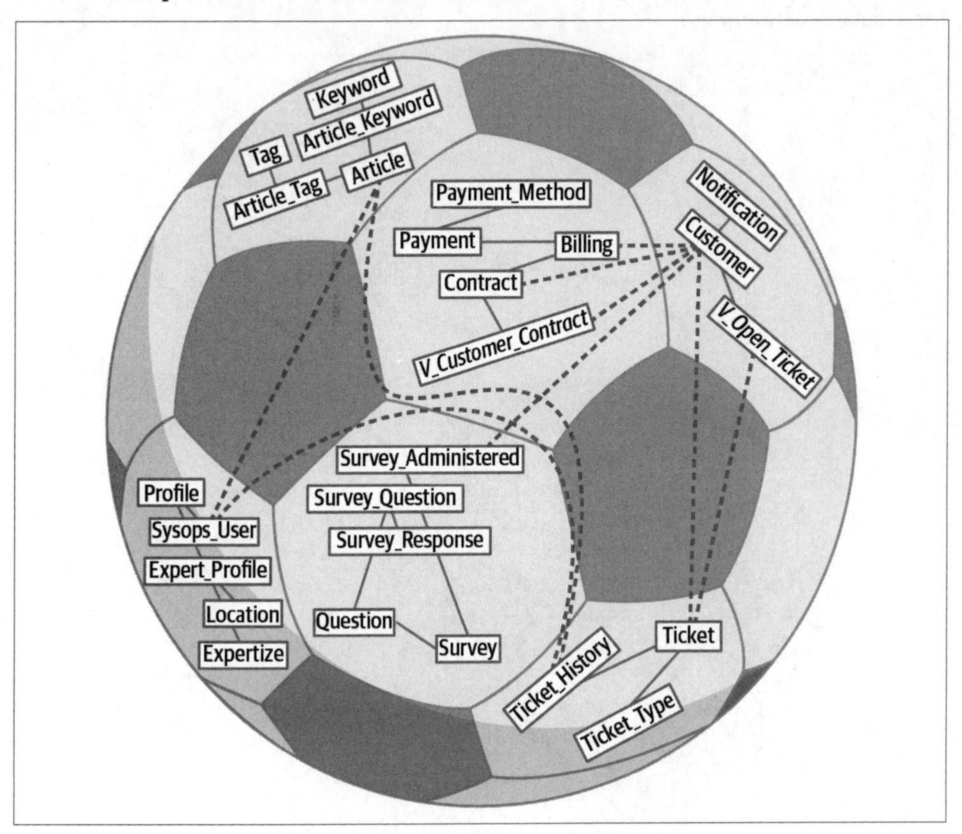

Figure 6-18. Database objects in a hexagon belong in a data domain

Visualizing the database this way allows the architect and database team to clearly see data domain boundaries and also the cross-domain dependencies (such as foreign keys, views, stored procedures, and so on) that need to be broken. Notice in Figure 6-18 that *within* each white hexagon, all data table dependencies and relationships can be preserved, but not *between* each white hexagon. For example, in the diagram notice that solid lines represent dependencies that are self-contained to the data domain, while the dotted lines cross data domains and must be removed when the data domains are extracted into separate schemas.

When extracting a data domain, these cross-domain dependencies must be removed. This means removing foreign-key constraints, views, triggers, functions, and stored procedures between data domains. Database teams can leverage the refactoring patterns found in the book *Refactoring Databases: Evolutionary Database Design*, by Scott Ambler and Pramod Sadalage (Addison-Wesley), to safely and iteratively remove these data dependencies.

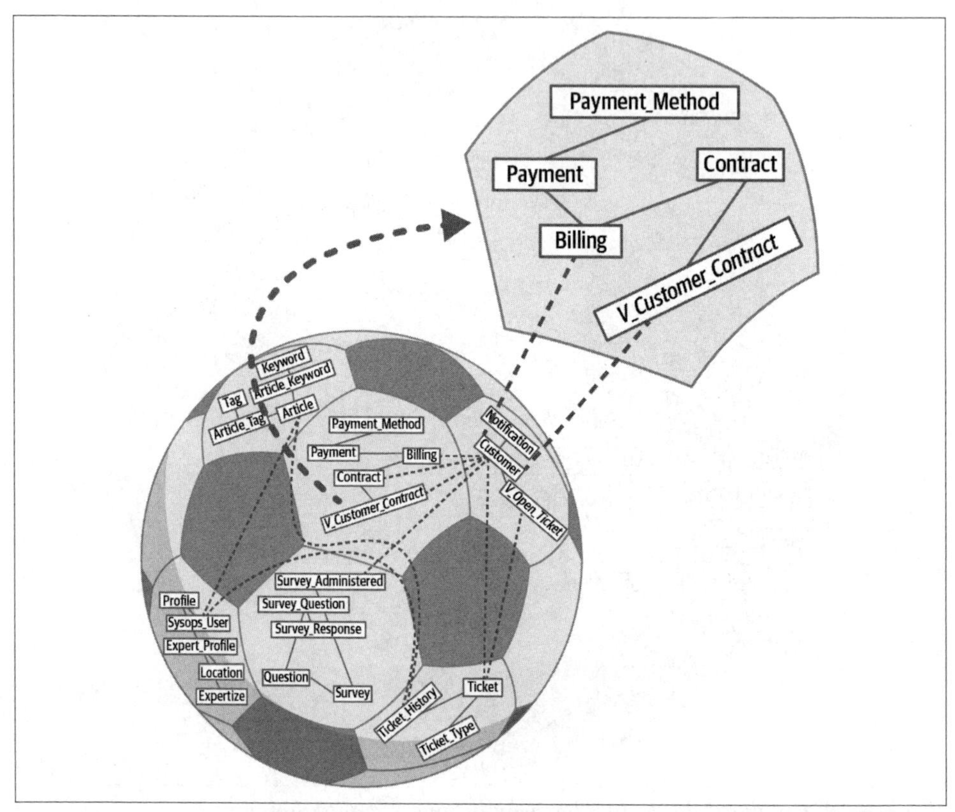

Figure 6-19. Tables belonging to data domains, extracted out, and connections that need to be broken

To illustrate the process of defining a data domain and removing cross-domain references, consider the diagram in Figure 6-19, where a data domain representing Payment is created. Since the `customer` table belongs to a different data domain than the `v_customer_contract`, the `customer` table must be removed from the view in the Payment domain. The original view `v_customer_contract` prior to defining the data domain is defined in Example 6-1.

Example 6-1. Database view to get open tickets for customer with cross-domain joins

```
CREATE VIEW [payment].[v_customer_contract]
  AS
SELECT
    customer.customer_id, customer.customer_name,
    contract.contract_start_date, contract.contract_duration,
    billing.billing_date, billing.billing_amount
FROM payment.contract AS contract
INNER JOIN customer.customer AS customer
    ON ( contract.customer_id = customer.customer_id )
INNER JOIN payment.billing AS billing
    ON ( contract.contract_id = billing.contract_id )
WHERE contract.auto_renewal = 0
```

Notice in the updated view shown in Example 6-2 that the join between `customer` and `payment` tables is removed, as is the column for the customer name (`customer.customer_name`).

Example 6-2. Database view to get open tickets in ticket domain for a given customer

```
CREATE VIEW [payment].[v_customer_contract]
  AS
SELECT
    billing.customer_id, contract.contract_start_date,
    contract.contract_duration, billing.billing_date,
    billing.billing_amount
FROM payment.contract AS contract
INNER JOIN payment.billing AS billing
    ON ( contract.contract_id = billing.contract_id )
WHERE contract.auto_renewal = 0
```

The bounded context rules for data domains apply just the same as individual tables—a service cannot talk to multiple data domains. Therefore, by removing this table from the view, the Payment service must now call the Customer service to get the customer name that it originally had from the view.

Once architects and database teams understand the concept of a data domain, they can apply the five-step process for decomposing a monolithic database. Those five steps are outlined in the following sections.

Step 1: Analyze Database and Create Data Domains

As illustrated in Figure 6-20, all services have access to all data in the database. This practice, known as the shared database (*https://oreil.ly/EFqtc*) integration style described by Gregor Hohpe and Bobby Woolf in their book *Enterprise Integration Patterns: Designing, Building, and Deploying Messaging Solutions* (Addison-Wesley), creates a tight coupling between data and the services accessing that data. As discussed in "Data Decomposition Drivers" on page 132, this tight coupling in the database makes change management very difficult.

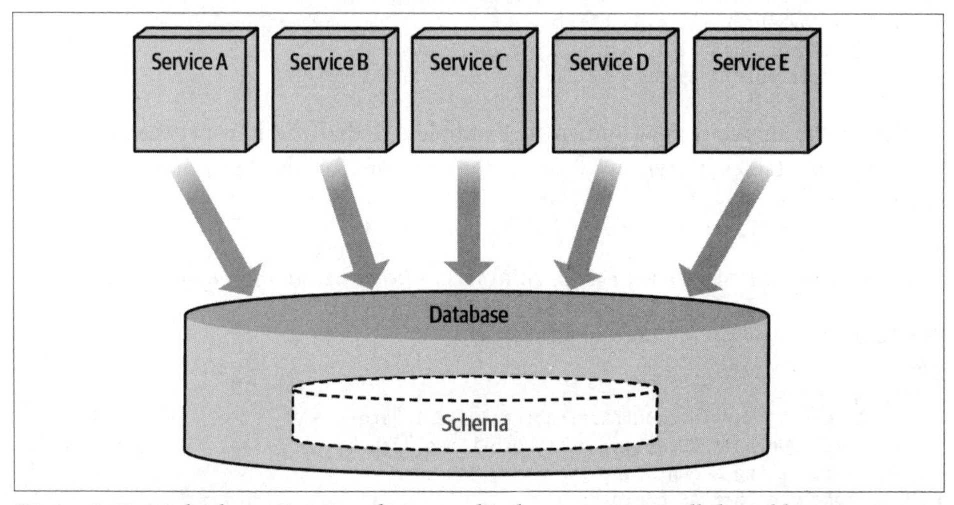

Figure 6-20. Multiple services use the same database, accessing all the tables necessary for read or write purposes

The first step in breaking apart a database is to identify specific domain groupings within the database. For example, as shown in Table 6-5, related tables are grouped together to help identify possible data domains.

Step 2: Assign Tables to Data Domains

The next step is to group tables along a specific bounded context, assigning tables that belong to a specific data domain into their own schema. A *schema* is a logical construct in database servers. A schema contain objects such as tables, views, functions, and so on. In some database servers, like Oracle, the schema is same as the user, while in other databases, like SQL Server, a schema is logical space for database objects where users have access to these schemas.

As illustrated in Figure 6-21, we have created schemas for each data domain and moved tables to the schemas to which they belong.

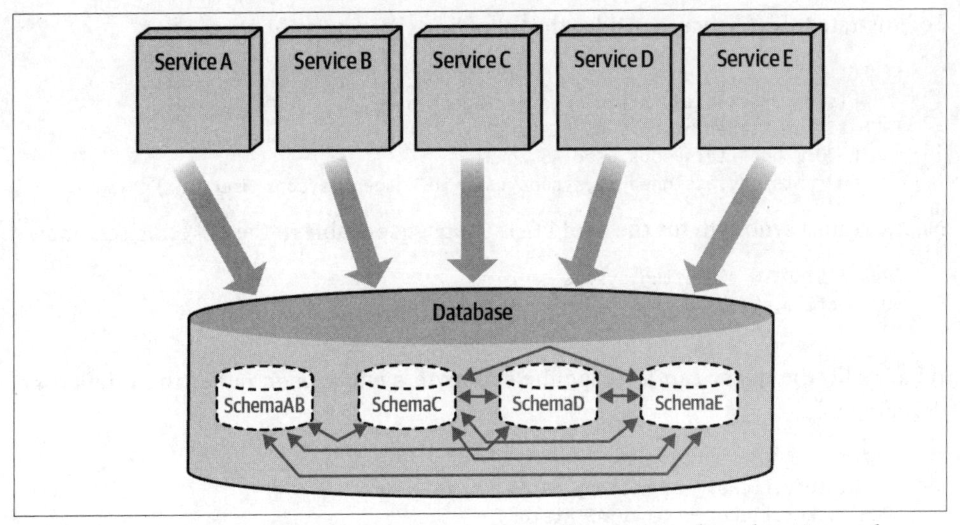

Figure 6-21. Services use the primary schema according to their data domain needs

When tables belonging to different data domains are tightly coupled and related to one another, data domains must necessarily be combined, creating a broader bounded context where multiple services own a specific data domain. Combining data domains is discussed in more detail in Chapter 9.

Data Domain Versus Database Schema

A *data domain* is an architectural concept, whereas a *schema* is a database construct that holds the database objects belonging to a particular data domain. While the relationship between a data domain and a schema is usually one to one, data domains can be mapped to one or more schemas, particularly when combining data domains because of tightly coupled data relationships. We will be referring to a data domain and schema to mean the same thing, and will be using the terms interchangeably.

To illustrate the assignment of tables to schemas, consider the Sysops Squad example where the `billing` table must be moved from its original schema to another data domain schema called `payment`:

```
ALTER SCHEMA payment TRANSFER sysops.billing;
```

Alternatively, a database team can create synonyms for tables that do not belong in their schema. *Synonyms* are database constructs, similar to *symlink*, that provide an alternate name for another database object that can exist in the same or different schema or server. While the idea of synonyms is to eliminate cross-schema queries, read or write privileges are needed to access them.

To illustrate this practice, consider the following cross-domain query:

```
SELECT
    history.ticket_id, history.notes, agent.name
FROM ticket.ticket_history AS history
INNER JOIN profile.sysops_user AS agent
    ON ( history.assigned_to_sysops_user_id = agent.sysops_user_id )
```

Next, create a synonym for the `profile.sysops_user` table in the ticketing schema:

```
CREATE SYNONYM ticketing.sysops_user
FOR profile.sysops_user;
GO
```

As a result, the query can leverage the synonym `sysops_user` rather than the cross-domain table:

```
SELECT
    history.ticket_id, history.notes, agent.name
FROM ticket.ticket_history AS history
INNER JOIN ticket.sysops_user AS agent
    ON ( history.assigned_to_sysops_user_id = agent.sysops_user_id )
```

Unfortunately, creating synonyms this way for tables that are accessed across schemas provides the application developers with coupling points. To form proper data domains, these coupling points need to be broken apart at some later time, therefore moving the integration points from the database layer to the application layer.

While synonyms do not really get rid of cross-schema queries, they do allow for easier dependency checking and code analysis, making it easier to split these later on.

Step 3: Separate Database Connections to Data Domains

In this step, the database connection logic within each service is refactored to ensure services connect to a specific schema and have read and write access to the tables belonging only to their data domain. This transition, illustrated in Figure 6-22, is the most difficult since all cross-schema access must be resolved at the service level.

Notice that the database configuration has been changed so that all data access is done strictly via services and their connected schemas. In this example, Service C communicates with Service D and not with SchemaD. There is no cross-schema access; all synonyms created in "Step 2: Assign Tables to Data Domains" on page 156 are removed.

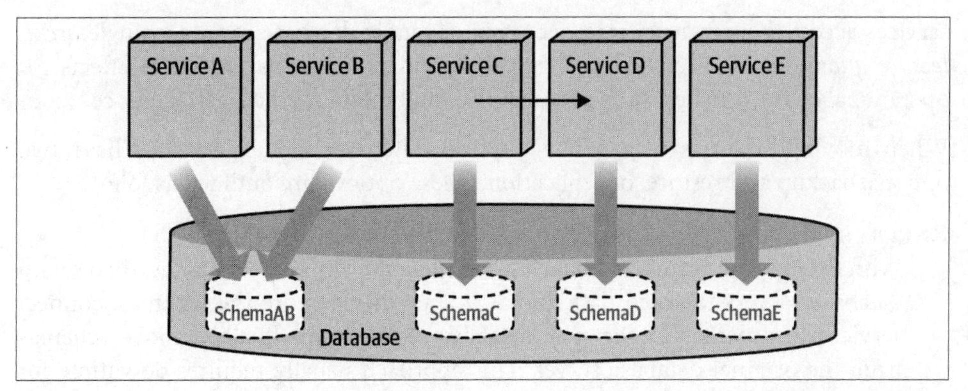

Figure 6-22. Move the cross-schema object access to the services, away from direct cross-schema access

When data from other domains is needed, do not reach into their databases. Instead, access it using the service that owns the data domain.

Upon completion of this step, the database is in a state of *data sovereignty per service*, which occurs when each service owns its own data. Data sovereignty per service is the nirvana state for a distributed architecture. Like all practices in architecture, it includes benefits and shortcomings:

Benefits

- Teams can change the database schema without worrying about affecting changes in other domains.
- Each service can use the database technology and database type best suitable for their use case.

Shortcomings

- Performance issues occur when services need access to large volumes of data.
- Referential integrity cannot be maintained in the database, resulting in the possibility of bad data quality.
- All database code (stored procedures, functions) that access tables belonging to other domains must be moved to the service layer.

Step 4: Move Schemas to Separate Database Servers

Once database teams have created and separated data domains, and have isolated services so that they access their own data, they can now move the data domains to separate physical databases. This is often a necessary step because even though services access their own schemas, accessing a single database creates a single *architecture quantum*, as discussed in Chapter 2, which might have adverse effects for operational characteristics, such as scalability, fault tolerance, and performance.

When moving schemas to separate physical databases, database teams have two options: backup and restore, or replication. These options are outlined as follows:

Backup and restore
: With this option, teams first back up each schema with data domains, then set up database servers for each data domain. They then restore the schemas, connect services to schemas in the new database servers, and finally remove schemas from the original database server. This approach usually requires downtime for the migration.

Replicate
: Using the replicate option, teams first set up database servers for each data domain. Next they replicate the schemas, switch connections over to the new database servers, and then remove the schemas from the original database server. While this approach avoids downtime, it does require more work to set up the replication and manage increased coordination.

Figure 6-23 shows an example of the replication option, where the database team sets up multiple database servers so that there is one database server for each data domain.

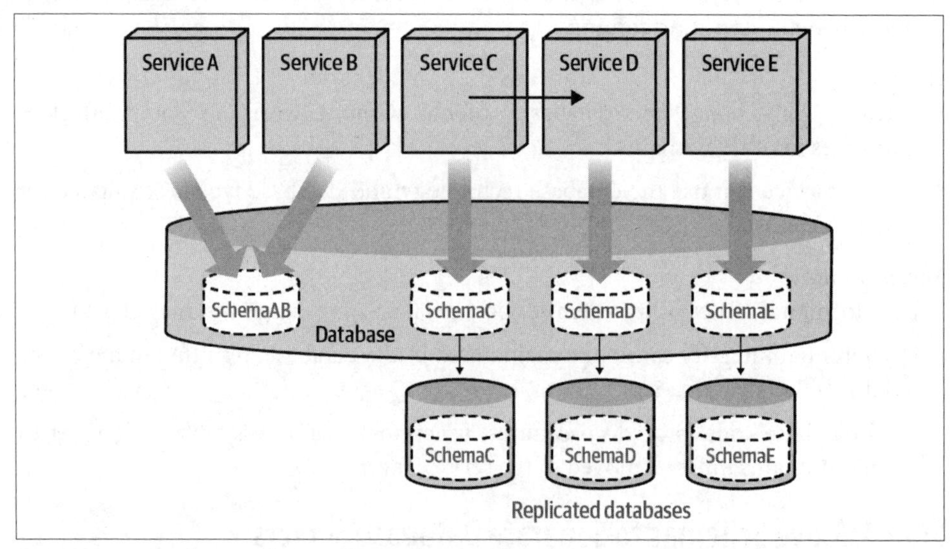

Figure 6-23. Replicate schemas (data domains) to their own database servers

Step 5: Switch Over to Independent Database Servers

Once the schemas are fully replicated, the service connections can be switched. The last step in getting the data domains and services to act as their own independent deployable units is to remove the connection to the old database servers and remove the schemas from the old database servers as well. The final state is seen in Figure 6-24.

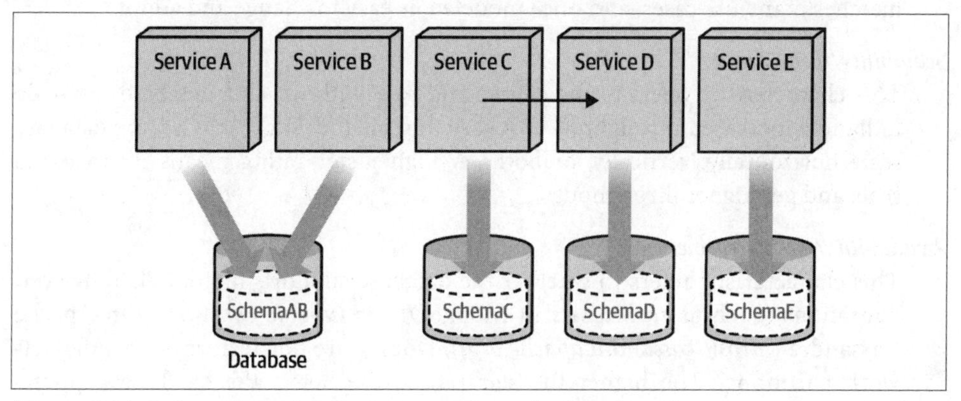

Figure 6-24. Independent database servers for each data domain

Once the database team has separated the data domains, isolated the database connections, and finally moved the data domains to their own database servers, they can optimize the individual database servers for availability and scalability. Teams can also analyze the data to determine the most appropriate database type to use, introducing polyglot database usage within the ecosystem.

Selecting a Database Type

Beginning around 2005, a revolution has occurred in database technologies. Unfortunately, the number of products that have emerged during this time have created a problem known as The Paradox of Choice (*https://oreil.ly/pBjGZ*). Having such a large number of products and choices means having more trade-off decisions to make. Given that each product is optimized for certain trade-offs, it rests on both software and data architects to pick the appropriate product with these trade-offs in mind as it relates to their problem space.

In this section, we introduce star ratings for the various database types, using the following characteristics in our analysis:

Ease-of-learning curve
: This characteristic refers to the ease with which new developers, data architects, data modelers, operational DBAs, and other users of the databases can learn and adopt. For example, it's assumed that most software developers understand SQL,

whereas something like Gremlin (a graph query language) may be a niche skill. The higher the star rating, the easier the learning curve. The lower the star rating, the harder the learning curve.

Ease of data modeling
: This characteristic refers to the ease with which data modelers can represent the domain in terms of a data model. A higher star rating means data modeling matches many use cases, and once modeled, is easy to change and adopt.

Scalability/throughput
: This characteristic refers to the degree and ease with which a database can scale to handle increased throughput. Is it easy to scale the database? Can the database scale horizontally, vertically, or both? A higher star rating means it's easier to scale and get higher throughput.

Availability/partition tolerance
: This characteristic refers to whether the database supports high availability configurations (such as *replica-sets* in MongoDB or *tunable consistency* in Apache Cassandra (*https://cassandra.apache.org*)). Does it provide features to handle network partitions? The higher the star rating, the better the database supports higher availability and/or better partition tolerance.

Consistency
: This characteristic refers to whether the database supports an "always consistent" paradigm. Does the database support ACID transactions, or does it lean toward BASE transactions with an eventual consistency model? Does it provide features to have tunable consistency models for different types of writes? The higher the star rating, the more consistency the database supports.

Programming language support, product maturity, SQL support, and community
: This characteristic refers to which (and how many) programming languages the database supports, how mature the database is, and the size of the database community. Can an organization easily hire people who know how to work with the database? Higher star ratings means there is better support, the product is mature, and it's easy to hire talent.

Read/write priority
: This characteristic refers to whether the database prioritizes reads over writes, or writes over reads, or if it is balanced in its approach. This is not a binary choice—rather, it's more of a scale toward which direction the database optimizes.

Relational Databases

Relational databases (also known as an RDBMS) have been the database of choice for more than three decades. There is significant value in their usage and the stability they provide, particularly within most business-related applications. These databases are known for the ubiquitous Structured Query Language (SQL) and the ACID properties they provide. The SQL interface they provide makes them a preferred choice for implementing different `read` models on top of the same `write` model. The star ratings for relational databases appear in Figure 6-25.

Rating subject	RDBMS databases (Oracle, SQL Server, Postgres, etc.)
Ease of learning	☆☆☆☆
Ease of data modeling	☆☆☆
Scalability/throughput	☆☆
Availability/partition tolerance	☆
Consistency	☆☆☆☆☆
Programming language support, product maturity, SQL support, community	☆☆☆☆
Read/write priority	Read ▲ Write

Figure 6-25. Relational databases rated for various adoption characteristics

Ease-of-learning curve
: Relational databases have been around for many years. They are commonly taught in schools, and mature documentation and tutorials exist. Therefore, they are much easier to learn than other database types.

Ease of data modeling
: Relational databases allow for flexible data modeling. They allow the modeling of key-value, document, graph-like structures, and they allow for changes in read patterns with addition of new indexes. Some models are really difficult to achieve, such as graph structures with arbitrary depth. Relational databases organize data into tables and rows (similar to spreadsheets), something that is natural for most database modelers.

Scalability/throughput
: Relational databases are generally vertically scaled using large machines. However, setup with replications and automated switchover are complex, requiring higher coordination and setup.

Availability/partition tolerance
: Relational databases favor consistency over availability and partition tolerance, discussed in "Table Split Technique" on page 254.

Consistency
: Relational databases have been dominant for years because of their support for ACID properties. The ACID features handle many concerns in concurrent systems and allow for developing applications without being concerned about lower-level details of concurrency and how the databases handle them.

Programming language support, product maturity, SQL support, and community
: Since relational databases have been around for many years, well-known design, implementation, and operational patterns can be applied to them, thus making them easy to adopt, develop, and integrate within an architecture. Many of the relational databases lack support for reactive stream APIs and similar new concepts; newer architectural concepts take longer to implement in well-established relational databases. Numerous programming language interfaces work with relational databases, and the community of users is large (although splintered among all the vendors).

Read/write priority
: In relational databases, the data model can be designed in such a way that either reads become more efficient or writes become more efficient. The same database can handle different types of workloads, allowing for balanced read-write priority. For example, not all use cases need ACID properties, especially in large data and traffic scenarios, or when really flexible schema is desired such as in survey administration. In these cases, other database types may be a better option.

MySQL (*https://www.mysql.com*), Oracle (*https://www.oracle.com*), Microsoft SQL Server (*https://oreil.ly/LP7jK*), and PostgreSQL (*https://www.postgresql.org*) are the most popular relational databases and can be run as standalone installations or are available as *Database as a Service* on major cloud provider platforms.

Aggregate Orientation

Aggregate orientation is the preference to operate on data that is related and has a complex data structure. Aggregate is a term originated in *Domain-Driven Design: Tackling Complexity in the Heart of Software* by Erik Evans. Think of `ticket` or `customer` with all its dependent tables in the Sysops Squad—they are aggregates. Like all practices in architecture, aggregate orientation includes benefits and shortcomings:

Benefits

- Enables easy distribution of data in clusters of servers, as the whole aggregate can be copied over to different servers.
- Improves read and write performance, as it reduces joins in the database.
- Reduces impedance mismatch between the application model and storage model.

Shortcomings

- It's difficult to arrive at proper aggregates, and changing aggregate boundaries is hard.
- Analyzing data across aggregates is difficult.

Key-Value Databases

Key-value databases are similar to a hash table (*https://oreil.ly/2FOQy*) data structure, something like tables in an RDBMS with an `ID` column as the key and a `blob` column as the value, which can consequently store any type of data. Key-value databases are part of a family known as NoSQL databases. In the book *NoSQL Distilled: A Brief Guide to the Emerging World of Polyglot Persistence* (Addison-Wesley), Pramod Sadalage (one of your authors) and Martin Fowler describe the rise of NoSQL databases and the motivations, usages, and trade-offs of using these types of databases, and is a good reference for further information on this database type.

Key-value databases are easiest to understand among the NoSQL databases. An application client can insert a key and a value, get a value for a known key, or delete a known key and its value. A key-value database does not know what's inside the value part, nor does it care what's inside, meaning that the database can query using the key and nothing else.

Unlike relational databases, key-value databases should be picked based on needs. There are persistent key-value databases like Amazon DynamoDB or Riak KV, nonpersistent databases like MemcacheDB, and other databases like Redis that can be configured to be persistent or not. Other relational database constructs like `joins`, `where`, and `order by` are not supported, but rather the operations `get`, `put`, and `delete`. The ratings for key-value databases appear in Figure 6-26.

Rating subject	Key value databases (Redis, DynamoDB, Riak, etc.)
Ease of learning	☆☆☆
Ease of data modeling	☆
Scalability/throughput	☆☆☆☆
Availability/partition tolerance	☆☆☆☆
Consistency	☆☆
Programming language support, product maturity, SQL support, community	☆☆☆
Read/write priority	Read ▲ Write

Figure 6-26. Key-value databases rated for various adoption characteristics

Ease-of-learning curve
: Key-value databases are easy to understand. Since they use "Aggregate Orientation" on page 165, it's important to design the aggregate properly because any change in the aggregate means rewriting all the data. Moving from relational databases to any of the NoSQL databases takes practice and unlearning familiar practices. For example, a developer cannot simply query "Get me all the keys."

Ease of data modeling
: Since key-value databases are aggregate oriented, they can use memory structures like arrays, maps, or any other type of data, including big blob. The data can be queried only by key or ID, which means the client should have access to the key outside of the database. Good examples of a key include `session_id`, `user_id`, and `order_id`.

Scalability/throughput
: Since key-value databases are indexed by key or ID, key lookups are very fast as there are no `joins` or `order by` operations. The value is fetched and returned to the client, which allows for easier scaling and higher throughput.

Availability/partition tolerance
: Since there are many types of key-value databases and each has different properties, even the same database can be configured to act in different ways either for an installation or for each read. For example, in Riak users can use *quorum* properties such as `all`, `one`, `quorum`, and `default`. When we use *one* quorum, the query can return `success` when any one node responds. When the `all` quorum

is used, all nodes have to respond for the query to return success. Each query can tune the partition tolerance and availability. Hence, assuming that all key-value stores are the same is a mistake.

Consistency
: During each write operation, we can apply configurations that are similar to applying quorum during read operations; these configurations provide what is known as *tunable consistency*. Higher consistency can be achieved by trading off latency. For a write to be highly consistent, all nodes have to respond, which reduces partition tolerance. Using a *majority quorum* is considered a good trade-off.

Programming language support, product maturity, SQL support, and community
: Key-value databases have good programming language support, and many open source databases have an active community to help learn and understand them. Since most databases have an *HTTP REST API*, they are much easier to interface with.

Read/write priority
: Since key-value databases are aggregate oriented, access to data via a key or ID is geared toward *read* priority. Key-value databases can be used for session storage, and can be used to cache user properties and preferences as well.

Sharding in Databases

The concept of partitioning is well-known in relational databases: the table data is partitioned into sets based on a *schema* on the same database server. *Sharding* is similar to partitioning, but data resides on different servers or nodes. Nodes collaborate to figure out where data exists or where data should be stored based on a sharding key. The word *shard* means horizontal partition (*https://oreil.ly/34AOj*) of data in a database.

Document Databases

Documents such as JSON or XML are the basis of document databases. Documents are human-readable, self-describing, hierarchical tree structures. Document databases are another type of NoSQL database, whose ratings appear in Figure 6-27. These databases understand the structure of the data and can index multiple attributes of the documents, allowing for better query flexibility.

Rating subject	Document databases (MongoDB, CouchDB, Marklogic, etc.)
Ease of learning	☆☆☆
Ease of data modeling	☆☆☆
Scalability/throughput	☆☆
Availability/partition tolerance	☆☆☆
Consistency	☆☆
Programming language support, product maturity, SQL support, community	☆☆☆
Read/write priority	Read ▲ Write

Figure 6-27. Document databases rated for various adoption characteristics

Ease-of-learning curve

Document databases are like key-value databases where the value is human readable. This makes learning the database much easier. Enterprises are used to dealing with documents, such as XML and JSON in different contexts, such as API payloads and JavaScript frontends.

Ease of data modeling

Just like key-value databases, data modeling involves modeling aggregates such as orders, tickets, and other domain objects. Document databases are forgiving when it comes to aggregate design, as the parts of the aggregate are queryable and can be indexed.

Scalability/throughput

Document databases are aggregate oriented and easy to scale. Complex indexing reduces scalability, and increased data size leads to a need for partitioning or sharding. Once sharding is introduced, it increases the complexity and also forces the selection of a sharding key.

Availability/partition tolerance

Like key-value databases, document databases can be configured for higher availability. The setup gets complicated when there are replicated clusters for sharded collections. The cloud providers are trying to make these setups more usable.

Consistency
Some document databases have started supporting ACID transactions within a collection, but this may not work in some edge cases. Just like key-value databases, document databases provide the ability to tune the read and write operations using the quorum mechanism.

Programming language support, product maturity, SQL support, and community
Document databases are the most popular of the NoSQL databases, with an active user community, numerous online learning tutorials, and many programming language drivers that allow for easier adoption.

Read/write priority
Document databases are aggregate oriented and have secondary indexes to query, so these databases are favoring read priority.

Schema-less Databases

One common theme in NoSQL databases is duplication of data and schema attribute names. No two entries have to be the same in terms of schema or attribute names. This introduces interesting change control dynamics and provides flexibility. The schema-less nature of the database is powerful, but it's important to understand that the data always has a schema even if it's implicit or defined elsewhere. The application needs to handle multiple versions of the schema returned by a database. The claim that NoSQL databases are entirely schema-less is misleading.

Column Family Databases

Column family databases, also known as *wide column databases* or *big table databases*, have rows with varying numbers of columns, where each column is a *name-value* pair. With columnar databases, the `name` is known as a *column-key*, the `value` is known as a *column-value*, and the primary key of a `row` is known as a *row key*. Column family databases are another type of NoSQL database that group related data that is accessed at the same time, and whose ratings appear in Figure 6-28.

Rating subject	Column family databases (Cassandra, Scylla, Druid, etc.)
Ease of learning	☆☆
Ease of data modeling	☆
Scalability/throughput	☆☆☆☆
Availability/partition tolerance	☆☆☆☆
Consistency	☆
Programming language support, product maturity, SQL support, community	☆☆
Read/write priority	Read ▲ Write

Figure 6-28. Column family databases rated for various adoption characteristics

Ease-of-learning curve
: Column family databases are difficult to understand. Since a collection of name-value pairs belong to a row, each row can have different name-value pairs. Some name-value pairs can have a map of columns and are known as *super columns*. Understanding how to use these takes practice and time.

Ease of data modeling
: Data modeling with column family databases takes some getting used to. Data needs to be arranged in groups of name-value pairs that have a single row identifier, and designing this row key takes multiple iterations. Some column family databases like Apache Cassandra have introduced a SQL-like query language known as Cassandra Query Language (CQL) that makes data modeling accessible.

Scalability/throughput
: All column family databases are highly scalable and suit use cases where high write or read throughput is needed. Column family databases scale horizontally for read and write operations.

Availability/partition tolerance
: Column family databases naturally operate in clusters, and when some nodes of the cluster are down, it is transparent to the client. The default replication factor is three, which means at least three copies of data are made, improving availability and partition tolerance. Similar to key-value and document databases, column family databases can tune writes and reads based on quorum needs.

Consistency
: Column family databases, like other NoSQL databases, follow the concept of tunable consistency. This means that, based on needs, each operation can decide how much consistency is desired. For example, in `high write` scenarios where some data loss can be tolerated, the write consistency level of `ANY` could be used, which means at least one node has accepted the write, while a consistency level of `ALL` means all nodes have to accept the write and respond success. Similar consistency levels can be applied to read operations. It's a trade-off—higher consistency levels reduce availability and partition tolerance.

Programming language support, product maturity, SQL support, and community
: Column family databases like Cassandra and Scylla have active communities, and the development of SQL-like interfaces has made the adoption of these databases easier.

Read/write priority
: Column family databases use the concepts of *SSTables*, *commit logs*, and *memtables*, and since the name-value pairs are populated when data is present, they can handle sparse data much better than relational databases. They are ideal for high write-volume scenarios.

All NoSQL databases are designed to understand aggregate orientation. Having aggregates improves read and write performance, and also allows for higher availability and partition tolerance when the databases are run as a cluster. The notion of CAP theorem is covered in "Table Split Technique" on page 254 at more length.

Graph Databases

Unlike relational databases, where relations are implied based on references, graph databases use nodes to store entities and their properties. These nodes are connected with *edges*, also known as *relationships*, which are explicit objects. Nodes are organized by relationships and allow for analysis of the connected data by traversing along specific edges.

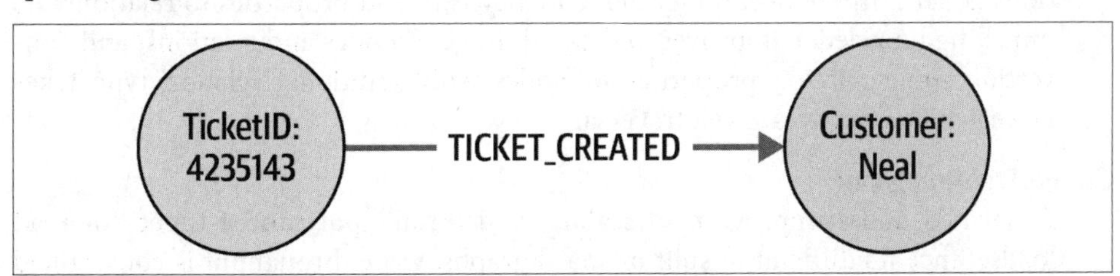

Figure 6-29. In graph databases, direction of the edge has significance when querying

The edges in graph databases have directional significance. In Figure 6-29, an edge of type `TICKET_CREATED` connecting a *ticket* node with ID `4235143` to a *customer* node with ID `Neal`. We can traverse from the ticket node via the *outgoing* edge `TICKET_CREATED` or the customer node via the *incoming* edge `TICKET_CREATED`. When the directions get mixed up, querying the graph becomes really difficult. The ratings for graph databases are illustrated in Figure 6-29.

Rating subject	Graph databases (Neo4J, Infinite Graph, Tigergraph, etc.)
Ease of learning	☆
Ease of data modeling	☆☆
Scalability/throughput	☆☆☆
Availability/partition tolerance	☆☆☆
Consistency	☆☆☆
Programming language support, product maturity, SQL support, community	☆☆
Read/write priority	Read ▲ Write

Figure 6-30. Graph databases rated for various adoption characteristics

Ease-of-learning curve
: Graph databases have a steep learning curve. Understanding how to use the nodes, relations, relation type, and properties takes time.

Ease of data modeling
: Understanding how to model the domains and convert them into nodes and relations is hard. In the beginning, the tendency is to add properties to relations. As modeling knowledge improves, increased usage of nodes and relations, and converting some relation properties to nodes with additional relation type takes place, which improves graph traversal.

Scalability/throughput
: Replicated nodes improve read scaling, and throughput can be tuned for read loads. Since it's difficult to split or shard graphs, write throughput is constrained with the type of graph database picked. Traversing the relationships is very fast, as the indexing and storage is persisted and not calculated at query time.

Availability/partition tolerance
: Some of the graph databases that have high partition tolerance and availability are distributed. Graph database clusters can use nodes that can be promoted as leaders when current leaders are unavailable.

Consistency
: Many graph databases support ACID transactions. Some graph databases, such as Neo4j (*https://neo4j.com*), support transactions, so that data is always consistent.

Programming language support, product maturity, SQL support, and community
: Graph databases have lots of support in the community. Many algorithms, like Dijkstra's algorithm (*https://oreil.ly/TFr1D*) or *node similarity*, are implemented in the database, reducing the need to write them from scratch. The language framework known as Gremlin works across many different databases, helping in the ease of use. Neo4J (*https://neo4j.com*) supports a query language known as Cypher, allowing developers to easily query the database.

Read/write priority
: In graph databases, data storage is optimized for relationship traversal as opposed to relational databases, where we have to query the relationships and derive them at query time. Graph databases are better for read-heavy scenarios.

Graph databases allow the same node to have various types of relationships. In the Sysops Squad example, a sample graph might look as follows: a `knowledge_base` was `created_by` user `sysops_user` and `knowledge_base` `used_by` `sysops_user`. Thus, the relationships `created_by` and `used_by` join the same nodes for different relationship types.

Changing Relationship Types

Changing relationship types is an expensive operation, since each relationship type has to be re-created. When this happens, both nodes connected by the edge have to be visited, the new edge created, and the old edge removed. Hence, edge type or relationship types have to be thought about carefully.

NewSQL Databases

Matthew Aslett first used the term *NewSQL* to define new databases that aimed to provide the scalability of NoSQL databases while supporting the features of relational databases like ACID. NewSQL databases use different types of storage mechanisms, and all of them support SQL.

NewSQL databases, whose ratings appear in Figure 6-31, improve upon relational databases by providing automated data partitioning or sharding, allowing for horizontal scaling and improved availability, while at the same time allowing an easy transition for developers to use the known paradigm of SQL and ACID.

Rating subject	New SQL databases (VoltDB, NuoDB, ClustrixDB, etc.)
Ease of learning	☆☆☆
Ease of data modeling	☆☆☆
Scalability/throughput	☆☆☆
Availability/partition tolerance	☆☆☆
Consistency	☆☆
Programming language support, product maturity, SQL support, community	☆☆
Read/write priority	Read ▲ Write

Figure 6-31. New SQL databases rated for various adoption characteristics

Ease-of-learning curve
: Since NewSQL databases are just like relational databases (with SQL interface, added features of horizontal scaling, ACID compliant), the learning curve is much easier. Some of them are available as only Database as a Service (DBaaS), which may make learning them more difficult.

Ease of data modeling
: Since NewSQL databases are like relational databases, data modeling is familiar to many and easier to pick up. The extra wrinkle is sharding design, allowing sharded data placement in geographically different locations.

Scalability/throughput
: NewSQL databases are designed to support horizontal scaling for distributed systems, allowing for multiple active nodes, unlike relational databases that have only one active leader, and the rest of the nodes are followers. The multiple active nodes allow NewSQL databases to be highly scalable and to have better throughput.

Availability/partition tolerance
: Because of the multiple active nodes design, the benefits to availability can be really high with greater partition tolerance. CockroachDB is a popular NewSQL database that survives disk, machine, and data center failures.

Consistency
: NewSQL databases support strongly consistent ACID transactions. The data is always consistent, and this allows for relational database users to easily transition to NewSQL databases.

Programming language support, product maturity, SQL support, and community
: There are many open source NewSQL databases, so learning them is accessible. Some of the databases also support wire-compatible protocols with existing relational databases, which allows them to replace relational databases without any compatibility problems.

Read/write priority
: NewSQL databases are used just like relational databases, with added benefits of indexing and distributing geographically either to improve read performance or write performance.

Cloud Native Databases

With increased cloud usage, cloud databases such as Snowflake (*https://snowflake.com*), Amazon Redshift (*https://aws.amazon.com/redshift*), Datomic (*https://datomic.com*), and Azure CosmosDB (*https://oreil.ly/Tvkx3*) have gained in popularity. These databases reduce operational burden, provide cost transparency, and are an easy way to experiment since no up-front investments are needed. Ratings for cloud native databases appear in Figure 6-32.

Rating subject	Cloud databases (Snowflake, Amazon Redshift, etc.)
Ease of learning	☆☆
Ease of data modeling	☆☆
Scalability/throughput	☆☆☆☆
Availability/partition tolerance	☆☆☆
Consistency	☆☆☆
Programming language support, product maturity, SQLsupport, community	☆☆
Read/write priority	Read ▲ Write

Figure 6-32. Cloud native databases rated for various adoption characteristics

Ease-of-learning curve
: Some cloud databases like AWS Redshift are like relational databases and therefore are easier to understand. Databases like Snowflake, which have a SQL interface but have different storage and compute mechanisms, require some practice. Datomic is totally different in terms of models and uses `immutable` atomic facts. Thus, the learning curve varies with each database offering.

Ease of data modeling
: Datomic does not have the concept of tables or the need to define attributes in advance. It is necessary to define properties of individual attributes, and entities can have any attribute. Snowflake and Redshift are used more for data warehousing type workloads. Understanding the type of modeling provided by the database is critical in selecting the database to use.

Scalability/throughput
: Since all these databases are cloud only, scaling them is relatively simple since resources can be allocated automatically for a price. In these decisions, the trade-off typically relates to price.

Availability/partition tolerance
: Databases in this category (such as Datomic) are highly available when deployed using *Production Topology*. They have no single point of failure and are supported by extensive caching. Snowflake, for example, replicates its databases across regions and accounts. Other databases in this category support higher availability with various options to configure. For example, Redshift runs in a single availability zone and would need to be run in multiple clusters to support higher availability.

Consistency
: Datomic supports ACID transactions using storage engines to store blocks in block storage. Other databases, like Snowflake and Redshift, support ACID transactions.

Programming language support, product maturity, SQL support, and community
: Many of these databases are new, and finding experienced help can be difficult. Experimenting with these databases requires a cloud account, which can create another barrier. While cloud native databases reduce operational workload on the operational DBAs, they do have a higher learning curve for developers. Datomic uses Clojure (*https://clojure.org*) in all its examples, and stored procedures are written with Clojure, so not knowing Clojure maybe a barrier to usage.

Read/write priority
: These databases can be used for both read-heavy or write-heavy loads. Snowflake and Redshift are geared more toward data warehouse type workloads, lending

them toward read priority, while Datomic can support both type of loads with different indexes such as EAVT (Entity, Attribute, Value, then Transaction) first.

Time-Series Databases

Given the trends, we see increased usage of IoT, microservices, self-driving cars, and observability, all of which have driven a phenomenal increase in time-series analytics. This trend has given rise to databases optimized for storing sequences of data points collected during a time window, enabling users to track changes over any duration of time. The ratings for this database type appear in Figure 6-33.

Rating subject	Time series databases (InfluxDB, TimescaleDB, etc.)
Ease of learning	☆
Ease of data modeling	☆☆
Scalability/throughput	☆☆☆☆
Availability/partition tolerance	☆☆
Consistency	☆☆☆
Programming language support, product maturity, SQL support, community	☆☆
Read/write priority	Read ▲ Write (closer to Read)

Figure 6-33. Time-series databases rated for various adoption characteristics

Ease-of-learning curve
: Understanding time-series data is often easy—every data point is attached to a timestamp, and data is almost always inserted and never updated or deleted. Understanding append-only operations takes some unlearning from other database usage, where errors in the data can be corrected with an update. InfluxDB, Kx, and TimeScale are some of the popular time-series databases.

Ease of data modeling
: The underlying concept with time-series databases is to analyze changes in data over time. For example, with the Sysops Squad example, changes done to a ticket object can be stored in a time-series database, where the `timestamp` of change and `ticket_id` are tagged. It's considered bad practice to add more than one piece of information in one tag. For example, `ticket_status=Open, ticket_id=374737` is better than `ticket_info=Open.374737`.

Scalability/throughput

Timescale is based on PostgreSQL and allows for standard scaling and throughput improvement patterns. Running InfluxDB in cluster mode by using *meta* nodes that manage metadata and *data* nodes that store actual data provides scaling and throughput improvements.

Availability/partition tolerance

Some databases like InfluxDB have better availability and partition tolerance options with configurations for meta and data nodes, along with *replication* factors.

Consistency

Time-series databases that use relational databases as their storage engine get ACID properties for consistency, while other databases can tune consistency using `consistency-level` of `any`, `one`, or `quorum`. Higher consistency-level configuration generally results in higher *consistency* and lower *availability*, so it's a trade-off that needs to be considered.

Programming language support, product maturity, SQL support, and community

Time-series databases have become popular lately, and there are many resources to learn from. Some of these databases, such as InfluxDB, provide a SQL-like query language known as InfluxQL.

Read/write priority

Time-series databases are append only and tend to be better suited for read-heavy workloads.

When using time-series databases, the database automatically attaches a timestamp to every datum creation, and the data contains tags or attributes of information. The data is queried based on some fact between specific time windows. Therefore, time-series databases are not general-purpose databases.

In summary of all the database types discussed in this section, Table 6-6 shows some popular database products for the database type.

Table 6-6. Summary of database types and products in the database type

Database type	Products
Relational	PostgreSQL, Oracle, Microsoft SQL
Key-value	Riak KV, Amazon DynamoDB, Redis
Document	MongoDB, Couchbase, AWS DocumentDB
Column family	Cassandra, Scylla, Amazon SimpleDB
Graph	Neo4j, Infinite Graph, Tiger Graph
NewSQL	VoltDB, ClustrixDB, SimpleStore (aka MemSQL)

Database type	Products
Cloud native	Snowflake, Datomic, Redshift
Time-series	InfluxDB, kdb+, Amazon Timestream

Sysops Squad Saga: Polyglot Databases

Thursday, December 16, 16:05

Now that the team had formed data domains from the monolithic Sysops Squad database, Devon noticed that the Survey data domain would be a great candidate for migrating from a traditional relational database to a document database using JSON. However, Dana, the head of data architecture, didn't agree and wanted to keep the tables as relational.

"I simply don't agree," said Dana. "The survey tables have always worked in the past as relational tables, so I see no reason to change things around. "

"Actually," said Skyler, "if you had originally talked with us about this when the system was first being developed, you would understand that from a user interface perspective, it's really hard to deal with relational data for something like a customer survey. So I disagree. It may work out good for you, but from a user interface development standpoint, dealing with relational data for the survey stuff has been a major pain point."

"See, so there you are," said Devon. "This is why we need to change it to a document database."

"You seem to forget that as the data architect for this company, I am the one who has ultimate responsibility for all these different databases. You can't just start adding different database types to the system," said Dana.

"But it would be a much better solution," said Devon.

"Sorry, but I'm not going to cause a disruptor on the database teams just so Skyler can have an easier job maintaining the user interface. Things don't work that way."

"Wait," said Skyler, "didn't we all agree that part of the problem of the current monolithic Sysops Squad application was that the development teams didn't work close enough with the database teams?"

"Yes," said Dana.

"Well then," said Skyler, "let's do that. Let's work together to figure this out."

"OK," said Dana, "but what I'm going to need from you and Devon is a good solid justification for introducing another type of database into the mix."

"You got it," said Devon. "We'll start working on that right away."

Devon and Skyler knew that a document database would be a much better solution for the customer survey data, but they weren't sure how to build the right justifications for Dana to agree to migrate the data. Skyler suggested that they meet with Addison to get some help, since both agreed that this was somewhat an architectural concern. Addison agreed to help, and set up a meeting with Parker (the Sysops Squad product owner) to validate whether there was any business justification to migrating the customer survey tables to a document database.

"Thanks for meeting with us, Parker," said Addison. "As I mentioned to you before, we are thinking of changing the way the customer survey data is stored, and have a few questions for you."

"Well," said Parker, "that was one of the reasons why I agreed to this meeting. You see, the customer survey part of the system has been a major pain point for the marketing department, as well as for me."

"Huh?" asked Skyler. "What do you mean?"

"How long does it take you to apply even the smallest of change requests to the customer surveys?" asked Parker.

"Well," said Devon, "it's not too bad from the database side. I mean, it's a matter of adding a new column for a new question or changing the answer type."

"Hold on," said Skyler. "Sorry, but for me it's a major change, even when you add an additional question. You have no idea how hard it is to query all of that relational data and render a customer survey in the user interface. So, my answer is, a *very long time*."

"Listen," said Parker. "We on the business side of things get very frustrated ourselves when even the simplest of changes take you literally days to do. It's simply not acceptable."

"I think I can help here," said Addison. "So Parker, what you're saying is that the customer survey changes frequently, and it is taking too long to make the changes?"

"Correct," said Parker. "The marketing department not only wants better flexibility in the customer surveys, but better response from the IT department as well. Many times they don't place change requests because they know it will just end in frustration and additional cost they didn't plan for."

"What if I were to tell you that the lack of flexibility and responsiveness to change requests has everything to do with the technology used to store customer surveys, and that by changing the way we store data, we could significantly improve flexibility as well as response time for change requests?" asked Addison.

"Then I would be the happiest person on Earth, as would the marketing department," said Parker.

"Devon and Skyler, I think we have our business justification," said Addison.

With the business justification established, Devon, Skyler, and Addison convinced Dana to use a document database. Now the team had to figure out the optimal structure for the customer survey data. The existing relational database tables are illustrated in Figure 6-34. Each customer survey consisted of two primary tables—a Survey table and a Question table, with a one-to-many relationship between the two tables.

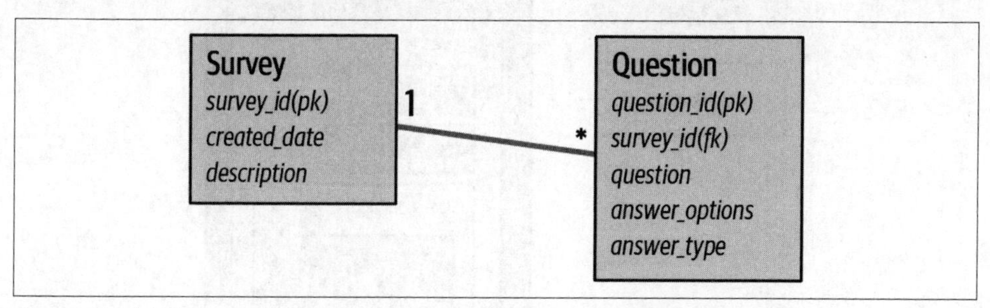

Figure 6-34. Tables and relationships in the sysops survey data domain

An example of the data contained in each table is shown Figure 6-35, where the Question table contains the question, the answer options, and the data type for the answer.

Survey

survey_id	created_date	description
19998	May 20 2022	Expert performance survey.
19999	May 20 2022	Service satisfaction survey.

Question

question_id	survey_id	question	answer_options	answer_type
50000	19999	Did the..	{Yes,No}	Boolean
50001	19999	Rate..	{1,2,3,4,5}	Option

Figure 6-35. Relational data in tables for survey and question in the survey data domain

"So, essentially we have two options for modeling the survey questions in a document database," said Devon. "A single aggregate document or one that is split."

"How to we know which one to use?" asked Skyler, happy that the development teams were now finally working with the database teams to arrive at a unified solution.

"I know," said Addison, "let's model both so we can visually see the trade-offs with each approach."

Devon showed the team that with the single aggregate option, as shown in Figure 6-36, with the corresponding source code listing in Example 6-3, both the survey data and all related question data were stored as one document. Therefore, the entire customer survey could be retrieved from the database by using a single **get** operation, making it easy for Skyler and others on the development team to work with the data.

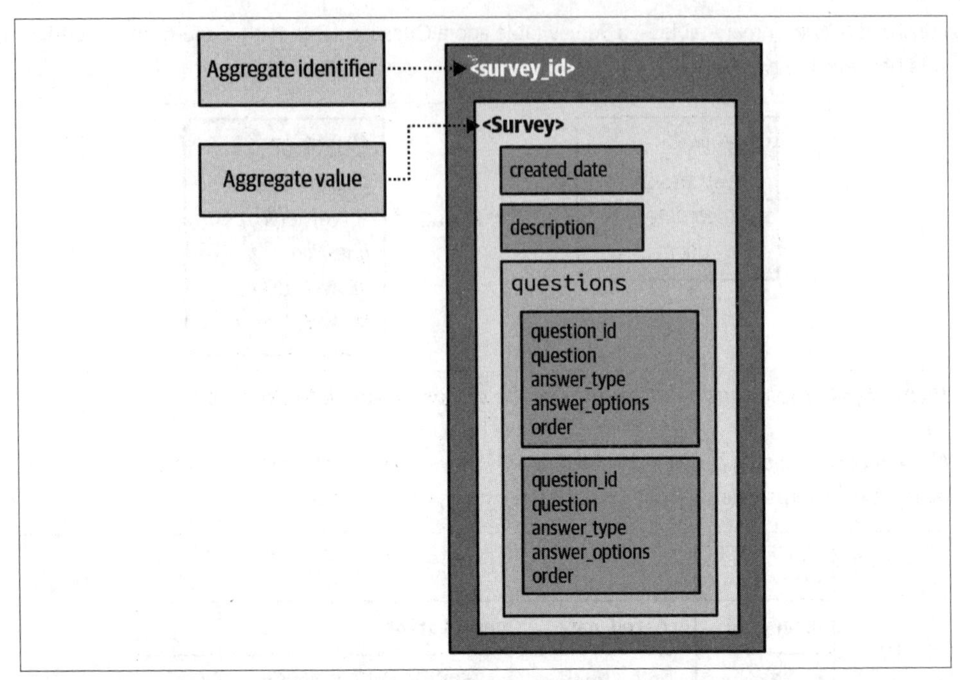

Figure 6-36. Survey model with single aggregate

Example 6-3. JSON document for single aggregate design with children embedded

```
# Survey aggregate with embedded questions
{
    "survey_id": "19999",
    "created_date": "Dec 28 2021",
    "description": "Survey to gauge customer...",
    "questions": [
        {
            "question_id": "50001",
            "question": "Rate the expert",
            "answer_type": "Option",
            "answer_options": "1,2,3,4,5",
            "order": "2"
        },
        {
            "question_id": "50000",
            "question": "Did the expert fix the problem?",
            "answer_type": "Boolean",
            "answer_options": "Yes,No",
            "order": "1"
        }
    ]
}
```

"I really like that approach," said Skyler. "Essentially, I wouldn't have to worry so much about aggregating things myself in the user interface, meaning I could simply render the document I retrieve on the web page."

"Yeah," said Devon, "but it would require additional work on the database side as questions would be replicated in each survey document. You know, the whole reuse argument. Here, let me show you the other approach."

Skyler explained that another way to think about aggregates was to split the *survey* and *question* model so that the questions could be operated on in an independent fashion, as shown in Figure 6-37, with the corresponding source code listing in Example 6-4. This would allow the same question to be used in multiple surveys, but would be harder to render and retrieve than the single aggregate.

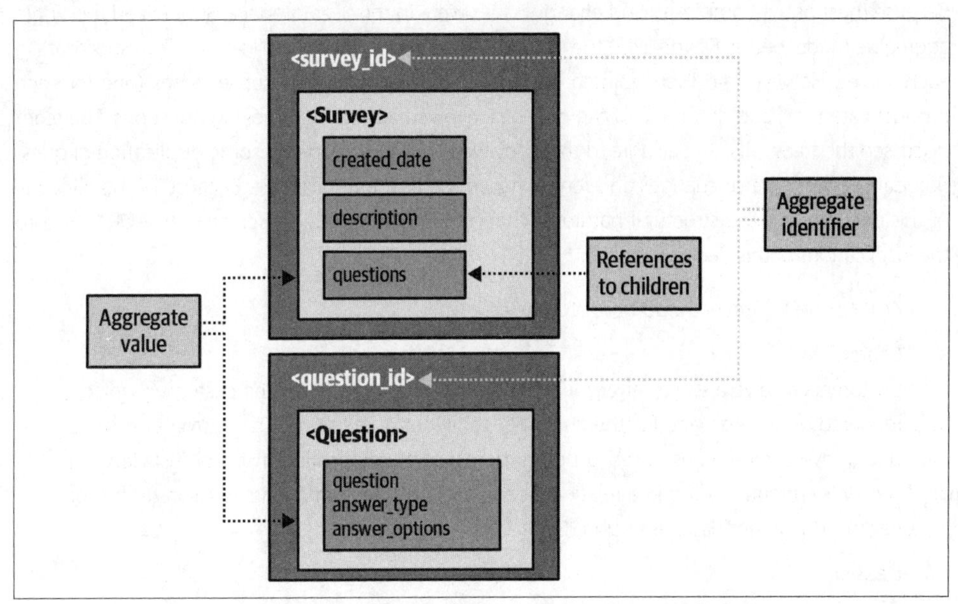

Figure 6-37. Survey model with multiple aggregates with references

Example 6-4. JSON document with aggregates split and parent document showing references to children

```
# Survey aggregate with references to Questions
{
    "survey_id": "19999",
    "created_date": "Dec 28",
    "description": "Survey to gauge customer...",
    "questions": [
        {"question_id": "50001", "order": "2"},
        {"question_id": "50000", "order": "1"}
    ]
```

```
}
# Question aggregate
{
    "question_id": "50001",
    "question": "Rate the expert",
    "answer_type": "Option",
    "answer_options": "1,2,3,4,5"
}
{
    "question_id": "50000",
    "question": "Did the expert fix the problem?",
    "answer_type": "Boolean",
    "answer_options": "Yes,No"
}
```

Because most of the complexity and change issues were in the user interface, Skyler liked the single aggregate model better. Devon liked the multiple aggregate to avoid duplication of question data in each survey. However, Addison pointed out that there were only five survey types (one for each product category), and that most of the changes involved adding or removing questions. The team discussed the trade-offs, and all agreed that they were willing to trade off some duplication of question data for the ease of changes and rendering on the user interface side. Because of the difficulty of this decision and the structural nature of changing the data, Addison created an ADR to record the justifications of this decision:

> *ADR: Use of Document Database for Customer Survey*
>
> *Context*
> Customers receive a survey after the work has been completed by the customer, which is rendered on a web page for the customer to fill out and submit. The customer receives one of five survey types based on the type of electronic product fixed or installed. The survey is currently stored in a relational database, but the team wants to migrate the survey to a document database using JSON.
>
> *Decision*
> We will use a document database for the customer survey.
>
> The Marketing Department requires more flexibility and timeliness for changes to the customer surveys. Moving to a document database would not only provide better flexibility, but also better timeliness for changes needed to the customer surveys.
>
> Using a document database would simplify the customer survey user interface and better facilitate changes to the surveys.
>
> *Consequences*
> Since we will be using a single aggregate, multiple documents would need to be changed when a common survey question is updated, added, or removed.
>
> Survey functionality will need to be shut down during the data migration from the relational database to the document database.

CHAPTER 7

Service Granularity

Thursday, October 14, 13:33

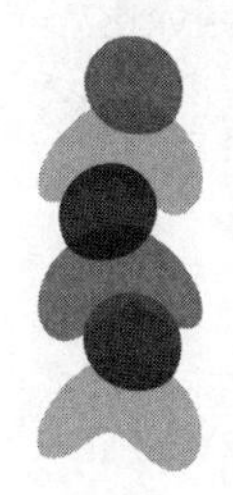

As the migration effort got underway, both Addison and Austen started getting overwhelmed with all of the decisions involved with breaking apart the domain services previously identified. The development team also had its own opinions, which made decision making for service granularity even more difficult.

"I'm still not sure what to do with the core ticketing functionality," said Addison. "I can't decide whether ticket creation, completion, expert assignment, and expert routing should be one, two, three, or even four services. Taylen is insisting on making everything fine-grained, but I'm not sure that's the right approach."

"Me neither," said Austen. "And I've got my own issues trying to figure out if the customer registration, profile management, and billing functionality should even be broken apart. And on top of all that, I've got another game this evening."

"You've always got a game to go to," said Addison. "Speaking of customer functionality, did you ever figure out if the customer login functionality is going to be a separate service?"

"No," said Austen, "I'm still working on that as well. Skyler says it should be separate, but won't give me a reason other than to say it's separate functionality."

"This is hard stuff," said Addison. "Do you think Logan can shed any light on this?"

"Good idea," said Austen, "This seat-of-the-pants analysis is really slowing things down."

Addison and Austen invited Taylen, the Sysops Squad tech lead, to the meeting with Logan so that all of them could be on the same page with regard to the service granularity issues they were facing.

"I'm telling you," said Taylen, "we need to break up the domain services into smaller services. They are simply too coarse-grained for microservices. From what I remember, *micro* means small. We are, after all, moving to microservices. What Addison and Austen are suggesting simply doesn't fit with the microservices model."

"Not every portion of an application has to be microservices," said Logan. "That's one of the biggest pitfalls of the microservices architecture style."

"If that's the case, then how do you determine what services should and shouldn't be broken apart?" asked Taylen.

"Let me ask you something, Taylen," said Logan. "What is your reason for wanting to make all of the services so small?"

"Single-responsibility principle," answered Taylen. "Look it up. That's what microservices is based on."

"I know what the single-responsibility principle is," said Logan. "And I also know how subjective it can be. Let's take our customer notification service as an example. We can notify our customers through SMS, email, and we even send out postal letters. So tell me everyone, one service or three services?"

"Three," immediately answered Taylen. "Each notification method is its own thing. That's what microservices is all about."

"One," answered Addison. "Notification itself is clearly a single responsibility."

"I'm not sure," answered Austen. "I can see it both ways. Should we just toss a coin?"

"This is exactly why we need help," sighed Addison.

"The key to getting service granularity right," said Logan, "is to remove opinion and gut feeling, and use granularity disintegrators and integrators to objectively analyze the trade-offs and form solid justifications for whether or not to break apart a service."

"What are granularity disintegrators and integrators?" asked Austen.

"Let me show you," said Logan.

Architects and developers frequently confuse the terms *modularity* and *granularity*, and in some cases even treat them to mean the same thing. Consider the following dictionary definitions of each of these terms:

Modularity
: Constructed with standardized units or dimensions for flexibility and variety in use.

Granularity
: Consisting of or appearing to consist of one of numerous particles forming a larger unit.

It's no wonder so much confusion exists between these terms! Although the terms have similar dictionary definitions, we want to distinguish between them because they mean different things within the context of software architecture. In our usage, *modularity* concerns breaking up systems into separate parts (see Chapter 3), whereas *granularity* deals with the *size* of those separate parts. Interestingly enough, most issues and challenges within distributed systems are typically not related to modularity, but rather granularity.

Determining the right level of granularity—the size of a service—is one of the many hard parts of software architecture that architects and development teams continually struggle with. Granularity is not defined by the number of classes or lines of code in a service, but rather what the service does—hence why it is so hard to get service granularity right.

Architects can leverage metrics to monitor and measure various aspects of a service to determine the appropriate level of service granularity. One such metric used to objectively measure the size of a service is to calculate the number of statements in a service. Every developer has a different coding style and technique, which is why the number of classes and number of lines of code are poor metrics to use to measure granularity. The number of statements, on the other hand, at least allows an architect or development team to objectively measure *what* the service is doing. Recall from Chapter 4 that a *statement* is a single complete action performed in the source code, usually terminated by a special character (such as a semicolon in languages such as Java, C, C++, C#, Go, JavaScript; or a newline in languages such as F#, Python, and Ruby).

Another metric to determine service granularity is to measure and track the number of *public* interfaces or operations exposed by a service. Granted, while there is still a bit of subjectiveness and variability with these two metrics, it's the closest thing we've come up with so far to objectively measure and assess service granularity.

Two opposing forces for service granularity are granularity disintegrators and granularity integrators. These opposing forces are illustrated in Figure 7-1. *Granularity disintegrators* address the question "When should I consider breaking apart a service into smaller parts?", whereas *Granularity integrators* address the question "When should I consider putting services back together?" One common mistake many development teams make is focusing too much on granularity disintegrators while ignoring granularity integrators. The secret of arriving at the appropriate level of granularity for a service is achieving an equilibrium between these two opposing forces.

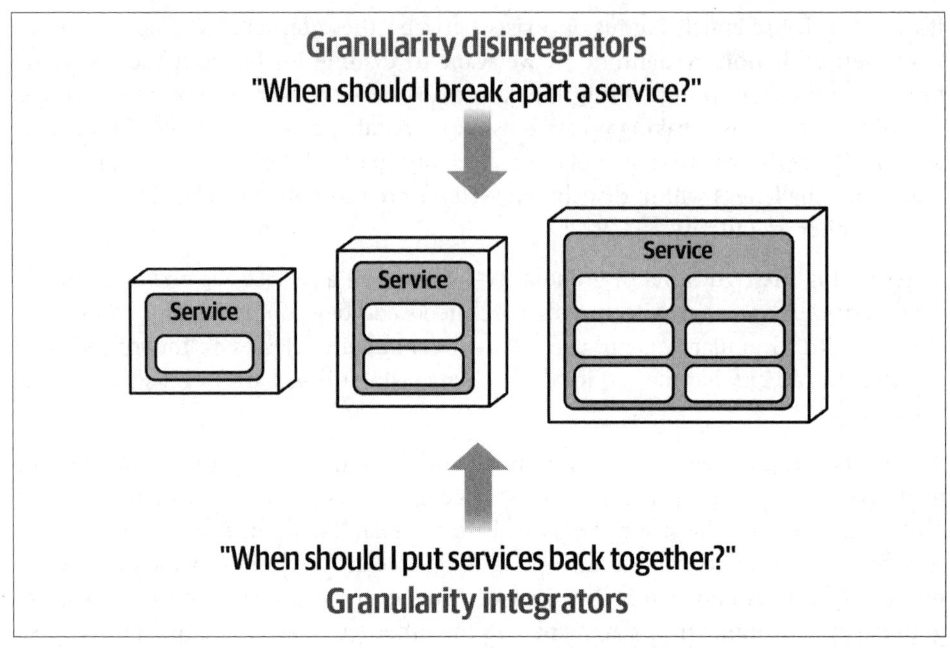

Figure 7-1. Service granularity depends on a balance of disintegrators and integrators

Granularity Disintegrators

Granularity disintegrators provide guidance and justification for when to break a service into smaller pieces. While the justification for breaking up a service may involve only a single driver, in most cases the justification will be based on multiple drivers. The six main drivers for granularity disintegration are as follows:

Service scope and function
: Is the service doing too many unrelated things?

Code volatility
: Are changes isolated to only one part of the service?

Scalability and throughput
: Do parts of the service need to scale differently?

Fault tolerance
: Are there errors that cause critical functions to fail within the service?

Security
: Do some parts of the service need higher security levels than others?

Extensibility
: Is the service always expanding to add new contexts?

The following sections detail each of these granularity disintegration drivers.

Service Scope and Function

The service scope and function is the first and most common driver for breaking up a single service into smaller ones, particularly with regard to microservices. There are two dimensions to consider when analyzing the service scope and function. The first dimension is *cohesion*: the degree and manner to which the operations of a particular service interrelate. The second dimension is the overall *size* of a component, measured usually in terms of the total number of statements summed from the classes that make up that service, the number of public entrypoints into the service, or both.

Consider a typical Notification Service that does three things: notifies a customer through SMS (Short Message Service (*https://oreil.ly/caVCG*)), email, or a printed postal letter that is mailed to the customer. Although it is very tempting to break this service into three separate single-purpose services (one for SMS, one for email, and one for postal letters) as illustrated in Figure 7-2, this alone is not enough to justify breaking the service apart because it already has relatively strong cohesion—all of these functions relate to one thing, notifying the customer. Because "single purpose" is left for individual opinion and interpretation, it is difficult to know whether to break apart this service or not.

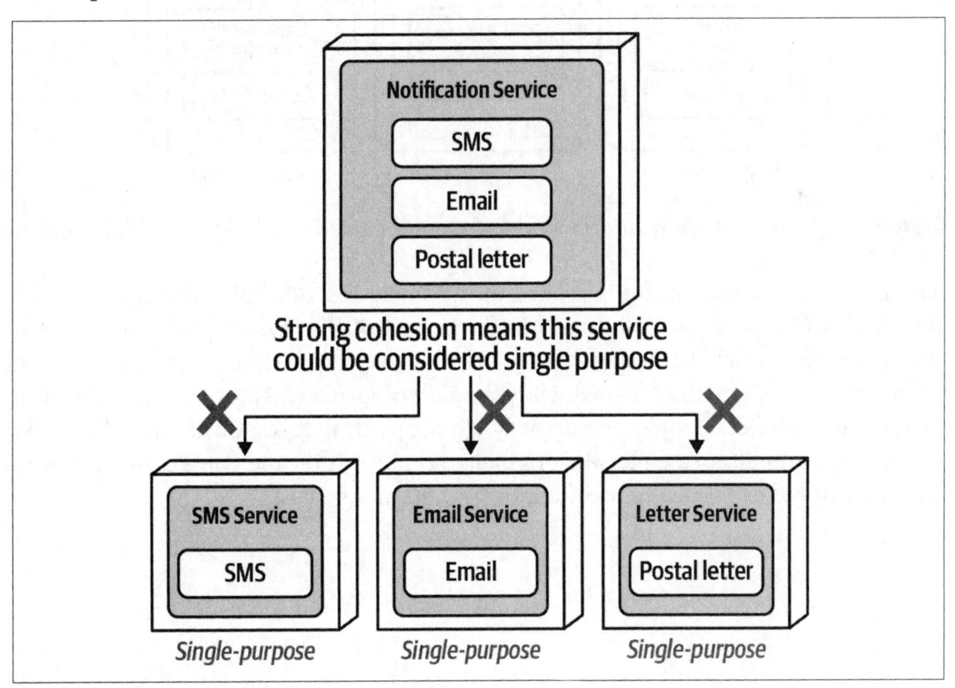

Figure 7-2. A service with relatively strong cohesion is not a good candidate for disintegration based on functionality alone

Now consider a single service that manages the customer profile information, customer preferences, and also customer comments made on the website. Unlike the previous Notification Service example, this service has relatively weak cohesion because these three functions relate to a broader scope—customer. This service is possibly doing too much, and hence should probably be broken into three separate services, as illustrated in Figure 7-3.

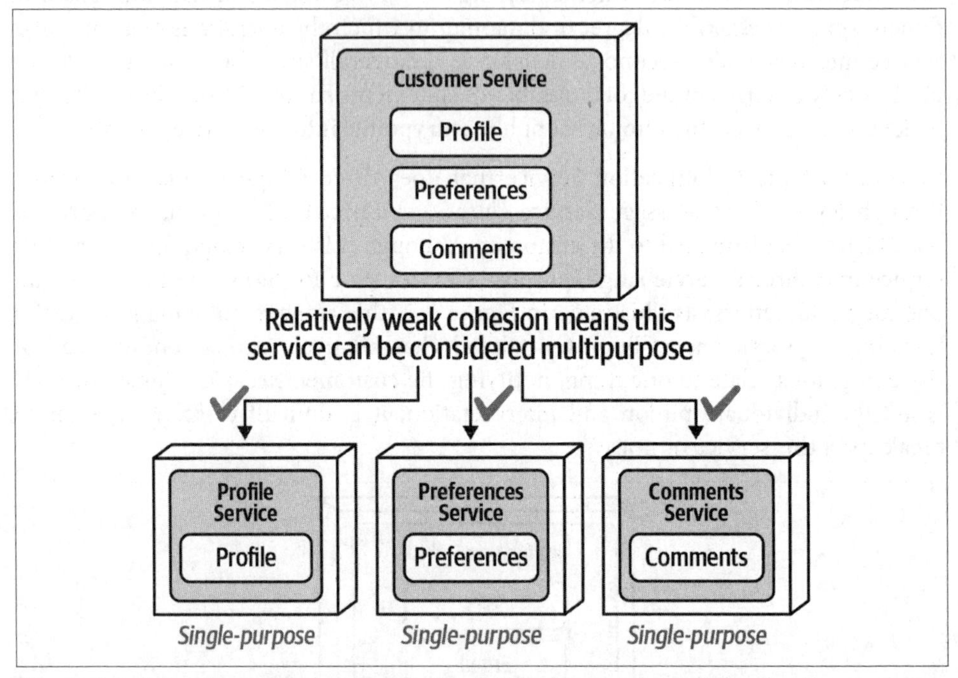

Figure 7-3. A service with relatively weak cohesion is a good candidate for disintegration

This granularity disintegrator is related to the single-responsibility principle (*https://oreil.ly/JZpcT*) coined by Robert C. Martin as part of his SOLID principles (*https://oreil.ly/r64Yw*), which states, "every class should have responsibility over a single part of that program's functionality, which it should encapsulate. All of that module, class or function's services should be narrowly aligned with that responsibility." While the single-responsibility principle was originally scoped within the context of classes, in later years it has expanded to include components and services.

Within the microservices architecture style, a *microservice* is defined as a single-purpose, separately deployed unit of software that does *one thing* really well. No wonder developers are so tempted to make services as small as possible without considering why they are doing so! The subjectiveness related to what is and isn't a single responsibility is where most developers get into trouble with regard to service granularity. While there are some metrics (such as LCOM (*https://oreil.ly/qOtdg*)) to measure cohesion, it is nevertheless highly subjective when it comes to services—is notifying the customer one single thing, or is notifying via email one single thing? For this reason, it is vital to understand other granularity disintegrators to determine the appropriate level of granularity.

Code Volatility

Code volatility--the rate at which the source code changes—is another good driver for breaking a service into smaller ones. This is also known as *volatility-based decomposition*. Objectively measuring the frequency of code changes in a service (easily done through standard facilities in any source code version-control system) can sometimes lead to a good justification for breaking apart a service. Consider the notification service example again from the prior section. Service scope (cohesion) alone was not enough to justify breaking the service apart. However, by applying change metrics, relevant information is revealed about the service:

- SMS notification functionality rate of change: every six months (avg)
- Email notification functionality rate of change: every six months (avg)
- Postal letter notification functionality rate of change: weekly (avg)

Notice that the postal letter functionality changes weekly (on average), whereas the SMS and email functionality rarely changes. As a single service, any change to the postal letter code would require the developer to test and redeploy the entire service, including SMS and email functionality. Depending on the deployment environment, this also might mean SMS and email functionality would not be available when the postal letter changes are deployed. Thus, as a single service, testing scope is increased and deployment risk is high. However, by breaking this service into two separate services (Electronic Notification and Postal Letter Notification), as illustrated in Figure 7-4, frequent changes are now isolated into a single, smaller service. This in turn means that the testing scope is significantly reduced, deployment risk is lower, and SMS and email functionality is not disrupted during a deployment of postal letter changes.

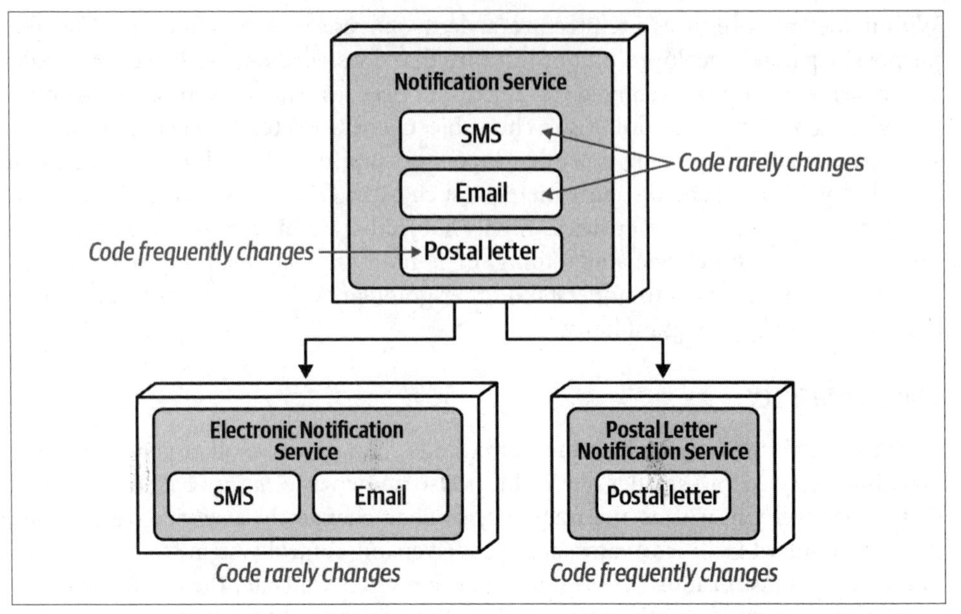

Figure 7-4. An area of high code change in a service is a good candidate for disintegration

Scalability and Throughput

Another driver for breaking up a service into separate smaller ones is *scalability* and *throughput*. The scalability demands of different functions of a service can be objectively measured to qualify whether a service should be broken apart. Consider once again the Notification Service example, where a single service notifies customers through SMS, email, and printed postal letter. Measuring the scalability demands of this single service reveals the following information:

- SMS notification: 220,000/minute
- Email notification: 500/minute
- Postal letter notification: 1/minute

Notice the extreme variation between sending out SMS notifications and postal letter notifications. As a single service, email and postal letter functionality must unnecessarily scale to meet the demands of SMS notifications, impacting cost and also elasticity in terms of mean time to startup (MTTS). Breaking the Notification Service into three separate services (SMS, Email, and Letter), as illustrated in Figure 7-5, allows each of these services to scale independently to meet their varying demands of throughput.

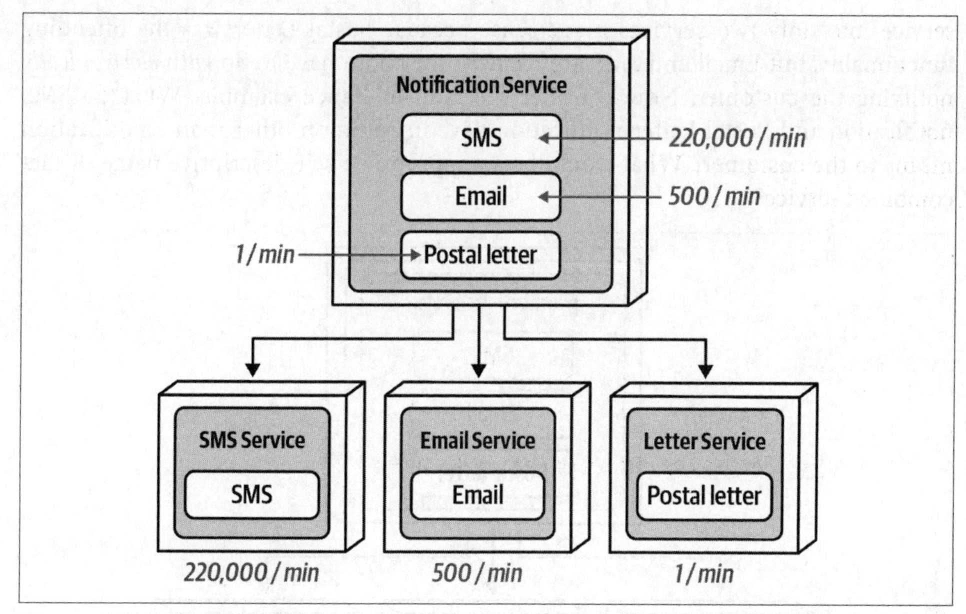

Figure 7-5. Differing scalability and throughput needs is a good disintegration driver

Fault Tolerance

Fault tolerance describes the ability of an application or functionality within a particular domain to continue to operate, even though a fatal crash occurs (such as an out-of-memory condition). Fault Tolerance is another good driver for granularity disintegration.

Consider the same consolidated Notification Service example that notifies customers through SMS, email, and postal letter (Figure 7-6). If the email functionality continues to have problems with out-of-memory conditions and fatally crashes, the entire service comes down, including SMS and postal letter processing.

Separating this single consolidated Notification Service into three separate services provides a level of fault tolerance for the domain of customer notification. Now, a fatal error in the functionality of the email service doesn't impact SMS or postal letters.

Notice in this example that the Notification Service is split into three separate services (SMS, Email, and Postal Letter), even though email functionality is the only issue with regard to frequent crashes (the other two are very stable). Since email functionality is the only issue, why not combine the SMS and postal letter functionality into a single service?

Consider the code volatility example from the prior section. In this case Postal Letter changes constantly, whereas the other two (SMS and Email) do not. Splitting this service into only two services made sense because Postal Letter was the offending functionality, but Email and SMS are *related*—they both have to do with *electronically* notifying the customer. Now consider the fault-tolerance example. What do SMS notification and Postal Letter notification have in common other than a notification means to the customer? What would be an appropriate self-descriptive name of that combined service?

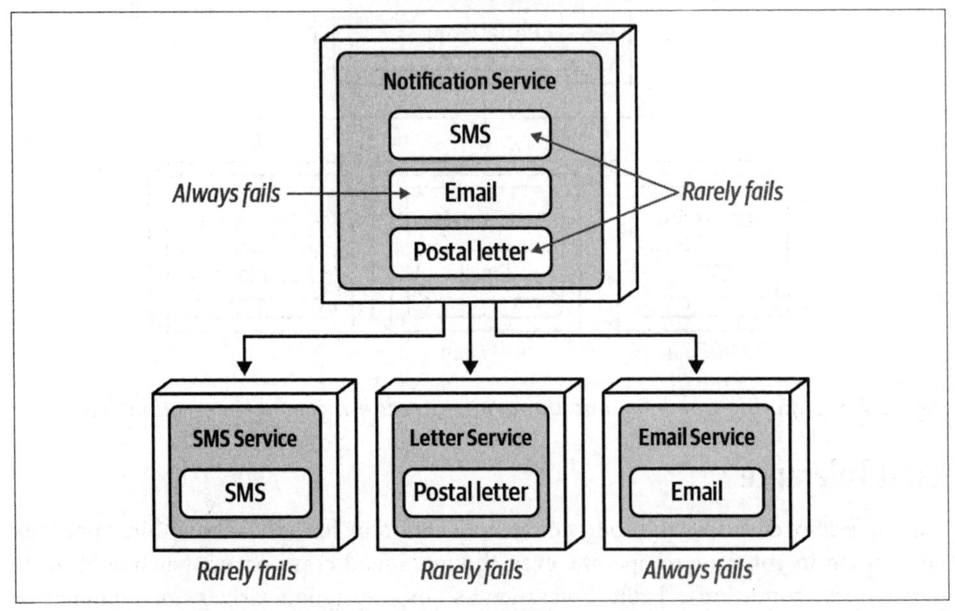

Figure 7-6. Fault tolerance and service availability are good disintegration drivers

Moving the email functionality to a separate service disrupts the overall *domain cohesion* because the resulting cohesion between SMS and postal letter functionality is weak. Consider what the likely service names would be: Email Service and...Other Notification Service? Email Service and...SMS-Letter Notification Service? Email Service and...Non-Email Service? This naming problem relates back to the service scope and function granularity disintegrator—if a service is too hard to name because it's doing multiple things, then consider breaking apart the service. The following disintegrations help in visualizing this important point:

- Notification Service → Email Service, Other Notification Service (poor name)
- Notification Service → Email Service, Non-Email Service (poor name)
- Notification Service → Email Service, SMS-Letter Service (poor name)
- Notification Service → Email Service, SMS Service, Letter Service (good names)

In this example, only the last disintegration makes sense, particularly considering the addition of another social media notification—where would that go? Whenever breaking apart a service, regardless of the disintegration driver, always check to see if strong cohesion can be formed with the "leftover" functionality.

Security

A common pitfall when securing sensitive data is to think only in terms of the storage of that data. For example, securing PCI (Payment Card Industry (*https://oreil.ly/Z5QRV*)) data from non-PCI data might be addressed through separate schemas or databases residing in different secure regions. What is sometimes missing from this practice, however, is also securing *how* that data is accessed.

Consider the example illustrated in Figure 7-7 that describes a Customer Profile Service containing two main functions: customer profile maintenance for adding, changing, or deleting basic profile information (name, address, and so on); and customer credit card maintenance for adding, removing, and updating credit card information.

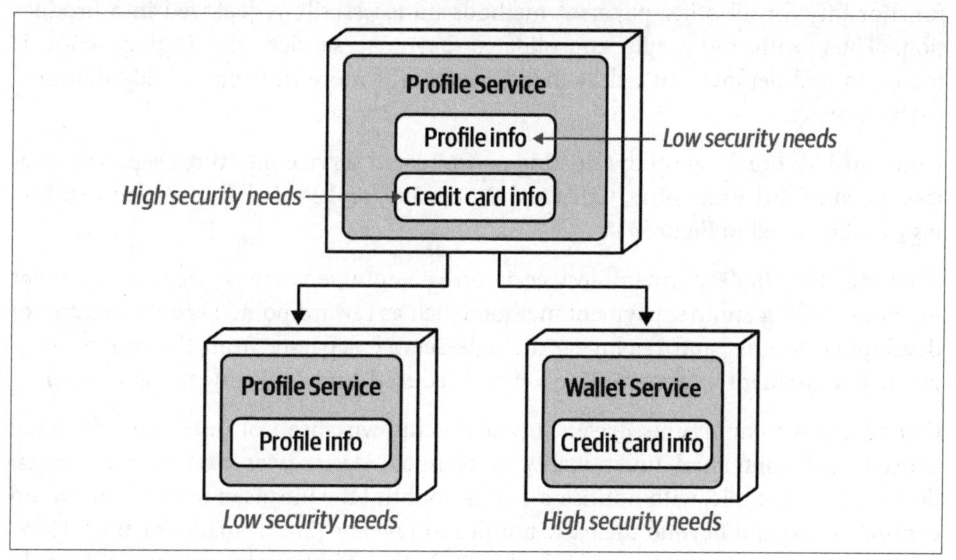

Figure 7-7. Security and data access are good disintegration drivers

While the credit card *data* may be protected, *access* to that data is at risk because the credit card functionality is joined together with the basic customer profile functionality. Although the API entry points into the consolidated customer profile service may differ, nevertheless there is risk that someone entering into the service to retrieve the customer name might also have access to credit card functionality. By breaking this service into two separate services, access to the *functionality* used to maintain credit card information can be made more secure because the set of credit card operations is going into only a single-purpose service.

Extensibility

Another primary driver for granularity disintegration is_ extensibility_—the ability to add additional functionality as the service context grows. Consider a payment service that manages payments and refunds through multiple payment methods, including credit cards, gift cards, and PayPal transactions. Suppose the company wants to start supporting other managed payment methods, such as reward points, store credit from returns; and other third-party payment services, such as ApplePay, SamsungPay, and so on. How easy is it to extend the payment service to add these additional payment methods?

These additional payment methods could certainly be added to a single consolidated payment service. However, every time a new payment method is added, the entire payment service would need to be tested (including other payment types), and the functionality for all other payment methods unnecessarily redeployed into production. Thus, with the single consolidated payment service, the testing scope is increased and deployment risk is higher, making it more difficult to add additional payment types.

Now consider breaking up the existing consolidated service into three separate services (Credit Card Processing, Gift Card Processing, and PayPal Transaction Processing), as illustrated in Figure 7-8.

Now that the single payment service is broken into separate services by payment methods, adding another payment method (such as reward points) is only a matter of developing, testing, and deploying a single service separate from the others. As a result, development is faster, testing scope is reduced, and deployment risk is lower.

Our advice is to apply this driver only if it is known ahead of time that additional consolidated contextual functionality is planned, desired, or part of the normal domain. For example, with notification, it is doubtful the means of notification would continually expand beyond the basic notification means (SMS, email, or letter). However, with payment processing, it is highly likely that additional payment types would be added in the future, and therefore separate services for each payment type would be warranted. Since it is often difficult to sometimes "guess" whether (and when) contextual functionality might expand (such as additional payment methods), our advice is to wait on this driver as a primary means of justifying a granularly disintegration until a pattern can be established or confirmation of continued extensibility can be confirmed.

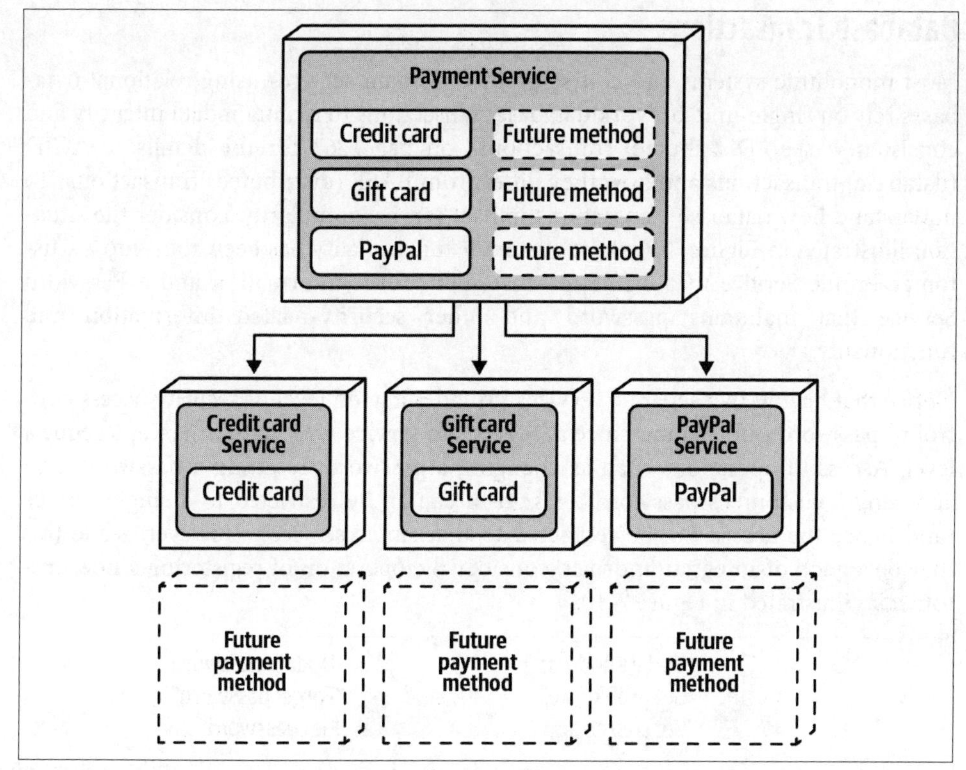

Figure 7-8. Planned extensibility is a good disintegration driver

Granularity Integrators

Whereas granularity disintegrators provide guidance and justification for when to break a service into smaller pieces, granularity integrators work in the opposite way—they provide guidance and justification for putting services back together (or not breaking apart a service in the first place). Analyzing the trade-offs between disintegration drivers and integration drivers is the secret to getting service granularity right. The four main drivers for granularity integration are as follows:

Database transactions
: Is an ACID transaction required between separate services?

Workflow and choreography
: Do services need to talk to one another? Shared code: Do services need to share codeamong one another? Database relationships: Although a service can be broken apart, can the data it uses be broken apart as well?

The following sections detail each of these granularity integration drivers.

Database Transactions

Most monolithic systems and course-grained domain services using relational databases rely on single-unit-of-work database transactions to maintain data integrity and consistency; see "Distributed Transactions" on page 263 for the details of ACID (database) transactions and how they differ from BASE (distributed) transactions. To understand how database transactions impact service granularity, consider the situation illustrated in Figure 7-9 where customer functionality has been split into a Customer Profile Service that maintains customer profile information and a Password Service that maintains password and other security-related information and functionality.

Notice that having two separate services provides a good level of security access control to password information since access is at a service level rather than at a request level. Access to operations such as changing a password, resetting a password, and accessing a customer's password for sign-in can all be restricted to a single service (and hence the access can be restricted to that single service). However, while this may be a good disintegration driver, consider the operation of registering a new customer, as illustrated in Figure 7-10.

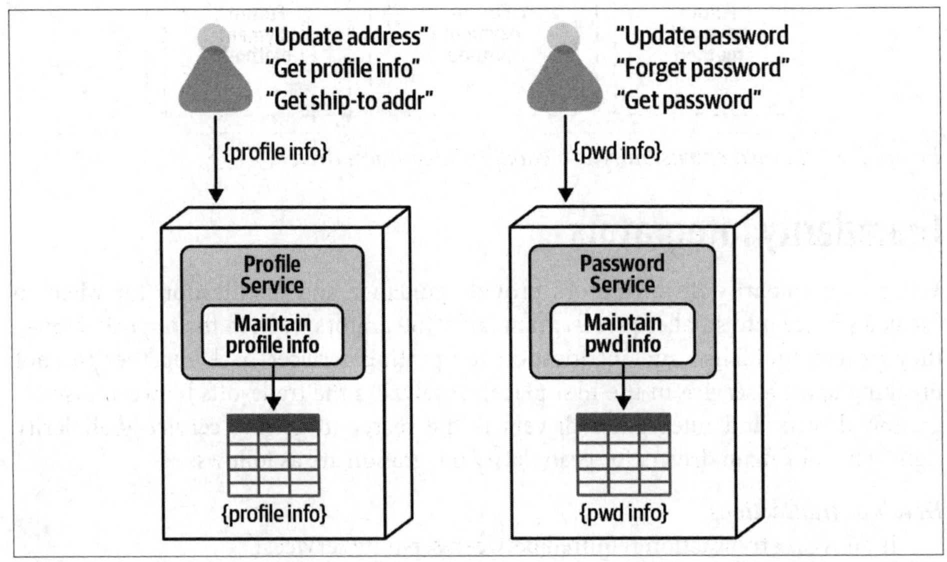

Figure 7-9. Separate services with atomic operations have better security access control

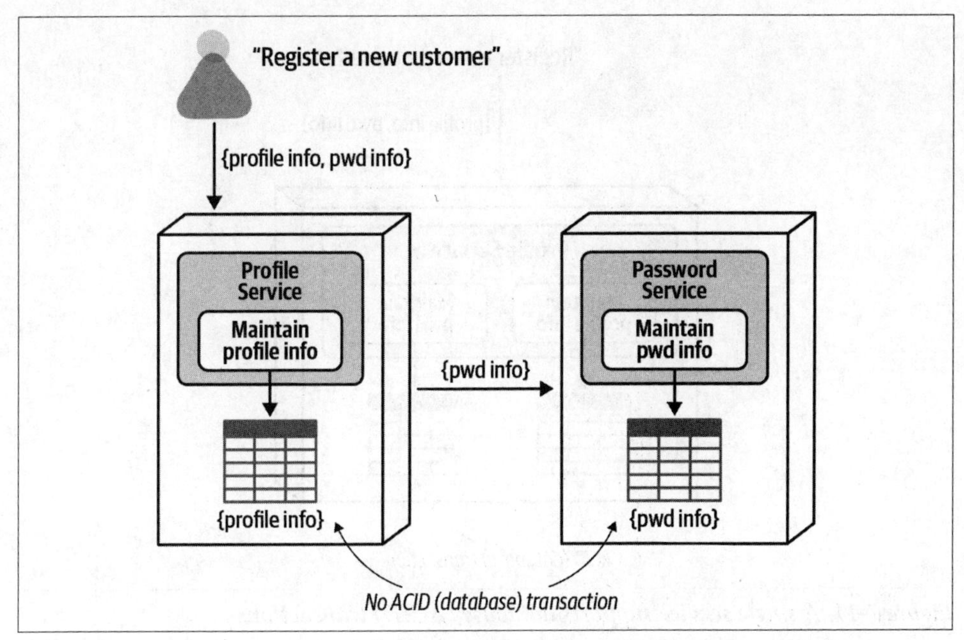

Figure 7-10. Separate services with combined operations do not support database (ACID) transactions

When registering a new customer, both profile and encrypted password information is passed into the Profile Service from a user interface screen. The Profile Service inserts the profile information into its corresponding database table, commits that work, and then passes the encrypted password information to the Password Service, which in turn inserts the password information into its corresponding database table and commits its own work.

While separating the services provides better security access control to the password information, the trade-off is that there is no ACID transaction for actions such as registering a new customer or unsubscribing (deleting) a customer from the system. If the password service fails during either of these operations, data is left in an inconsistent state, resulting in complex error handling (which is also error prone) to reverse the original profile insert or take other corrective action (see "Transactional Saga Patterns" on page 324 for the details of eventual consistency and error handling within distributed transactions). Thus, if having a single-unit-of-work ACID transaction is required from a business perspective, these services should be consolidated into a single service, as illustrated in Figure 7-11.

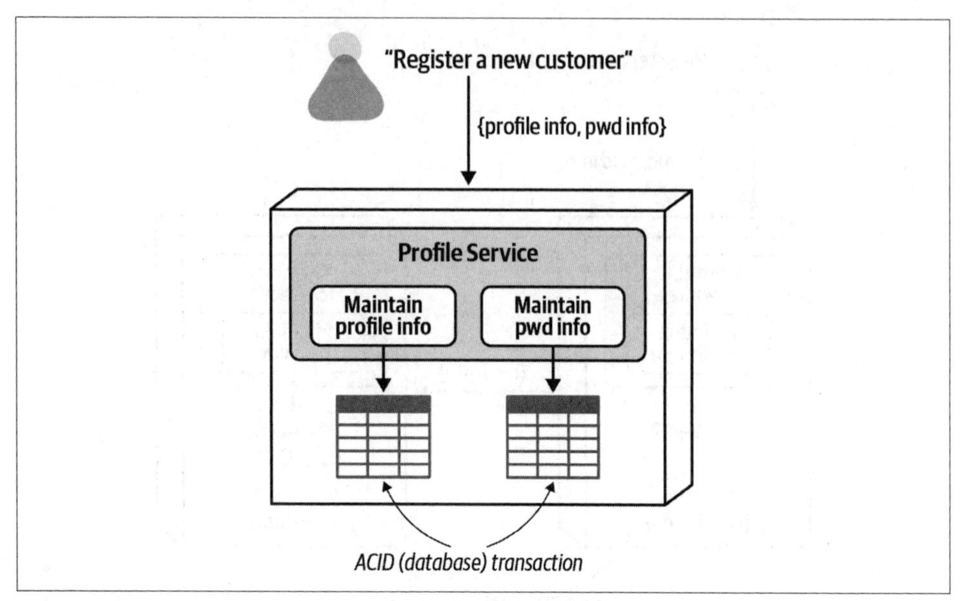

Figure 7-11. A single service supports database (ACID) transactions

Workflow and Choreography

Another common granularity integrator is *workflow* and *choreography*--services talking to one another (also sometimes referred to as *interservice* communication or *east-west* communications). Communication between services is fairly common and in many cases necessary in highly distributed architectures like microservices. However, as services move toward a finer level of granularity based on the disintegration factors outlined in the previous section, service communication can increase to a point where negative impacts start to occur.

Issues with overall fault tolerance is the first impact of too much synchronous inter-service communication. Consider the diagram in Figure 7-12: Service A communicates with services B and C, Service B communicates with Service C, Service D communicates with Service E, and finally Service E communicates with Service C. In this case, if Service C goes down, all other services become nonoperational because of a transitive dependency with Service C, creating an issue with overall fault tolerance, availability, and reliability.

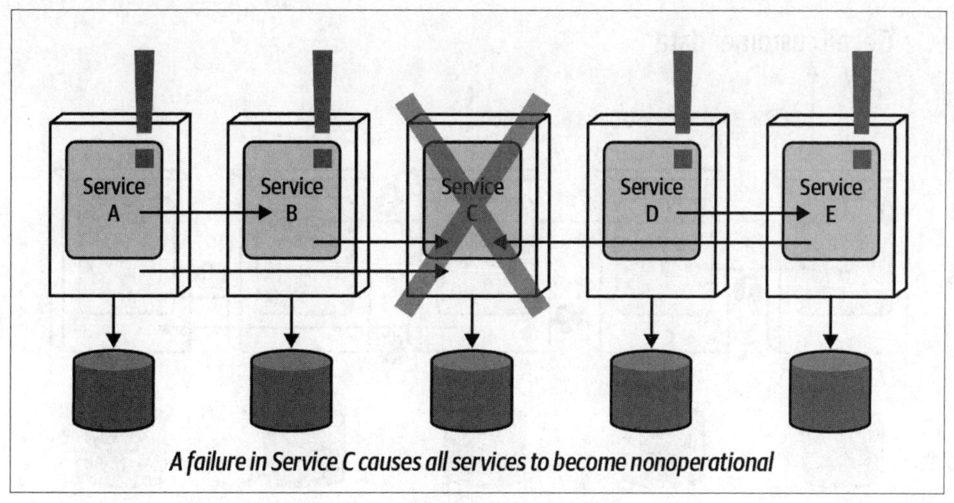

Figure 7-12. Too much workflow impacts fault tolerance

Interestingly enough, fault tolerance is one of the granularity disintegration drivers from the previous section—yet when those services need to talk to one another, nothing is really gained from a fault-tolerance perspective. When breaking apart services, always check to see if the functionalities are tightly coupled and dependent on one another. If it is, then overall fault tolerance from a business request standpoint won't be achieved, and it might be best to consider keeping the services together.

Overall performance and responsiveness is another driver for granularity integration (putting services back together). Consider the scenario in Figure 7-13: a large customer service is split into five separate services (services A through E). While each of these services has its own collection of cohesive atomic requests, retrieving all of the customer information collectively from a single API request into a single user interface screen involves five separate hops when using choreography (see Chapter 11 for an alternative solution to this problem using orchestration). Assuming 300 ms in network and security latency per request, this single request would incur an additional 1500 ms just in latency alone! Consolidating all of these services into a single service would remove the latency, therefore increasing overall performance and responsiveness.

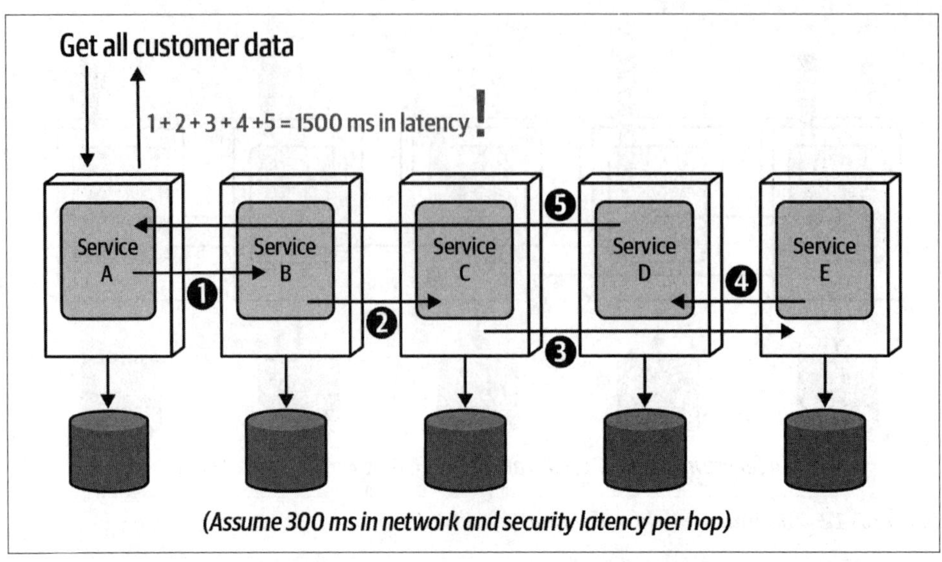

Figure 7-13. Too much workflow impacts overall performance and responsiveness

In terms of overall performance, the trade-off for this integration driver is balancing the need to break apart a service with the corresponding performance loss if those services need to communicate with one another. A good rule of thumb is to take into consideration the number of requests that require multiple services to communicate with one another, also taking into account the criticality of those requests requiring interservice communication. For example, if 30% of the requests require a workflow between services to complete the request and 70% are purely atomic (dedicated to only one service without the need for any additional communication), then it might be OK to keep the services separate. However, if the percentages are reversed, then consider putting them back together again. This assumes, of course, that overall performance matters. There's more leeway in the case of backend functionality where an end user isn't waiting for the request to complete.

The other performance consideration is with regard to the criticality of the request requiring workflow. Consider the previous example, where 30% of the requests require a workflow between services to complete the request, and 70% are purely atomic. If a critical request that requires extremely fast response time is part of that 30%, then it might be wise to put the services back together, even though 70% of the requests are purely atomic.

Overall reliability and data integrity are also impacted with increased service communication. Consider the example in Figure 7-14: customer information is separated into five separate customer services. In this case, adding a new customer to the system involves the coordination of all five customer services. However, as explained in a previous section, each of these services has its own database transaction. Notice in

Figure 7-14 that services A, B, and C have all committed part of the customer data, but Service D fails.

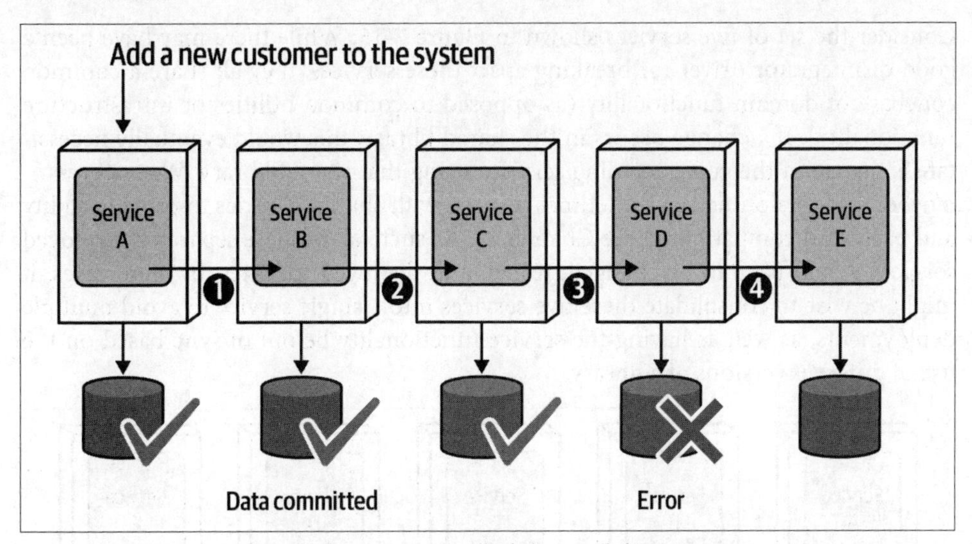

Figure 7-14. Too much workflow impacts reliability and data integrity

This creates a data consistency and data integrity issue because part of the customer data has already been committed, and may have already been acted upon through a retrieval of that information from another process or even a message sent out from one of those services broadcasting an action based on that data. In either case, that data would either have to be rolled back through compensating transactions or marked with a specific state to know where the transaction left off in order to restart it. This is very messy situation, one we describe in detail in "Transactional Saga Patterns" on page 324. If data integrity and data consistency are important or critical to an operation, it might be wise to consider putting those services back together.

Shared Code

Shared source code is a common (and necessary) practice in software development. Functions like logging, security, utilities, formatters, converters, extractors, and so on are all good examples of shared code. However, things can get complicated when dealing with shared code in a distributed architecture and can sometimes influence service granularity.

Shared code is often contained in a shared library, such as a JAR file in the Java Ecosystem, a GEM in the Ruby environment, or a DLL in the .NET environment, and is typically bound to a service at compile time. While we dive into code reuse patterns in detail in Chapter 8, here we illustrate only how shared code can sometimes influence service granularity and can become a granularity integrator (putting services back together).

Consider the set of five services shown in Figure 7-15. While there may have been a good disintegrator driver for breaking apart these services, they all share a common codebase of domain functionality (as opposed to common utilities or infrastructure functionality). If a change occurs in the shared library, this would eventually necessitate a change in the corresponding services using that shared library. We say *eventually* because versioning can sometimes be used with shared libraries to provide agility and backward compatibility (see Chapter 8). As such, all of these separately deployed services would have to be changed, tested, and deployed together. In these cases, it might be wise to consolidate these five services into a single service to avoid multiple deployments, as well as having the service functionality be out of sync based on the use of different versions of a library.

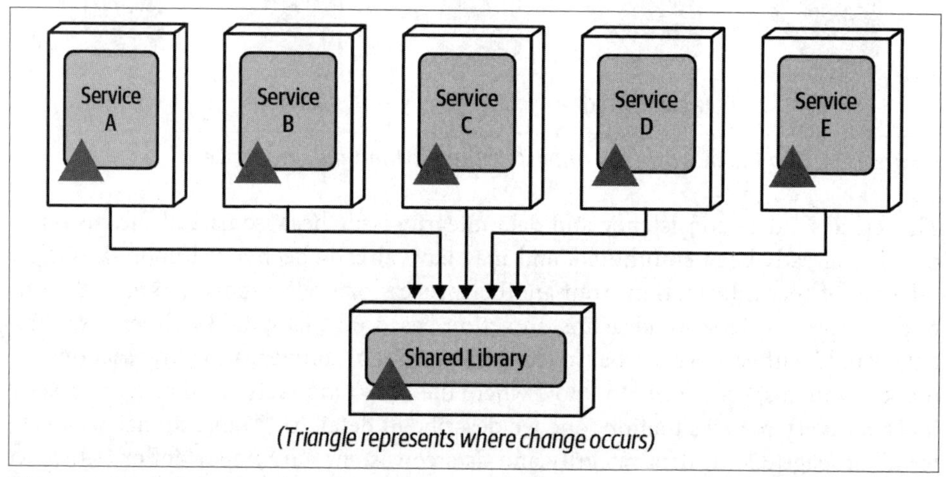

Figure 7-15. A change in shared code requires a coordinated change to all services

Not all uses of shared code drive granularity integration. For example, infrastructure-related cross-cutting functionality such as logging, auditing, authentication, authorization, and monitoring that all services use is *not* a good driver for putting services back together or even moving back to a monolithic architecture. Some of the guidelines for considering shared code as a granularity integrator are as follows:

Specific shared domain functionality
: Shared domain functionality is shared code that contains business logic (as opposed to infrastructure-related cross-cutting functionality). Our recommendation is to consider this factor as a possible granularity integrator if the percentage of shared domain code is relatively high. For example, suppose the common (shared) code for a group of customer-related functionality (profile maintenance, preference maintenance, and adding or removing comments) makes up over 40%

of the collective codebase. Breaking up the collective functionality into separate services would mean that almost half of the source code is in a shared library used only by those three services. In this example it might be wise to consider keeping the collective customer-related functionality in a single consolidated service along with the shared code (particularly if the shared code changes frequently, as discussed next).

Frequent shared code changes
Regardless of the size of the shared library, frequent changes to shared functionality require frequent coordinated changes to the services using that shared domain functionality. While versioning can sometimes be used to help mitigate coordinated changes, eventually services using that shared functionality will need to adopt the latest version. If the shared code changes frequently, it might be wise to consider consolidating the services using that shared code to help mitigate the complex change coordination of multiple deployment units.

Defects that cannot be versioned
While versioning can help mitigate coordinated changes and allow for backward compatibility and agility (the ability to respond quickly to change), at times certain business functionality must be applied to all services at the same time (such as a defect or a change in business rules). If this happens frequently, it might be time to consider putting services back together to simplify the changes.

Data Relationships

Another trade-off in the balance between granularity disintegrators and integrators is the relationship between the data that a single consolidated service uses as opposed to the data that separate services would use. This integrator driver assumes that the data resulting from breaking apart a service is not shared, but rather formed into tight bounded contexts within each service to facilitate change control and support overall availability and reliability.

Consider the example in Figure 7-16: a single consolidated service has three functions (A, B, and C) and corresponding data table relationships. The solid lines pointing to the tables represent writes to the tables (hence data ownership), and the dotted lines pointing away from the tables represent read-only access to the table. Performing a mapping operation between the functions and the tables reveals the results shown in Table 7-1, where *owner* implies writes (and corresponding reads) and *access* implies read-only access to a table not owned by that function.

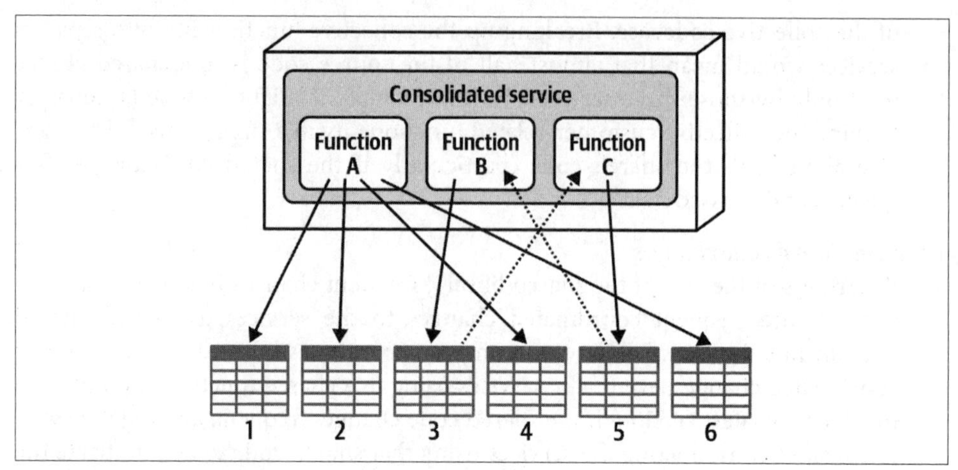

Figure 7-16. The database table relationships of a consolidated service

Table 7-1. Function-to-table mapping

Function	Table 1	Table 2	Table 3	Table 4	Table 5	Table 6
A	owner	owner		owner		owner
B			owner		access	
C			access		owner	

Assume that based on some of the disintegration drivers outlined in the prior section, this service was broken into three separate services (one for each of the functions in the consolidated service); see Figure 7-17. However, breaking apart the single consolidated service into three separate services now requires the corresponding data tables to be associated with each service in a bounded context.

Notice at the top of Figure 7-17 that Service A owns tables 1, 2, 4, and 6 as part of its bounded context; Service B owns table 3; and Service C owns table 5. However, notice in the diagram that every operation in Service B requires access to data in table 5 (owned by Service C), and every operation in Service C requires access to data in table 3 (owned by Service B). Because of the bounded context, Service B cannot simply reach out and directly query table 5, nor can Service C directly query table 3.

To better understand the bounded context and why Service C cannot simply access table 3, say Service B (which owns table 3) decides to make a change to its business rules that requires a column to be removed from table 3. Doing so would break Service C and any other services using table 3. This is why the bounded context concept is so important in highly distributed architectures like microservices. To resolve this issue, Service B would have to *ask* Service C for its data, and Service C would have to ask Service B for its data, resulting in back-and-forth interservice communication between these services, as illustrated at the bottom of Figure 7-17.

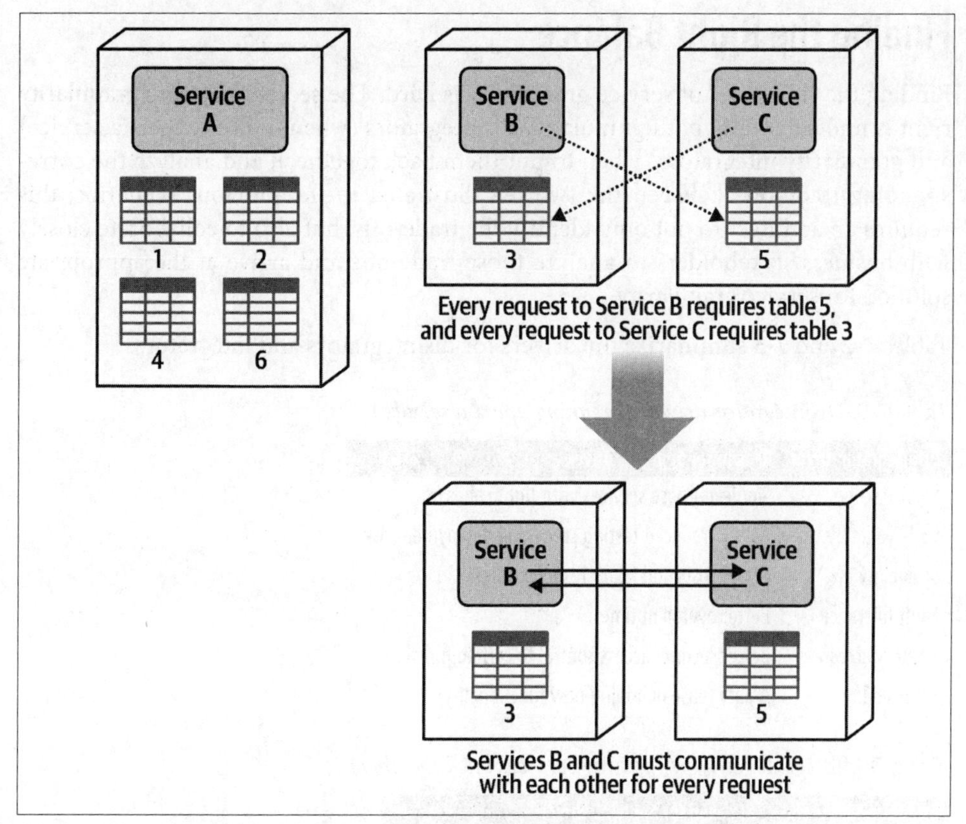

Figure 7-17. Database table relationships impact service granularity

Based on the dependency of the data between services B and C, it would be wise to consolidate those services into a single service to avoid the latency, fault tolerance, and scalability issues associated with the interservice communication between these services, demonstrating that relationships between tables can influence service granularity. We've saved this granularity integration driver for last because it is the one granularity integration driver with the fewest number of trade-offs. While occasionally a migration from a monolithic system requires a refactoring of the way data is organized, in most cases it isn't feasible to reorganize database table entity relationships for the sake of breaking apart a service. We dive into the details about breaking apart data in Chapter 6.

Finding the Right Balance

Finding the right level of service granularity is hard. The secret to getting granularity right is understanding both granularity disintegrators (when to break apart a service) and granularity integrators (when to put them back together), and analyze the corresponding trade-offs between the two. As illustrated in the previous scenarios, this requires an architect to not only identify the trade-offs, but also to collaborate closely with business stakeholders to analyze those trade-offs and arrive at the appropriate solution for service granularity.

Tables 7-2 and 7-3 summarize the drivers for disintegrators and integrators.

Table 7-2. Disintegrator drivers (breaking apart a service)

Disintegrator driver	Reason for applying driver
Service scope	Single-purpose services with tight cohesion
Code volatility	Agility (reduced testing scope and deployment risk)
Scalability	Lower costs and faster responsiveness
Fault tolerance	Better overall uptime
Security access	Better security access control to certain functions
Extensibility	Agility (ease of adding new functionality)

Table 7-3. Integrator drivers (putting services back together)

Integrator driver	Reason for applying driver
Database transactions	Data integrity and consistency
Workflow	Fault tolerance, performance, and reliability
Shared code	Maintainability
Data relationships	Data integrity and correctness

Architects can use the drivers in these tables to form trade-off statements that can then be discussed and resolved by collaborating with a product owner or business sponsor.

Example 1:

> **Architect**: "We want to break apart our service to isolate frequent code changes, but in doing so we won't be able to maintain a database transaction. Which is more important based on our business needs—better *overall agility* (maintainability, testability, and deployability), which translates to faster time-to-market, or stronger *data integrity and consistency*?"
>
> **Project Sponsor**: "Based on our business needs, I'd rather sacrifice a little bit slower time-to-market to have better data integrity and consistency, so let's leave it as a single service for right now."

Example 2:

> **Architect**: "We need to keep the service together to support a database transaction between two operations to ensure data consistency, but that means sensitive functionality in the combined single service will be less secure. Which is more important based on our business needs—better *data consistency* or better *security*?"
>
> **Project Sponsor**: "Our CIO has been through some rough situations with regard to security and protecting sensitive data, and it's on the forefront of their mind and part of almost every discussion. In this case, it's more important to secure sensitive data, so let's keep the services separate and work out how we can mitigate some of the issues with data consistency."

Example 3:

> **Architect**: "We need to break apart our payment service to provide better extensibility for adding new payment methods, but that means we will have increased workflow that will impact the responsiveness when multiple payment types are used for an order (which happens frequently). Which is more important based on our business needs—better extensibility within the payment processing, hence better *agility and overall time-to-market*, or better *responsiveness* for making a payment?"
>
> **Project Sponsor**: "Given that I see us adding only two, maybe three more payment types over the next couple of years, I'd rather have us focus on the overall responsiveness since the customer must wait for payment processing to be complete before the order ID is issued."

Sysops Squad Saga: Ticket Assignment Granularity

`Monday, October 25 11:08`

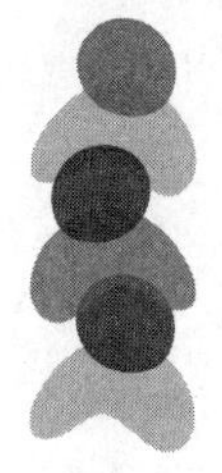

Once a trouble ticket has been created by a customer and accepted by the system, it must be assigned to a Sysops Squad expert based on their skill set, location, and availability. Ticket assignment involves two main components—a Ticket Assignment component that determines which consultant should be assigned the job, and the Ticket Routing component that locates the Sysops Squad expert, forwards the ticket to the expert's mobile device (via a custom Sysops Squad mobile app), and notifies the expert via an SMS text message that a new ticket has been assigned.

The Sysops Squad development team was having trouble deciding whether these two components (assignment and routing) should be implemented as a single consolidated service or two separate services, as illustrated in Figure 7-18. The development team consulted with Addison (one of the Sysops Squad architects) to help decide which option it should go with.

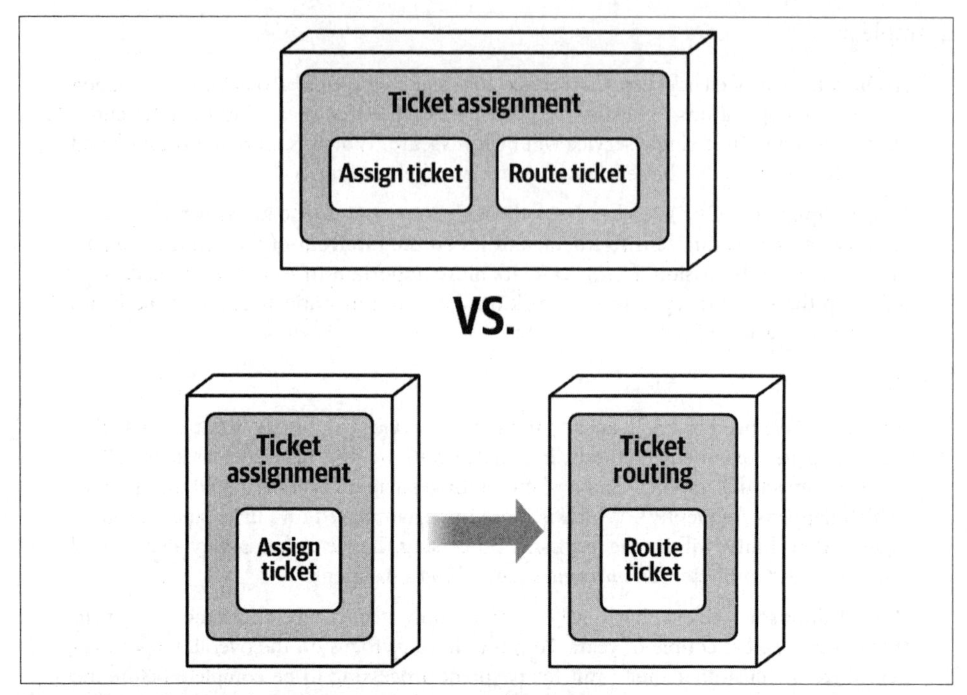

Figure 7-18. Options for ticket assignment and routing

"So you see," said Taylen, "the ticket assignment algorithms are very complex, and therefore should be isolated from the ticket routing functionality. That way, when those algorithms change, I don't have to worry about all of the routing functionality."

"Yes, but how much change is there to those assignment algorithms?" asked Addison. "And how much change do we anticipate in the future?"

"I apply changes to those algorithms at least two to three times a month. I read about volatility-based decomposition, and this situation fits it perfectly," said Taylen.

"But if we separated the assignment and routing functionality into two services, there would need to be constant communication between them," said Skyler. "Furthermore, assignment and routing are really one function, not two."

"No," said Taylen, "they are two separate functions."

"Hold on," said Addison. "I see what Skyler means. Think about it a minute. Once an expert is found that is available within a certain period of time, the ticket is immediately routed to that expert. If no expert is available, the ticket goes back in the queue and waits until an expert can be found."

"Yes, that's right," said Taylen.

"See," said Skyler, "you cannot make a ticket assignment without routing it to the expert. So the two functions are one."

"No, no, no," said Taylen. "You don't understand. If an expert is seen to be available within a certain amount of time, then that expert is assigned. Period. Routing is just a transport thing."

"What happens in the current functionality if a ticket can't be routed to the expert?" asked Addison.

"Then another expert is selected," said Taylen.

"OK, so think about it a minute, Taylen," said Addison. "If assignment and routing are two separate services, then the routing service would have to then communicate back to the assignment service, letting it know that the expert cannot be located and to pick another one. That's a lot of coordination between the two services."

"Yes, but they are still *two* separate functions, not one as Skyler is suggesting," said Taylen.

"I have an idea," said Addison. "Can we all agree that the assignment and routing are two separate activities, but are tightly bound synchronously to each other? Meaning, one function cannot exist without the other?"

"Yes," both Taylen and Skyler replied.

"In that case," said Addison, "let's analyze the trade-offs. Which is more important—isolating the assignment functionality for change control purposes, or combining assignment and routing into a single service for better performance, error handling, and workflow control?"

"Well," said Taylen, "when you put it that way, obviously the single service. But I still want to isolate the assignment code."

"OK," said Addison, "in that case, how about we make three distinct architectural components in the single service. We can delineate assignment, routing, and shared code with separate namespaces in the code. Would that help?"

"Yeah," said Taylen, "that would work. OK, you both win. Let's go with a single service then."

"Taylen," said Addison, "it's not about winning, it's about analyzing the trade-offs to arrive at the most appropriate solution; that's all."

With everyone agreeing to a single service for assignment and routing, Addison wrote the following architecture decision record (ADR) for this decision:

> *ADR: Consolidated Service for Ticket Assignment and Routing*
>
> *Context*
>
> Once a ticket is created and accepted by the system, it must be assigned to an expert and then routed to that expert's mobile device. This can be done through a single consolidated ticket assignment service or separate services for ticket assignment and ticket routing.

Decision
We will create a single consolidated ticket assignment service for the assignment and routing functions of the ticket.

Tickets are immediately routed to the Sysops Squad expert once they are assigned, so these two operations are tightly bound and dependent each other.

Both functions must scale the same, so there are no throughput differences between these services, nor is back-pressure (*https://oreil.ly/Vhjmv*) needed between these functions.

Since both functions are fully dependent on each other, fault tolerance is not a driver for breaking these functions apart.

Making these functions separate services would require workflow between them, resulting in performance, fault tolerance, and possible reliability issues.

Consequences
Changes to the assignment algorithm (which occur on a regular basis) and changes to the routing mechanism (infrequent change) would require testing and deployment of both functions, resulting in increased testing scope and deployment risk.

Sysops Squad Saga: Customer Registration Granularity

`Friday January 14, 13:15`

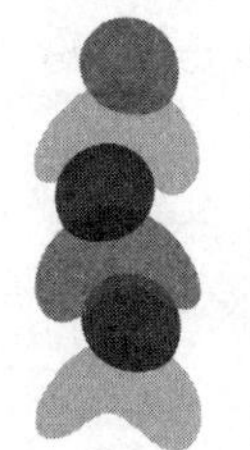

Customers must register with the system to gain access to the Sysops Squad support plan. During registration, customers must provide profile information (name, address, business name if applicable, and so on), credit card information (which is billed on a monthly basis), password and security question information, and a list of products purchased they would like to have covered under the Sysops Squad support plan.

Some members of the development team insisted that this should be a single consolidated Customer Service containing all of the customer information, yet other members of the team disagreed and thought that there should be a separate service for each of these functions (a Profile service, Credit Card service, Password service, and a Supported Product service). Skyler, having prior experience in PCI and PII data, thought that the credit card and password information should be a separate service from the rest, and hence only two services (a Profile service containing profile and product information and a separate Customer Secure service containing credit card and password information). These three options are illustrated in Figure 7-19.

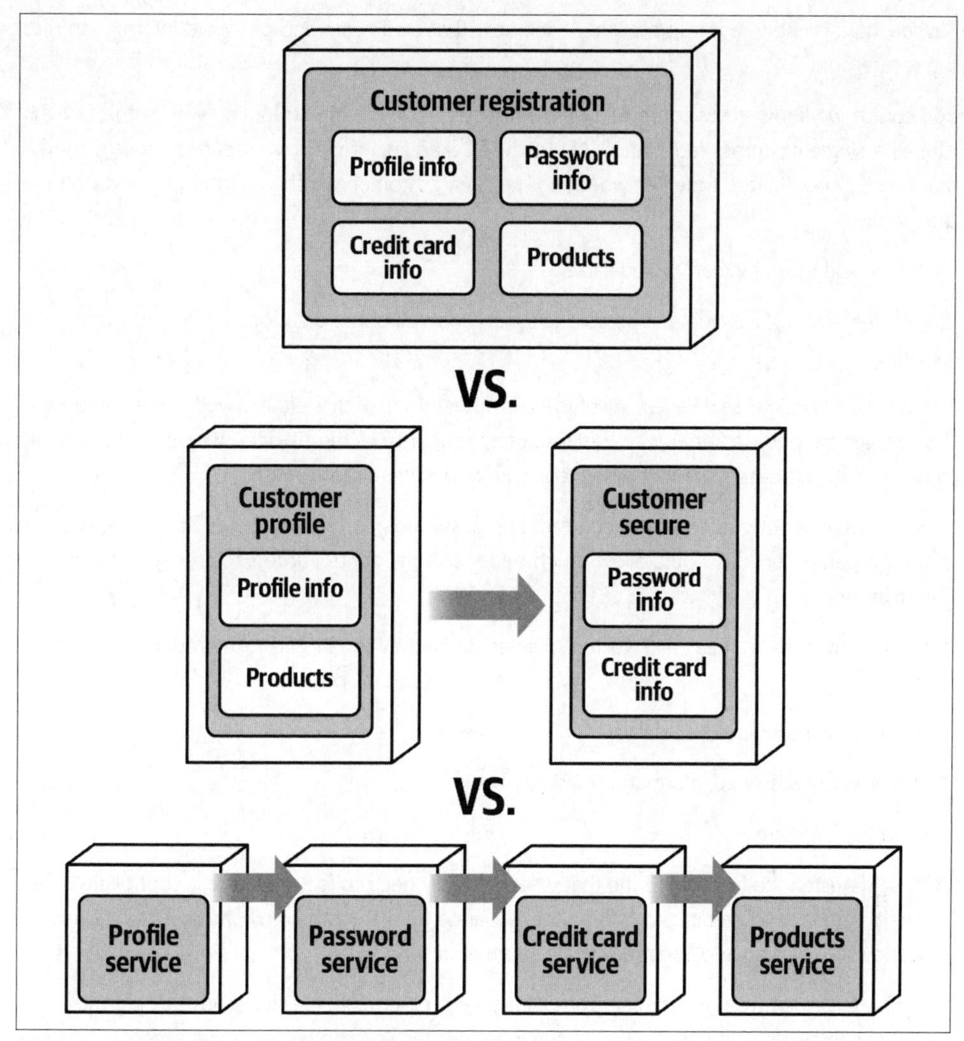

Figure 7-19. Options for customer registration

Because Addison was busy with the core ticketing functionality, the development team asked for Austen's help in resolving this granularity issue. Anticipating this will not be an easy decision, particularly since it involved security, Austen scheduled a meeting with Parker, (the product owner), and Sam, the Penultimate Electronics security expert to discuss these options.

"OK, so what can we do for you?" asked Parker.

"Well," said Austen, "we are struggling with how many services to create for registering customers and maintaining customer-related information, You see, there are four main pieces of data we are dealing with here: profile info, credit card info, password info, and purchased product info."

"Whoa, hold on now," interrupted Sam. "You know that credit card and password information *must* be secure, right?"

"Of course we know it has to be secure," said Austen. "What we're struggling with is the fact that there's a single customer registration API to the backend, so if we have separate services they all have to be coordinated together when registering a customer, which would require a distributed transaction."

"What do you mean by that?" asked Parker.

"Well," said Austen, "we wouldn't be able to synchronize all of the data together as one atomic unit of work."

"That's not an option," said Parker. "All of the customer information is either saved in the database, or it's not. Let me put it another way. We absolutely *cannot* have the situation where we have a customer record without a corresponding credit card or password record. *Ever.*"

"OK, but what about securing the credit card and password information?" asked Sam. "Seems to me, having separate services would allow much better security control access to that type of sensitive information."

"I think I may have an idea." said Austen. "The credit card information is tokenized in the database, right?"

"Tokenized *and* encrypted," said Sam.

"Great. And the password information?" asked Austen.

"The same," said Sam.

"OK," said Austen, "so it seems to me that what we really need to focus on here is controlling access to the password and credit card information separate from the other customer-related requests—you know, like getting and updating profile information, and so on."

"I think I see where you are coming from with your problem," said Parker. "You're telling me that if you separate all of this functionality into separate services, you can better secure access to sensitive data, but you cannot guarantee my all-or-nothing requirement. Am I right?"

"Exactly. That's the trade-off," said Austen.

"Hold on," said Sam. "Are you using the Tortoise security libraries to secure the API calls?"

"Yes. We use those libraries not only at the API layer, but also within each service to control access through the service mesh. So essentially it's a double-check," said Austen.

"Hmmm," said Sam. "OK, I'm good with a single service providing you use the Tortoise security framework."

"Me too, providing we can still have the all-or-nothing customer registration process," said Parker.

"Then I think we are all in agreement that the all-or-nothing customer registration is an absolute requirement and we will maintain multilevel security access using Tortoise," said Austen.

"Agreed," said Parker.

"Agreed," said Sam.

Parker noticed how Austen handled the meeting by facilitating the conversation rather than controlling it. This was an important lesson as an architect in identifying, understanding, and negotiating trade-offs. Parker also better understood the difference between design versus architecture in that security can be controlled through *design* (use of a custom library with special encryption) rather than *architecture* (breaking up functionality into separate deployment units).

Based on the conversation with Parker and Sam, Austen made the decision that customer-related functionality would be managed through a single consolidated domain service (rather than separately deployed services) and wrote the following ADR for this decision:

> *ADR: Consolidated Service for Customer-Related Functionality*
>
> *Context*
> Customers must register with the system to gain access to the Sysops Squad support plan. During registration, customers must provide profile information, credit card information, password information, and products purchased. This can be done through a single consolidated customer service, a separate service for each of these functions, or a separate service for sensitive and nonsensitive data.
>
> *Decision*
> We will create a single consolidated customer service for profile, credit card, password, and products supported.
>
> Customer registration and unsubscribe functionality *requires* a single atomic unit of work. A single service would support ACID transactions to meet this requirement, whereas separate services would not.
>
> Use of the Tortoise security libraries in the API layer and the service mesh will mitigate security access risk to sensitive information.
>
> *Consequences*
> We will require the Tortoise security library to ensure security access in both the API gateway and the service mesh.
>
> Because it's a single service, changes to source code for profile info, credit card, password, or products purchased will increase testing scope and increase deployment risk.
>
> The combined functionality (profile, credit card, password, and products purchased) will have to scale as one unit.
>
> The trade-off discussed in a meeting with the product owner and security expert is *transactionality* versus *security*. Breaking the customer functionality into separate services

provides better security access, but doesn't support the "all-or-nothing" database transaction required for customer registration or unsubscribing. However, the security concerns are mitigated through the use the custom Tortoise security library.

PART II

Putting Things Back Together

> Attempting to divide a cohesive module would only result in increased coupling and decreased readability.
>
> —Larry Constantine

Once a system is broken apart, architects often find it necessary to stitch it back together to make it work as one cohesive unit. As Larry Constantine so eloquently infers in the preceding quote, it's not quite as easy as it sounds, with lots of trade-offs involved when breaking things apart.

In this second part of this book, we discuss various techniques for overcoming some of the hard challenges associated with distributed architectures, including managing service communication, contracts, distributed workflows, distributed transactions, data ownership, data access, and analytical data.

Part I was about *structure*; Part II is about *communication*. Once an architect understands the structure and the decisions that lead to it, it's time to think about how the structural parts interact with each other.

CHAPTER 8

Reuse Patterns

`Wednesday, February 2, 15:15`

As the development team members worked on breaking apart the domain services, they started running into disagreements about what to do with all the shared code and shared functionality. Taylen, upset with what Skyler was doing with regard to the shared code, walked over to Skyler's desk.

"What in the world are you doing?" asked Taylen.

"I'm moving all of the shared code to a new workspace so we can create a shared DLL from it," replied Skyler.

"A *single* shared DLL?"

"That's what I was planning," said Skyler. "Most of the services will need this stuff anyway, so I'm going to create a single DLL that all the services can use."

"That's the worst idea I've ever heard," said Taylen. "Everyone knows you should have multiple shared libraries in a distributed architecture!"

"Not in my opinion," said Sydney. "Seems to me it's much easier to manage a single shared library DLL rather than dozens of them."

"Given that I'm the tech lead for this application, I want you to split that functionality into separate shared libraries."

"OK, OK, I suppose I can move the all of the authorization into its own separate DLL if that would make you happy," said Skyler.

"What?" said Taylen. "The authorization code has to be a shared service, you know——*not* in a shared library.""

"No," said Skyler. "That code should be in a shared DLL."

"What's all the shouting about over there?" asked Addison.

"Taylen wants the authorization functionality to be in a shared service. That's just crazy. I think it should go in the common shared DLL," said Skyler.

"No way," said Taylen. "It's got to be in its own separate shared service."

"And," said Skyler, "Taylen is insisting on having multiple shared libraries for the shared functionality rather than a single shared library."

"Tell you what," said Addison. "Let's go over the trade-offs of shared library granularity, and also go over the trade-offs between a shared library and a shared service to see if we can resolve these issues in a more reasonable and thoughtful manner."

Code reuse is a normal part of software development. Common business domain functionality, such as formatters, calculators, validators, and auditing, are typically shared across multiple components, as is common infrastructure functionality, such as security, logging, and metrics gathering. In most monolithic architectures, code reuse is rarely given a second thought—it's a matter of simply importing or auto-injecting shared class files. However, in distributed architectures, as shown in Figure 8-1, things get a bit more complicated, as questions arise about how to deal with shared functionality.

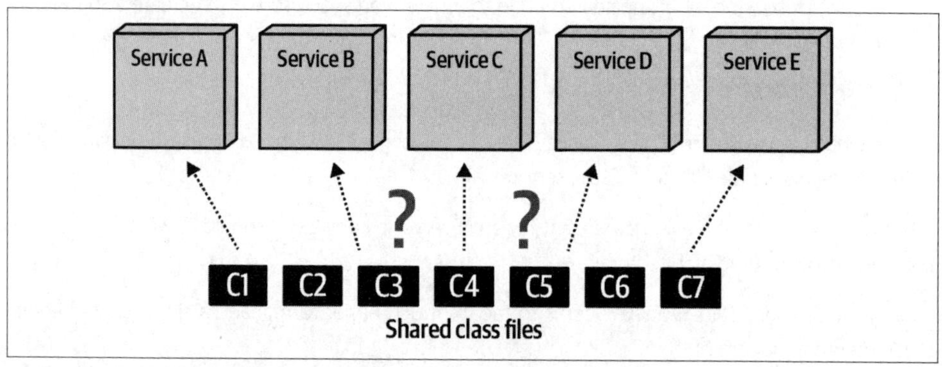

Figure 8-1. Code reuse is a hard part of distributed architecture

Frequently within highly distributed architectures like microservices and serverless environments, phrases like "reuse is abuse!" and "share nothing!" are touted by architects in an attempt to reduce the amount of shared code within these types of architectures. Architects in these environments have even been found to offer countering advice to the famous DRY principle (*https://oreil.ly/dTVrX*) (Don't repeat yourself) by using an opposing acronym called WET (Write every time or Write everything twice).

While developers should try to limit the amount of code reuse within distributed architectures, it is nevertheless a fact of life in software development and must be addressed, particularly in distributed architectures. In this chapter, we introduce several techniques for managing code reuse within a distributed architecture, including replicating code, shared libraries, shared services, and sidecars within a service mesh. For each of these options, we also discuss the pros, cons, and trade-offs of each approach.

Code Replication

In *code replication*, shared code is copied into each service (or more specifically, each service source code repository), as shown in Figure 8-2, thereby avoiding code sharing altogether. While it might sound crazy, this technique became popular in the early days of microservices when a lot of confusion and misunderstanding arose about the *bounded context* concept, hence the drive to create a "share nothing architecture." In theory, code replication seemed like a good approach at that time to reduce code sharing, but in practice it quickly fell apart.

Figure 8-2. With replication, shared functionality is copied into each service

While code replication isn't used much today, it nevertheless is still a valid technique for addressing code reuse across multiple distributed services. This technique should be approached with extreme caution for the obvious reason that if a bug is found in the code or an important change to the code is needed, it would be very difficult and time-consuming to update all services containing the replicated code.

At times, however, this technique can prove useful, particularly for highly static one-off code that most (or all) services need. For example, consider the Java code in Example 8-1 and the corresponding C# code in Example 8-2 that identifies the class in the service that represents the service entry point (usually the restful API class within a service).

Example 8-1. Source code defining a service entry point annotation (Java)

```
@Retention(RetentionPolicy.RUNTIME)
@Target(ElementType.TYPE)
public @interface ServiceEntrypoint {}

/* Usage:
@ServiceEntrypoint
public class PaymentServiceAPI {
   ...
}
*/
```

Example 8-2. Source code defining a service entry point attribute (C#)

```
[AttributeUsage(AttributeTargets.Class)]
class ServiceEntrypoint : Attribute {}

/* Usage:
[ServiceEntrypoint]
class PaymentServiceAPI {
   ...
}
*/
```

Note that the source code in Example 8-1 actually contains no functionality whatsoever. The annotation is simply a marker (or tag) used to identify a particular class as representing the service entry point. However, this simple annotation is very useful for placing other metadata annotations about a particular service, including the service type, domain, bounded context, and so on; see Chapter 89 in *97 Things Every Java Programmer Should Know* by Kevlin Henney and Trisha Gee (O'Reilly) for a description of these metadata custom annotations.

This kind of source code makes a good candidate for replication because it's static and doesn't contain any bugs (and most likely will not in the future). If this were a unique one-off class, it might be worth copying it into each service code repository rather than creating a shared library for it. That said, we generally encourage investigating the other code-sharing techniques presented in this chapter before opting for the code replication technique.

While the replication technique preserves the bounded context, it does make it difficult to apply changes if the code ever does need to be modified. Table 8-1 lists the various trade-offs associated with this technique.

Trade-Offs

Table 8-1. Trade-offs for the code replication technique

Advantages	Disadvantages
Preserves the bounded context	Difficult to apply code changes
No code sharing	Code inconsistency across services
	No versioning capabilities across services

When to Use

The replication technique is a good approach when developers have simple static code (like annotations, attributes, simple common utilities, and so on) that is either a one-off class or code that is unlikely to ever change because of defects or functional changes. However, as mentioned earlier, we encourage exploring other code-reuse options before embracing the code replication technique.

When migrating from a monolithic architecture to a distributed one, we've also found that the replication technique can sometimes work for common static utility classes. For example, by replicating a `Utility.cs` C# class to all services, each service can now remove (or enhance) the `Utility.cs` class to suit its particular needs, therefore eliminating unnecessary code and allowing the utility class to evolve for each specific context (similar to the tactical forking technique described in Chapter 3). Again, the risk with this technique is that a defect or change is very difficult to propagate to all services because the code is duplicated for each service.

Shared Library

One of the most common techniques for sharing code is to use a shared library. A shared library is an external artifact (such as a JAR file, DLL, and so on) containing source code that is used by multiple services which is typically bound to the service at compile time (see Figure 8-3). Although the shared library technique seems simple and straightforward, it has its share of complexities and trade-offs, not the least of which is shared library granularity and versioning.

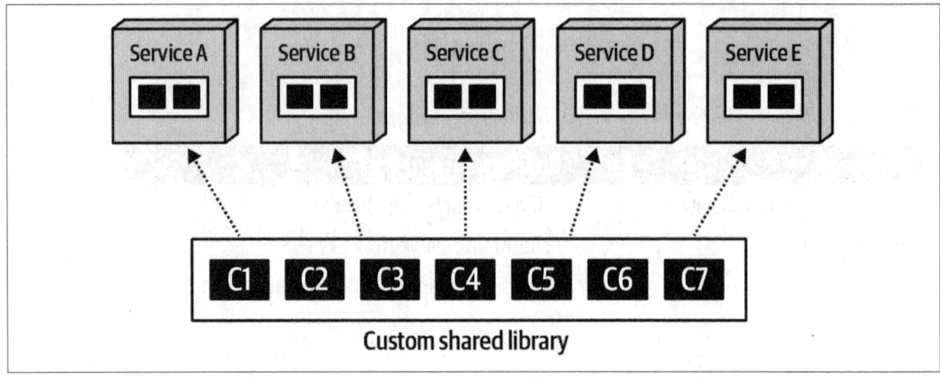

Figure 8-3. With the shared library technique, common code is consolidated and shared at compile time

Dependency Management and Change Control

Similar to service granularity (discussed in Chapter 7), there are trade-offs associated with the granularity of a shared library. The two opposing forces that form trade-offs with shared libraries are dependency management and change control.

Consider the coarse-grained shared library illustrated in Figure 8-4. Note that while the dependency management is relatively straightforward (each service uses the single shared library), change control is not. If a change occurs to any of the class files in the coarse-grained shared library, *every* service, whether it cares about the change or not, must eventually adopt the change because of a version deprecation of the shared library. This forces unnecessary retesting and redeployment of all the services using that library, therefore significantly increasing the overall testing scope of a shared library change.

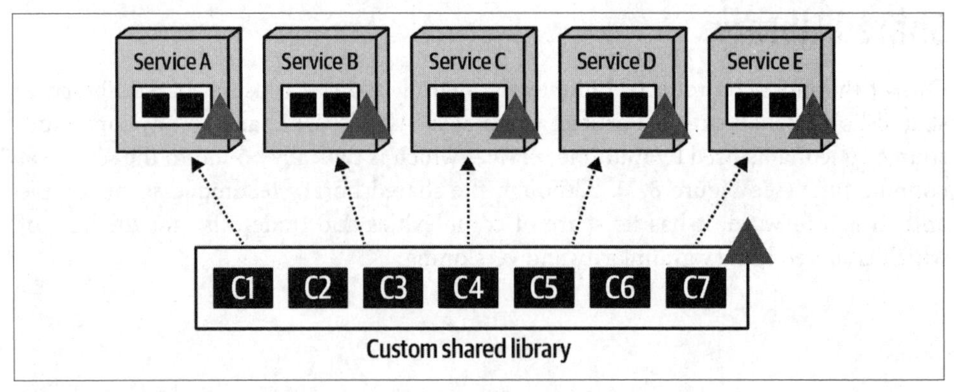

Figure 8-4. Changes to coarse-grained shared libraries impact multiple services but keep dependencies low

Breaking shared code into smaller functionality-based shared libraries (such as security, formatters, annotations, calculators, and so on) is better for change control and overall maintainability, but unfortunately creates a mess in terms of dependency management. As shown in Figure 8-5, a change in shared class C7 impacts only Service D and Service E, but managing the dependency matrix between shared libraries and services quickly starts looking like a big ball of distributed mud (or what some people refer to as a *distributed monolith*).

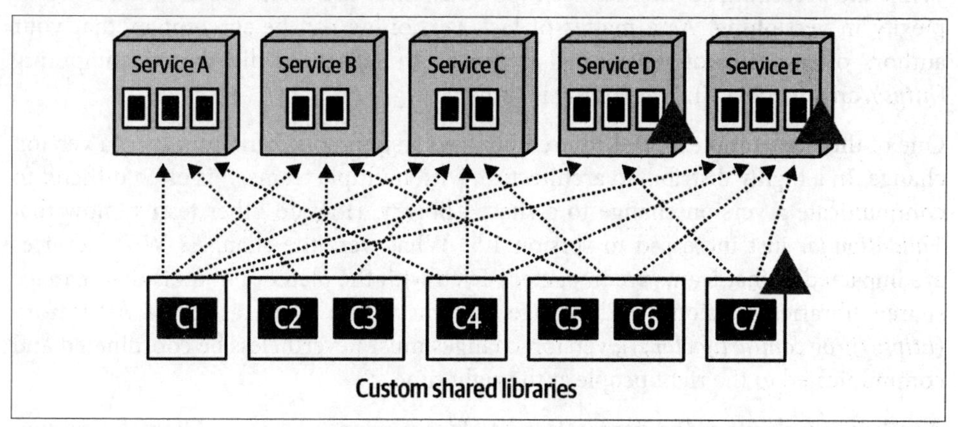

Figure 8-5. Changes to fine-grained shared libraries impact fewer services but increase dependencies

The choice of shared library granularity may not matter much with only a few services, but as the number of services increases, so do the issues associated with change control and dependency management. Just imagine a system with 200 services and 40 shared libraries—it would quickly become overly complex and unmaintainable.

Given these trade-offs of change control and dependency management, our advice is to generally avoid large, coarse-grained shared libraries and strive for smaller, functionally partitioned libraries whenever possible, thus favoring change control over dependency management. For example, carving off relatively static functionality such as formatters and security (authentication and authorization) into their own shared libraries isolates this static code, therefore reducing the testing scope and unnecessary version deprecation deployments for other shared functionality.

Versioning Strategies

Our general advice about shared library versioning is to *always use versioning!* Versioning your shared libraries provides not only backward compatibility, but also a high level of agility—the ability to respond quickly to change.

To illustrate this point, consider a shared library containing common field validation rules called *Validation.jar* that is used by 10 services. Suppose one of those services needs an immediate change to one of the validation rules. By versioning the *Validation.jar* file, the service needing the change can immediately incorporate the new *Validation.jar* version and be deployed to production right away, without any impact to the other 9 services. Without versioning, all 10 services would have to be tested and redeployed when making the shared library change, thereby increasing the amount of time and coordination for the shared library change (hence less agility).

While the preceding advice may seem obvious, there are trade-offs and hidden complexity in versioning. As a matter of fact, versioning can be so complex that your authors often think of versioning as the ninth fallacy of distributed computing (*https://oreil.ly/a9ADS*): "versioning is simple"

One of the first complexities of shared library versioning is communicating a version change. In a highly distributed architecture with multiple teams, it is often difficult to communicate a version change to a shared library. How do other teams know that *Validation.jar* just increased to version 1.5? What were the changes? What services are impacted? What teams are impacted? Even with the plethora of tools that manage shared libraries, versions, and change documentation (such as JFrog Artifactory (*https://jfrog.com/artifactory*)), version changes must nevertheless be coordinated and communicated to the right people at the right time.

Another complexity is the deprecation of older versions of a shared library—removing those versions no longer supported after a certain date. Deprecation strategies range from *custom* (for individual shared libraries) all the way to *global* (for all shared libraries). And, not surprisingly, trade-offs are involved with both approaches.

Assigning a custom deprecation strategy to each shared library is usually the desired approach because libraries change at different rates. For example, if a *Security.jar* shared library doesn't change often, maintaining only two or three versions is a reasonable strategy. However, if the *Calculators.jar* shared library changes weekly, maintaining only two or three versions means that all services using that shared library will be incorporating a newer version on a monthly (or even weekly) basis—causing a lot of unnecessary frequent retesting and redeployment. Therefore, maintaining 10 versions of *Calculators.jar* would be a much more reasonable strategy because of the frequency of change. The trade-off of this approach, however, is that someone must maintain and track the deprecation for *each shared library*. This can sometimes be a daunting task and is definitely not for the faint of heart.

Because change is variable among the various shared libraries, the global deprecation strategy, while simpler, is a less effective approach. The global deprecation strategy dictates that *all* shared libraries, regardless of the rate of change, will not support more than a certain number of backward versions (for example, four). While this is easy to maintain and govern, it can cause significant *churn*--the constant retesting and redeploying of services—just to maintain compatibility with the latest version of a frequently changed shared library. This can drive teams crazy and significantly reduce overall team velocity and productivity.

Regardless of the deprecation strategy used, serious defects or breaking changes to shared code invalidate any sort of deprecation strategy, causing *all* services to adopt the latest version of a shared library at once (or within a very short period of time). This is another reason we recommend keeping shared libraries as fine-grained as appropriate and avoid the coarse-grained *SharedStuff.jar* type of libraries containing all the shared functionality in the system.

One last word of advice regarding versioning: avoid the use of the `LATEST` version when specifying which version of a library a service requires. It has been our experience that services using the `LATEST` version experience issues when doing quick fixes or emergency hot deployments into production, because something in the `LATEST` version might be incompatible with the service, therefore causing additional development and testing effort for the team to release the service into production.

While the shared library technique allows changes to be versioned (therefore providing a good level of agility for shared code changes), dependency management can be difficult and messy. Table 8-2 lists various trade-offs associated with this technique.

Trade-Offs

Table 8-2. Trade-offs for the shared library technique

Advantages	Disadvantages
Ability to version changes	Dependencies can be difficult to manage
Shared code is compile-based, reducing runtime errors	Code duplication in heterogeneous codebases
Good agility for code shared code changes	Version deprecation can be difficult
	Version communication can be difficult

When To Use

The shared library technique is a good approach for homogeneous environments where shared code change is low to moderate. The ability to version (although sometimes complex) allows for good levels of agility when making shared code changes. Because shared libraries are usually bound to the service at compile time, operational characteristics such as performance, scalability, and fault tolerance are not impacted, and the risk of breaking other services with a change to common code is low because of versioning.

Shared Service

The primary alternative to using a shared library for common functionality is to use a shared service instead. The *shared service* technique, illustrated in Figure 8-6, avoids reuse by placing shared functionality in a separately deployed service.

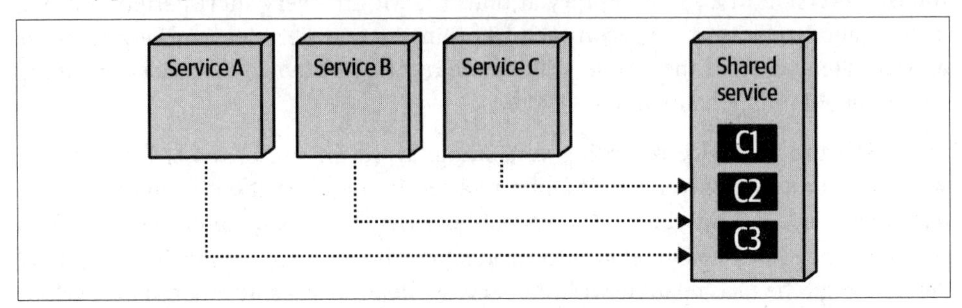

Figure 8-6. With the shared service technique, common functionality is made available at runtime through separate services

One distinguishing factor about the shared service technique is that the shared code must be in the form of *composition*, not *inheritance*. While there is a lot of debate about the use of composition over inheritance from a source code *design* standpoint (see the Thoughtworks article "Composition vs. Inheritance: How to Choose" (*https://oreil.ly/LMmZH*) and Martin Fowler's article "Designed Inheritance" (*https://oreil.ly/bW8CH*)), *architecturally* composition versus inheritance matters when choosing a code-reuse technique, particularly with the shared services technique.

Back in the day, shared services were a common approach to address shared functionality within a distributed architecture. Changes to shared functionality no longer require redeployment of services; rather, since changes are isolated to a separate service, they can be deployed without redeploying other services needing the shared functionality. However, like everything in software architecture, many trade-offs are associated with using shared services, including change risk, performance, scalability, and fault tolerance.

Change Risk

Changing shared functionality using the shared service technique turns out to be a double-edged sword. As illustrated in Figure 8-7, changing shared functionality is simply a matter of modifying the shared code contained in a separate service (such as a discount calculator), redeploying the service, and voila—the changes are now available to all services, without having to retest and redeploy any other service needing that shared functionality.

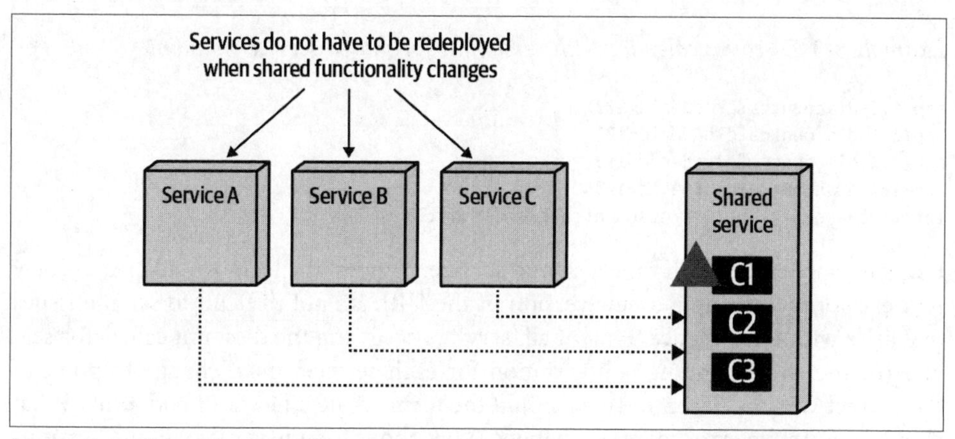

Figure 8-7. Shared functionality changes are isolated to only the shared service

If only life were that simple! The problem, of course, is that a change to a shared service is a *runtime* change, as opposed to a *compile-based* change with the shared library technique. As a result, a "simple" change in a shared service can effectively bring down an entire system, as illustrated in Figure 8-8.

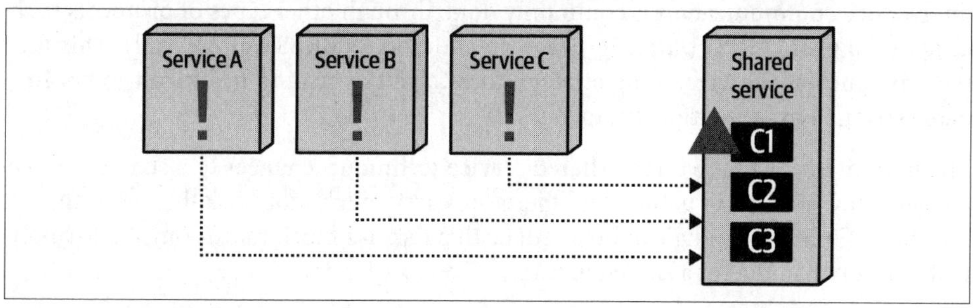

Figure 8-8. Changes to a shared service can break other services at runtime

This necessarily brings to the forefront the topic of versioning. In the shared library technique, versioning is managed through compile-time bindings, significantly reducing risk associated with a change in a shared library. However, how does one version a simple shared service change?

The immediate response, of course, is to use API endpoint versioning—in other words, create a new endpoint containing each shared service change, as shown in Example 8-3.

Example 8-3. Discount calendar with versioning for shared service endpoint

```
app/1.0/discountcalc?orderid=123
app/1.1/discountcalc?orderid=123
app/1.2/discountcalc?orderid=123
app/1.3/discountcalc?orderid=123
latest change -> app/1.4/discountcalc?orderid=123
```

Using this approach, each time a shared service changes, the team would create a new API endpoint containing a new version of the URI. It's not difficult to see the issues that arise with this practice. First of all, services accessing the discount calculator service (or the corresponding configuration for each service) must change to point to the correct version. Second, when should the team create a new API endpoint? What about for a simple error message change? What about for a new calculation? Versioning starts to become largely subjective at this point, *and* the services using the shared service must still change to point to the correct endpoint.

Another problem with API endpoint versioning is that it assumes all access to the shared service is through a RESTful API call going through a gateway or via point-to-point communication. However, in some cases, access to a shared service through interservice communication is commonly done through other types of protocols such as messaging and gRPC (*https://grpc.io*) (in addition to a RESTful API call). This further complicates the versioning strategy for a change, making it difficult to coordinate versions across multiple protocols.

The bottom line is that with the shared service technique, changes to a shared service are generally runtime in nature, and therefore carry much more risk than with shared libraries. While versioning can help reduce this risk, it's much more complex to apply and manage than that of a shared library.

Performance

Because services requiring the shared functionality must make an interservice call to a shared service, performance is impacted because of network latency (and security latency, assuming the endpoints to the shared service are secure). This trade-off, shown in Figure 8-9, does not exist with the shared library technique when accessing shared code.

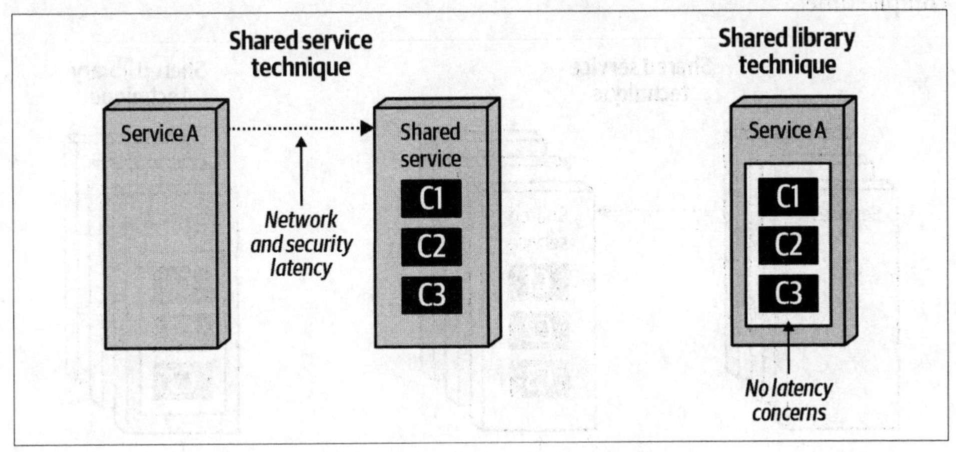

Figure 8-9. Shared service introduces network and security latency

Use of gRPC can help mitigate some of the performance issues by significantly reducing network latency, as can the use of asynchronous protocols like messaging. With messaging, the service needing the shared functionality can issue a request through a request queue, perform other work, and once needed, can retrieve the results through a separate reply queue using a correlation ID (see *Java Message Service*, Second Edition by Mark Richards et al. (O'Reilly) for more information about messaging techniques).

Scalability

Another drawback of the shared service technique is that the shared service must scale as services using the shared service scale. This can sometimes be a mess to manage, particularly with multiple services concurrently accessing the same shared service. However, as illustrated in Figure 8-10, the shared library technique does not have this issue because the shared functionality is contained within the service at compile time.

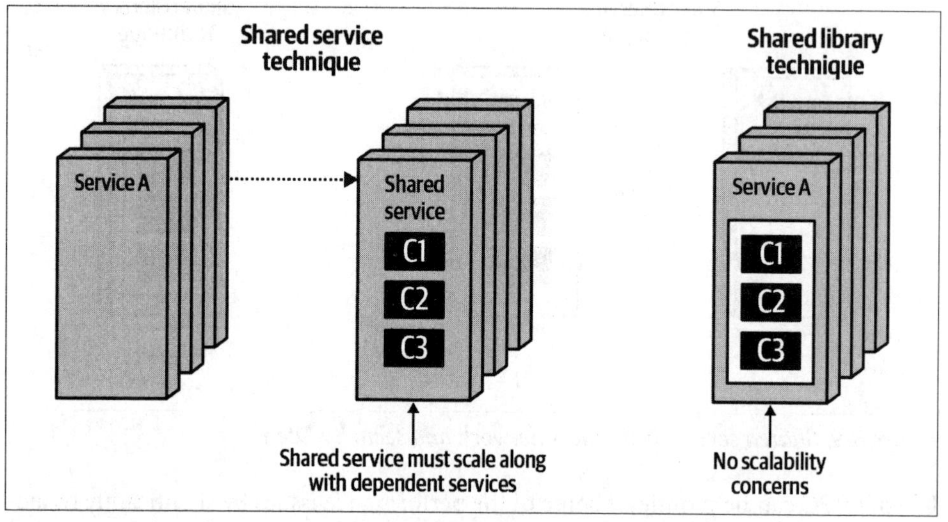

Figure 8-10. Shared services must scale as dependent services scale

Fault Tolerance

While fault-tolerance issues can usually be mitigated through multiple instances of a service, nevertheless it is a trade-off to consider when using the shared service technique. As illustrated in Figure 8-11, if the shared service becomes unavailable, services requiring the shared functionality are rendered nonoperational until the shared service is available. The shared library technique does not have this issue since the shared functionality is contained in the service at compile time, and therefore accessed through standard method or function calls.

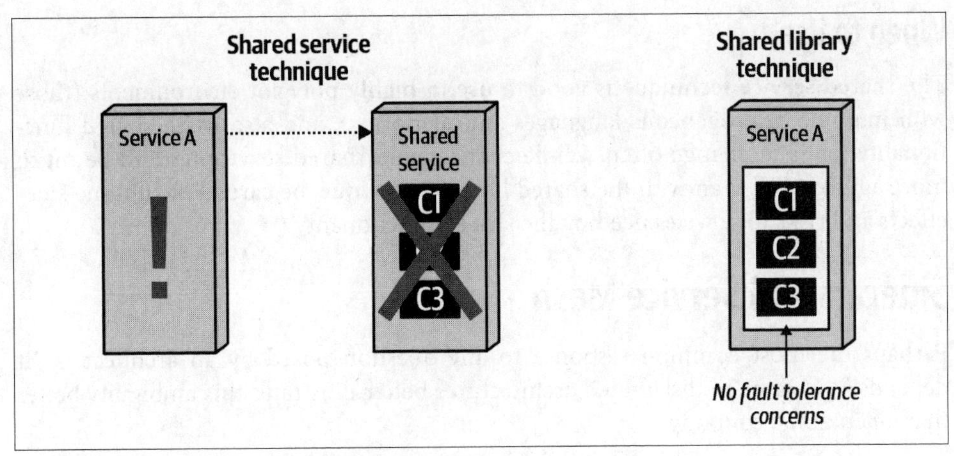

Figure 8-11. Shared services introduce fault-tolerance issues

While the shared service technique preserves the bounded context and is good for shared code that changes frequently, operational characteristics such as performance, scalability, and availability suffer. Table 8-3 lists the various trade-offs associated with this technique.

Trade-Offs

Table 8-3. Trade-offs for the shared service technique

Advantages	Disadvantages
Good for high code volatility	Versioning changes can be difficult
No code duplication in heterogeneous codebases	Performance is impacted due to latency
Preserves the bounded context	Fault tolerance and availability issues due to service dependency
No static code sharing	Scalability and throughput issues due to service dependency
	Increased risk due to runtime changes

When to Use

The shared service technique is good to use in highly polyglot environments (those with multiple heterogeneous languages and platforms), and also when shared functionality tends to change often. While changes in a shared service tend to be much more agile overall than with the shared library technique, be careful of runtime side-effects and risks to services needing the shared functionality.

Sidecars and Service Mesh

Perhaps the most common response to any question posed by an architect is "It depends!" No issue in distributed architectures better illustrates this ambiguity better than operational coupling.

One of the design goals of microservices architectures is a high degree of decoupling, often manifested in the advice "Duplication is preferable to coupling." For example, let's say that two Sysops Squad services need to pass customer information, yet the domain-driven design bounded context insists that implementation details remain private to the service. Thus, a common solution allows each service its own internal representation of entities such as `Customer`, passing that information in loosely coupled ways such as name-value pairs in JSON. Notice that this allows each service to change its internal representation at will, including the technology stack, without breaking the integration. Architects generally frown on duplicating code because it causes synchronization issues, semantic drift, and a host of other issues, but sometimes forces exist that are worse than the problems of duplication, and coupling in microservices often fits that bill. Thus, in microservices architecture, the answer to the question of "should we duplicate or couple to some capability?" is likely *duplicate*, whereas in another architecture style such as a service-based architecture, the correct answer is likely *couple*. It depends!

When designing microservices, architects have resigned themselves to the reality of implementation duplication to preserve decoupling. But what about the type of capabilities that *benefit* from high coupling? For example, consider common operational capabilities such as monitoring, logging, authentication and authorization, circuit breakers, and a host of other operational abilities that each service should have. But allowing each team to manage these dependencies often descends into chaos. For example, consider a company like Penultimate Electronics trying to standardize on a common monitoring solution to make it easier to operationalize the various services. Yet if each team is responsible for implementing monitoring for their service, how can the operations team be sure they did? Also, what about issues such as unified upgrades? If the monitoring tool needs to upgrade across the organization, how can teams coordinate that?

The common solution that has emerged in the microservices ecosystem over the last few years solves this problem in an elegant way, by using the Sidecar pattern. This pattern is based on a much earlier architecture pattern defined by Alistair Cockburn, known as the *hexagonal architecture*, illustrated in Figure 8-12.

In this Hexagonal pattern, what we would now call the domain logic resides in the center of the hexagon, which is surrounded by ports and adaptors to other parts of the ecosystem (in fact, this pattern is alternately known as the *Ports and Adaptors Pattern*). While predating microservices by a number of years, this pattern has similarities to modern microservices, with one significant difference: data fidelity. The hexagonal architecture treated the database as just another adaptor that can be plugged in, but one of the insights from DDD suggests that data schemas and transactionality should be inside the interior—like microservices.

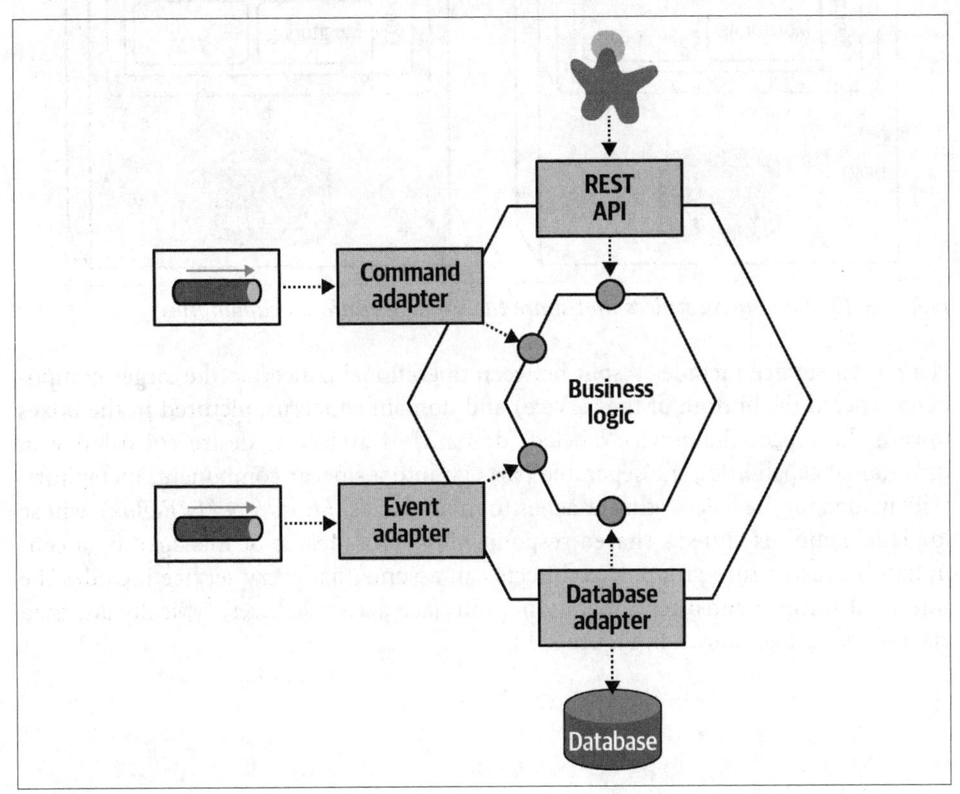

Figure 8-12. The Hexagonal pattern separated domain logic from technical coupling

The *Sidecar pattern* leverages the same concept as hexagonal architecture in that it decouples the domain logic from the technical (infrastructure) logic. For example, consider two microservices, as shown in Figure 8-13.

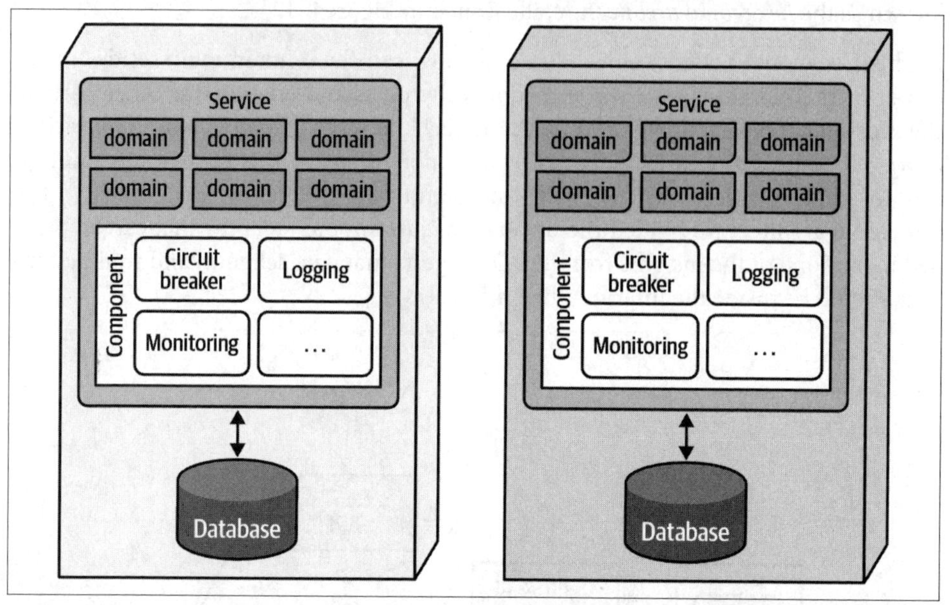

Figure 8-13. Two microservices that share the same operational capabilities

Here, each service includes a split between operational concerns (the larger components toward the bottom of the service) and domain concerns, pictured in the boxes toward the top of the service labeled "domain." If architects desire consistency in operational capabilities, the separable parts go into a sidecar component, metaphorically named for the sidecar that attaches to motorcycles (*https://oreil.ly/EcBuk*), whose implementation is either a shared responsibility across teams or managed by a centralized infrastructure group. If architects can assume that every service includes the sidecar, it forms a consistent operational interface across services, typically attached via a service plane, shown in Figure 8-14.

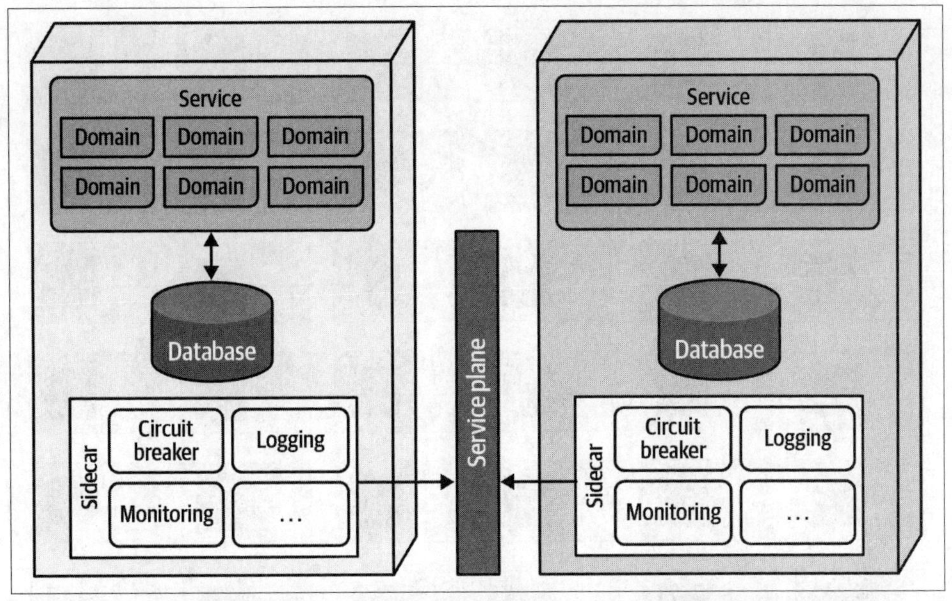

Figure 8-14. When each microservice includes a common component, architects can establish links between them for consistent control

If architects and operations can safely assume that every service includes the sidecar component (governed by fitness functions), it forms a service mesh, illustrated in Figure 8-15. The boxes to the right of each service all interconnect, forming a "mesh."

Having a mesh allows architects and DevOps to create dashboards, control operational characteristics such as scale, and a host of other capabilities.

The Sidecar pattern allows governance groups like enterprise architects a reasonable restraint over too many polyglot environments: one of the advantages of microservices is a reliance on integration rather than a common platform, allowing teams to choose the correct level of complexity and capabilities on a service-by-service basis. However, as the number of platforms proliferates, unified governance becomes more difficult. Therefore, teams often use the consistency of the service mesh as a driver to support infrastructure and other cross-cutting concerns across multiple heterogeneous platforms. For example, without a service mesh, if enterprise architects want to unify around a common monitoring solution, then teams must build a sidecar per platform that supports that solution.

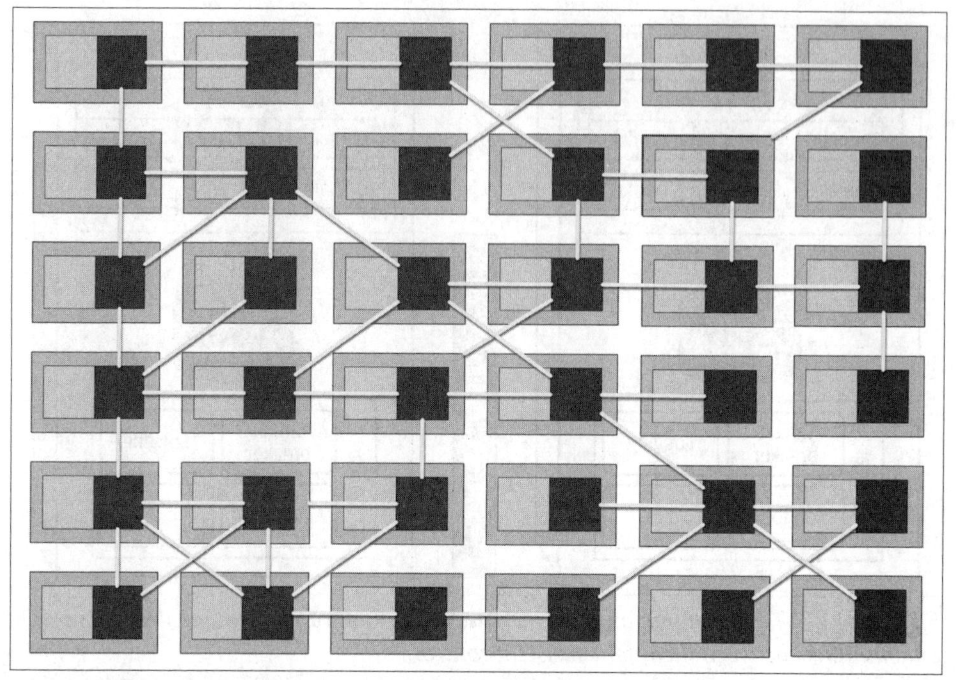

Figure 8-15. A service mesh is an operational link among services

The Sidecar pattern represents not only a way to decouple operational capabilities from domains—it's an *orthogonal reuse* pattern to address a specific kind of coupling (see "Orthogonal Coupling" on page 238). Often, architectural solutions require several types of coupling, such as our current example of domain versus operational coupling. An orthogonal reuse pattern presents a way to reuse some aspect counter to one or more seams in the architecture. For example, microservices architectures are organized around domains, but operational coupling requires cutting across those domains. A sidecar allows an architect to isolate those concerns in a cross-cutting, but consistent, layer through the architecture.

Orthogonal Coupling

In mathematics, two lines are *orthogonal* if they intersect at right angles, which also implies independence. In software architecture, two parts of an architecture may be *orthogonally coupled*: two distinct purposes that must still intersect to form a complete solution. The obvious example from this chapter is an operational concern such as monitoring, which is necessary but independent from domain behavior, like catalog checkout. Recognizing orthogonal coupling allows architects to find intersection points that cause the least entanglement between concerns.

While the Sidecar pattern offers a nice abstraction, it has trade-offs like all other architectural approaches, shown in Table 8-4.

Trade-Offs

Table 8-4. Trade-offs for the Sidecar pattern / service mesh technique

Advantages	Disadvantages
Offers a consistent way to create isolated coupling	Must implement a sidecar per platform
Allows consistent infrastructure coordination	Sidecar component may grow large/ complex
Ownership per team, centralized, or some combination	

When to Use

The Sidecar pattern and service mesh offer a clean way to spread some sort of cross-cutting concern across a distributed architecture, and can be used by more than just operational coupling (see Chapter 14). It offers an architectural equivalent to the Decorator Design Pattern (*https://oreil.ly/4hYmI*) from the Gang of Four *Design Patterns* book (Addison Wesley)—it allows an architect to "decorate" behavior across a distributed architecture independent of the normal connectivity.

Sysops Squad Saga: Common Infrastructure Logic

`Thursday, February 10, 10:34`

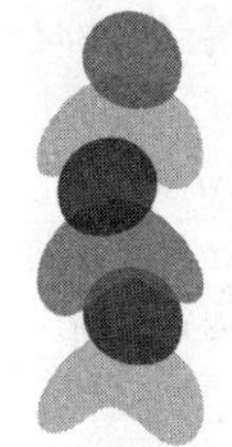

Sydney peeped into Taylen's office on a foggy morning. "Hey, are you using the shared Message Dispatch library?"

Taylen replied, "Yes, we're trying to consolidate on that to get some consistency on message resolution."

Sydney said, "OK, but now we're getting double log messages—it looks like the library writes to the logs, but our service also writes to the log. Is that as it should be?"

"No," Taylen replied. "We definitely don't want duplicate log entries. That just makes everything confusing. We should ask Addison about that."

Consequently, Sydney and Taylen darkened Addison's door. "Hey, do you have a minute?"

Addison replied, "Always for you—what's up?"

Sydney said, "We've been consolidating a bunch of our duplicated code into shared libraries, and that's working well—we're getting better at identifying the parts that rarely change. But, now we've hit the problem that brings us here—who is supposed to be writing log messages? Libraries, services, or something else? And, how can we make that consistent?"

Addison said, "We've bumped into operational shared behavior. Logging is just one of them. What about monitoring, service discovery, circuit breakers, even some of our utility functions, like the `JSONtoXML` library that a few teams are sharing? We need a better way to handle this to prevent issues. That's why we're in the process of implementing a service mesh with this common behavior in a sidecar component."

Sydney said, "I've read about sidecars and service mesh—it's a way to share things across a bunch of microservices, right?"

Addison said, "Sort of, but not all kinds of *things*. The intent of the service mesh and sidecar is to consolidate operational coupling, not domain coupling. For example, just like in our case, we want consistency for logging and monitoring across all our services, but don't want each team to have to worry about that. If we consolidate logging code into the common sidecar that every service implements, we can enforce consistency."

Taylen asked, "Who owns the shared library? Shared responsibility across all the teams?"

Addison replied, "We thought about that, but we have enough teams now; we've built a shared infrastructure team that is going to manage and maintain the sidecar component. They have built the deployment pipeline to automatically test the sidecar once it's been bound into the service with a set of fitness functions."

Sydney said, "So if we need to share libraries between services, just ask them to put it in the sidecar?"

Addison said, "Be careful—the sidecar isn't meant to be used for just anything, only operational coupling."

"I'm not sure what that distinction is," Taylen said.

"Operational coupling includes the things we've been discussing—logging, monitoring, service discovery, authentication and authorization, and so on. Basically, it covers all the plumbing parts of the infrastructure that have no domain responsibility. But you should never put domain shared components, like the `Address` or `Customer` class, in the sidecar."

Sydney asked, "But why? What if I need the same class definition in two services? Won't putting it in the sidecar make it available to both?"

Addison replied, "Yes, but now you are increasing coupling in exactly the way we try to avoid in microservices. In most architectures, a single implementation of that service would be shared across the teams that need it. However, in microservices, that creates a coupling point, tying several services together in an undesirable way—if one team changes the shared code, every team must coordinate with that change. However, the architects could decide to put the shared library in the sidecar—it is, after all a technical capability. Neither answer is unambiguously correct, making this an architect decision and worthy of trade-off analysis. For example, if the `Address` class changes and both services rely on it, they must both change—the definition of coupling. We handle those issues with contracts. The other issue concerns size: we don't want the sidecar to become the biggest part of the architecture. For example, consider the `JSONtoXML` library we were discussing before. How many teams use that?"

Taylen said, "Well, any team that has to integrate with the mainframe system for anything—probably 5 out of, what, 16 or 17 teams?"

Addison said, "Perfect. OK, what's the trade-off of putting the `JSONtoXML` in the sidecar?"

Sydney answered, "Well, that means that every team automatically has the library and doesn't have to wire it in through dependencies."

"And the bad side?" asked Addison.

"Well, adding it to the sidecar makes it bigger, but not by much—it's a small library." said Sydney.

"That's the key trade-off for shared utility code—how many teams need it versus how much overhead does it add to every service, particularly ones that don't need it."

"And if less than one-half the teams use it, it's probably not worth the overhead," Sydney said.

"Right! So, for now, we'll leave that out of the sidecar and perhaps reassess in the future," said Addison.

> *ADR: Using a Sidecar for Operational Coupling*
>
> *Context*
> Each service in our microservices architecture requires common and consistent operational behavior; leaving that responsibility to each team introduces inconsistencies and coordination issues.
>
> *Decision*
> We will use a sidecar component in conjunction with a service mesh to consolidate shared operational coupling.
>
> The shared infrastructure team will own and maintain the sidecar for service teams; service teams act as their customers. The following services will be provided by the sidecar:
>
> - Monitoring
> - Logging
> - Service discovery
> - Authentication
> - Authorization

Consequences

Teams should not add domain classes to the sidecar, which encourages inappropriate coupling.

Teams work with the shared infrastructure team to place shared, *operational* libraries in the sidecar if enough teams require it.

Code Reuse: When Does It Add Value?

Many architects fail to properly assess trade-offs when they encounter some situations, which isn't necessarily a deficiency—many trade-offs become obvious only after the fact.

Reuse is one of the most abused abstractions, because the general view in organizations is that *reuse* represents a laudable goal that teams should strive for. However, failing to evaluate all the trade-offs associated with reuse can lead to serious problems within architecture.

The danger of too much reuse was one of the lessons many architects learned from the early 20th century trend of orchestration-driven service-oriented architecture, where one of the primary goals for many organizations was to maximize reuse.

Consider the scenario from an insurance company, illustrated in Figure 8-16.

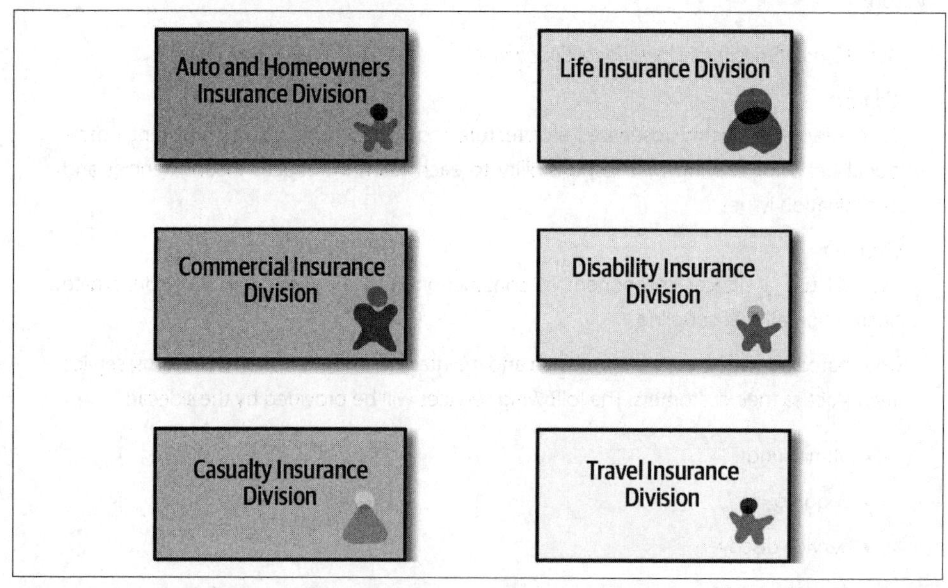

Figure 8-16. Each domain within a large insurance company has a view of the customer

Each division in the company has some aspect of customers it cares about. Years ago, architects were instructed to keep an eye out for this type of commonality; once discovered, the goal was to consolidate the organizational view of customer into a single service, shown in Figure 8-17.

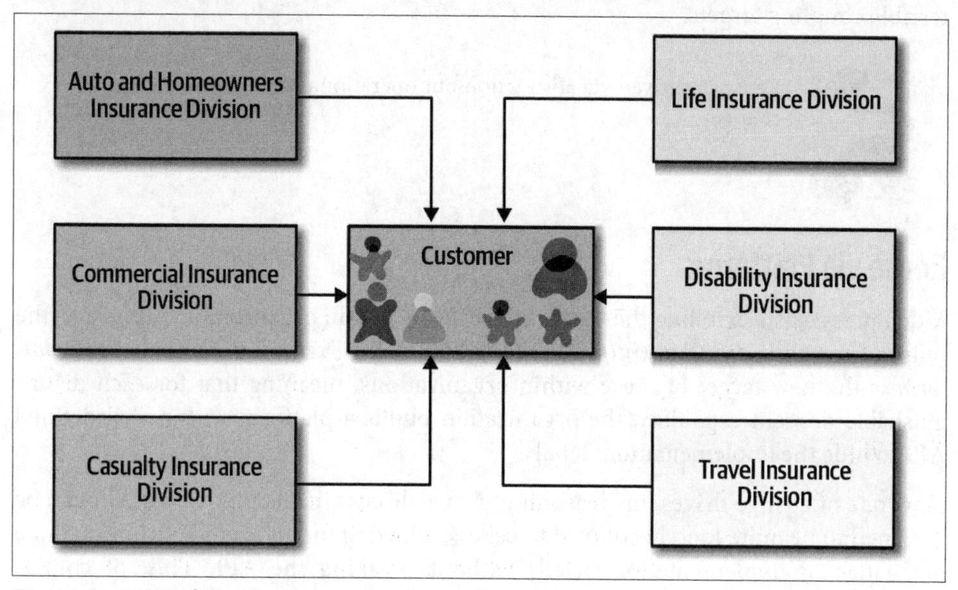

Figure 8-17. Unifying on a centralized Customer service

While the picture in Figure 8-17 may seem logical, it's an architectural disaster for two reasons. First, if all institutional information about a key entity like Customer must reside in a single place, that entity must be complex enough to handle any domain and scenario, making it difficult to use for simple things.

Secondly, though, it creates *brittleness* within the architecture. If every domain that needs customer information must get it from a single place, when that place changes, everything breaks. For example, in our example, what happens when `CustomerService` needs to add new capabilities on behalf of one of the domains? That change could potentially impact every other domain, requiring coordination and testing to ensure that the change hasn't "rippled" throughout the architecture.

What architects failed to realize is that reuse has two important aspects; they got the first one correct: abstraction. The way architects and developers discover candidates for reuse is via abstraction. However, the second consideration is the one that determines utility and value: *rate of change*.

Observing that some reuse causes brittleness begs the question about how that kind of reuse differs from the kinds we clearly benefit from. Consider things that everyone successfully reuses: operating systems, open source frameworks and libraries, and so on. What distinguishes those from assets that project teams build? The answer is *slow rate of change*. We benefit from technical coupling, like operating systems and external frameworks, because they have a well-understood rate of change and update cadence. Internal domain capabilities or quick-changing technical frameworks make terrible coupling targets.

Reuse is derived via abstraction but operationalized by slow rate of change.

Reuse via Platforms

Much press exists extolling the virtue of platforms within organizations, almost to the point of *semantic diffusion* (*https://oreil.ly/oYla7*). However, most agree that the *platform* is the new target of reuse within organizations, meaning that for each distinguishable domain capability, the organization builds a platform with a well-defined API to hide the implementation details.

Slow rate of change drives this reasoning. As we discuss in Chapter 13, an API can be designed to be quite loosely coupled to callers, allowing for an aggressive internal rate of change of implementation details without breaking the API. This, of course, doesn't protect the organization from changes to the semantics of the information it must pass between domains, but by careful design of encapsulation and contracts, architects can limit the amount of breaking change and brittleness in integration architecture.

Sysops Squad Saga: Shared Domain Functionality

`Tuesday, February 8, 12:50`

With Addison's approval, the development team had decided to split the core ticketing functionality into three separate services: a customer-facing Ticket Creation service, a Ticket Assignment service, and a Ticket Completion service. However, all three services used common database logic (queries and updates) and shared a set of database tables in the ticketing data domain.

Taylen wanted to create a shared data service that would contain the common database logic, thus forming a database abstraction layer, as shown in Figure 8-18.

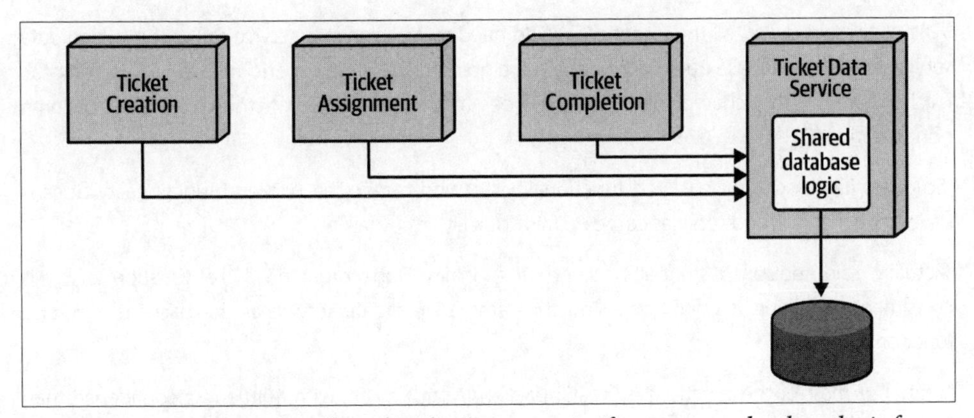

Figure 8-18. Option using a shared Ticket Data service for common database logic for the Sysops Squad ticketing services

Skyler hated the idea and wanted to use a single shared library (DLL) that each service would include as part of the build and deployment, as illustrated in Figure 8-19.

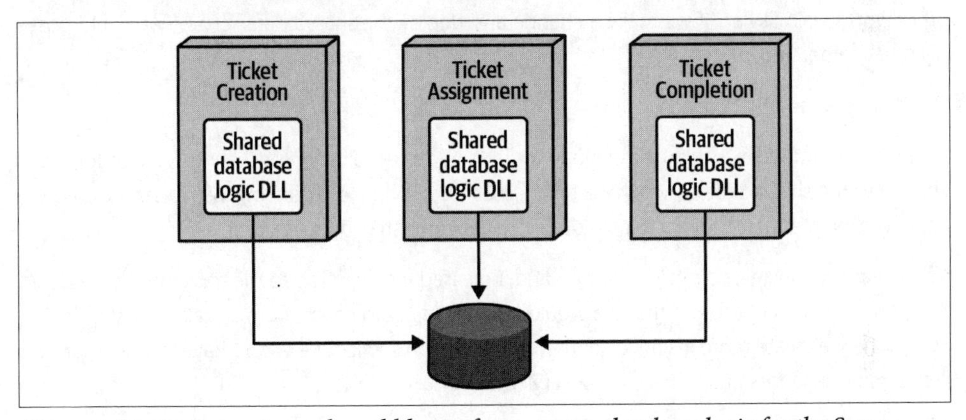

Figure 8-19. Option using a shared library for common database logic for the Sysops Squad ticketing services

Both developers met with Addison to resolve this roadblock.

"So, Addison, what is your opinion? Should the shared database logic be in a shared data service or a shared library?" asked Taylen.

"It's not about opinions," said Addison. "It's about analyzing the trade-offs to arrive at the most appropriate solution for the core shared ticketing database functionality. Let's do a hypothesis-based approach and hypothesize that the most appropriate solution is to use the shared data service."

"Hold on," said Skyler. "It's simply not a good architectural solution for this problem."

"Why?" asked Addison, prompting Skyler to start thinking in terms of trade-offs.

"First of all," said Skyler, "all three services would need to make an interservice call to the shared data service for *every* database query or update. We're going take a serious performance hit if we do that. Furthermore, if the shared data service goes down, all three of those services become nonoperational."

"So?" said Taylen. "It's all backend functionality, so who cares? The backend functionality doesn't have to be that fast, and services come up fairly quickly if they fail."

"Actually," said Addison, "it's not all backend functionality. Don't forget, the Ticket Creation service is customer facing, and it would be using the same shared data service as the backend ticketing functionality."

"Yeah, but *most* of the functionality is still backend," said Taylen, with a little less confidence than before.

"So far," said Addison, "it looks like the trade-off for using the shared data service is performance and fault tolerance for the ticketing services."

"Let's also not forget that any changes made to the shared data service are runtime changes. In other words," said Skyler, "if we make a change and deploy the shared data service, we could possibly break something."

"That's why we test," said Taylen.

"Yeah, but if you want to reduce risk you would have to test all of the ticketing services for every change to the shared data service, which increases testing time significantly. With a shared DLL, we could version the shared library to provide backward compatibility," said Skyler.

"OK, we will add increased risk for changes and increased testing effort to the trade-offs as well," said Addison. "Also, let's not forget that we would have extra coordination from a scalability standpoint. Every time we create more instances of the ticket creation service, we would have to make sure we create more instances of the shared data service as well."

"Let's not keep focusing so much on the negatives." said Taylen. "How about the positives of using a shared data service?"

"OK," said Addison, "let's talk about the benefits of using a shared data service."

"Data abstraction, of course," said Taylen. "The services wouldn't have to worry about any database logic. All they would have to do is make a remote service call to the shared data service."

"Any other benefits?" asked Addison.

"Well," said Taylen, "I was going to say centralized connection pooling, but we would need multiple instances anyway to support the customer ticket creation service. It would help, but it's not a major game changer since there are only three services without a lot of instances of each service. However, change control would be so much easier with a shared data service. We wouldn't have to redeploy any of the ticketing services for database logic changes."

"Let's take a look at those shared class files in the repository and see historically how much change there really is for that code," said Addison.

Addison, Taylen, and Skyler all looked at the repository history for the shared data logic class files.

"Hmm..." said Taylen, "I thought there were a lot more changes to that code than what is showing up in the repo. OK, so I guess the changes are fairly minimal for the shared database logic after all."

Through the conversation of discussing trade-offs, Taylen started to realize that the negatives of a shared service seemed to outweigh the positives, and there was no real compelling justification for putting the shared database logic in a shared service. Taylen agreed to put the shared database logic in a shared DLL, and Addison wrote an ADR for this architecture decision:

> *ADR: Use of a Shared Library for Common Ticketing Database Logic*
>
> *Context*
> The ticketing functionality is broken into three services: Ticket Creation, Ticket Assignment, and Ticket Completion. All three services use common code for the bulk of the database queries and update statements. The two options are to use a shared library or create a shared data service.
>
> *Decision*
> We will use a shared library for the common ticketing database logic.
>
> Using a shared library will improve performance, scalability, and fault tolerance of the customer-facing Ticket Creation service, as well as for the Ticket Assignment service.
>
> We found that the common database logic code does not change much and is therefore fairly stable code. Furthermore, change is less risky for the common database logic because services would need to be tested and redeployed. If changes are needed, we will apply versioning where appropriate so that not all services need to be redeployed when the common database logic changes.
>
> Using a shared library reduces service coupling and eliminates additional service dependencies, HTTP traffic, and overall bandwidth.
>
> *Consequences*
> Changes to the common database logic in the shared DLL will require the ticketing services to be tested and deployed, therefore reducing overall agility for common database logic for the ticketing functionality.
>
> Service instances will need to manage their own database connection pool.

CHAPTER 9

Data Ownership and Distributed Transactions

`Friday, December 10 09:12`

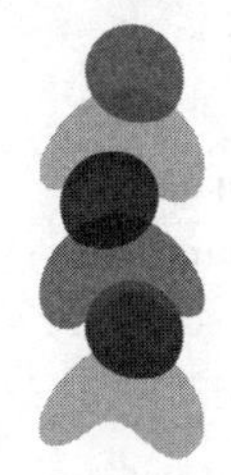

While the database team worked on decomposing the monolithic Sysops Squad database, the Sysops Squad development team, along with Addison, the Sysops Squad architect, started to work on forming bounded contexts between the services and the data, assigning table ownership to services in the process.

"Why did you add the expert profile table to the bounded context of the Ticket Assignment service?" asked Addison.

"Because," said Sydney, "the ticket assignment relies on that table for the assignment algorithms. It constantly queries that table to get the expert's location and skills information."

"But it only does queries to the expert table," said Addison. "The User Maintenance service contains the functionality to perform database updates to maintain that information. Therefore, it seems to me the expert profile table should be owned by the User Maintenance service and put within that bounded context."

"I disagree," said Sydney. "We simply cannot afford for the assignment service to make remote calls to the User Maintenance service for every query it needs. It simply won't work."

"In that case, how to you see updates occurring to the table when an expert acquires a new skill or changes their service location? And what about when we hire a new expert?" asked Addison. "How would that work?"

"Simple," said Sydney. "The User Maintenance service can still access the expert table. All it would need to do is connect to a different database. What's the big deal about that?"

"Don't you remember what Dana said earlier? It's OK for multiple services to connect to the same database schema, but it's not OK for a service to connect to multiple databases or schemas. Dana said that was a no-go and would not allow that to happen," said Addison.

"Oh, right, I forgot about that rule. So what do we do?" asked Sydney. "We have one service that needs to do occasional updates, and an entirely different service in an entirely different domain to do frequent reads from the table."

"I don't know what the right answer is," said Addison. "Clearly this is going to require more collaboration between the database team and us to figure these things out. Let me see if Dana can provide any advice on this."

Once data is pulled apart, it must be stitched back together to make the system work. This means figuring out which services own what data, how to manage distributed transactions, and how services can access data they need (but no longer own). In this chapter, we explore the ownership and transactional aspects of putting distributed data back together.

Assigning Data Ownership

After breaking apart data within a distributed architecture, an architect must determine which services own what data. Unfortunately, assigning data ownership to a service is not as easy as it sounds, and becomes yet another hard part of software architecture.

The general rule of thumb for assigning table ownership states that services that perform write operations to a table own that table. While this general rule of thumb works well for single ownership (only one service ever writes to a table), it gets messy when teams have joint ownership (multiple services do writes to the same table) or even worse, common ownership (most or all services write to the table).

The general rule of thumb for data ownership is that the service that performs write operations to a table is the owner of that table. However, joint ownership makes this simple rule complex!

To illustrate some of the complexities with data ownership, consider the example illustrated in Figure 9-1 showing three services: a Wishlist Service that manages all of the customer wish lists, a Catalog Service that maintains the product catalog, and an Inventory Service that maintains the inventory and restocking functionality for all products in the product catalog.

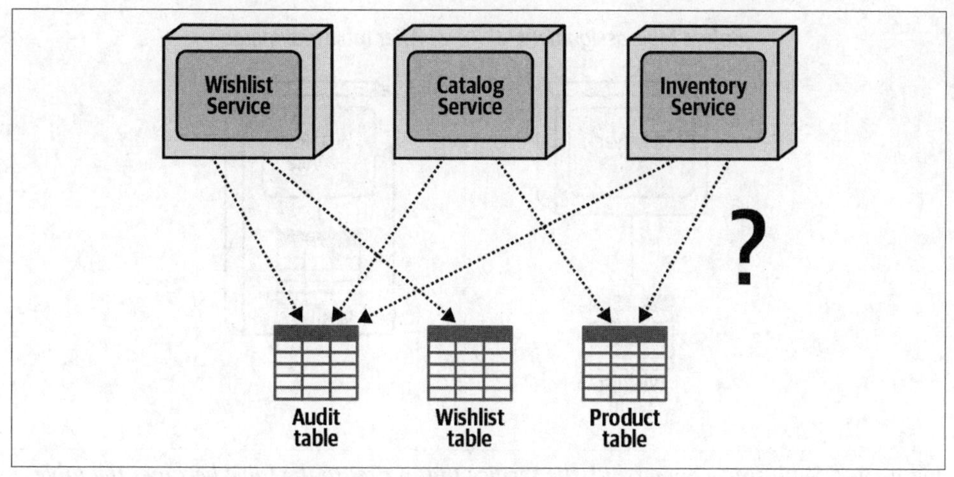

Figure 9-1. Once data is broken apart, tables must be assigned to services that own them

To further complicate matters, notice that the Wishlist Service writes to both the Audit table and the Wishlist table, the Catalog Service writes to the Audit table and the Product table, and the Inventory Service writes to the Audit table and the Product table. Suddenly, this simple real-world example makes assigning data ownership a complex and confusing task.

In this chapter, we unravel this complexity by discussing the three scenarios encountered when assigning data ownership to services (single ownership, common ownership, and joint ownership), and exploring techniques for resolving these scenarios, using Figure 9-1 as a common reference point.

Single Ownership Scenario

Single table ownership occurs when only one service writes to a table. This is the most straightforward of the data ownership scenarios and is relatively easy to resolve. Referring back to Figure 9-1, notice that the Wishlist table has only a single service that writes to it—the Wishlist Service.

In this scenario, it is clear that the Wishlist Service should be the owner of the Wishlist table (regardless of other services that need read-only access to the Wishlist table), see Figure 9-2. Notice that on the right side of this diagram, the Wishlist table becomes part of the bounded context of the Wishlist Service. This diagramming technique is an effective way to indicate table ownership and the bounded context formed between the service and its corresponding data.

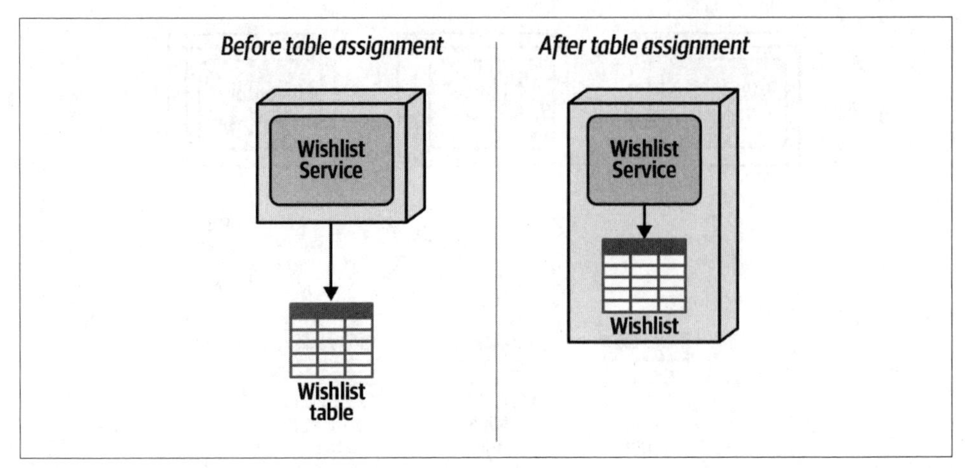

Figure 9-2. With single ownership, the service that writes to the table becomes the table owner

Because of the simplicity of this scenario, we recommend addressing single table ownership relationships first to clear the playing field in order to better address the more complicated scenarios that arise: common ownership and joint ownership.

Common Ownership Scenario

Common table ownership occurs when most (or all) of the services need to write to the same table. For example, Figure 9-1 shows that all services (Wishlist, Catalog, and Inventory) need to write to the Audit table to record the action performed by the user. Since all services need to write to the table, it's difficult to determine who should actually own the Audit table. While this simple example includes only three services, imagine a more realistic example where potentially hundreds (or even thousands) of services must write to the same Audit table.

The solution of simply putting the Audit table in a shared database or shared schema that is used by all services unfortunately reintroduces all of the data-sharing issues described at the beginning of Chapter 6, including change control, connection starvation, scalability, and fault tolerance. Therefore, another solution is needed to solve common data ownership.

A popular technique for addressing common table ownership is to assign a dedicated single service as the primary (and only) owner of that data, meaning only one service is responsible for writing data to the table. Other services needing to perform write actions would send information to the dedicated service, which would then perform the actual write operation on the table.

If no information or acknowledgment is needed by services sending the data, services can use persisted queues for asynchronous fire-and-forget messaging. Alternatively, if information needs to be returned to the caller based on a write action (such as returning a confirmation number or database key), services can use something like REST, gRPC, or request-reply messaging (pseudosynchronous) for a synchronous call.

Coming back to the Audit table example, notice in Figure 9-3 that the architect created a new Audit Service and assigned it ownership of the Audit table, meaning it is the only service that performs read or write actions on the table. In this example, since no return information is needed, the architect used asynchronous fire-and-forget messaging with a persistent queue so that the Wishlist Service, Catalog Service, and Inventory Service don't need to wait for the audit record to be written to the table. Making the queue persistent (meaning the message is stored on disk by the broker) provides guaranteed delivery in the event of a service or broker failure and helps ensure that no messages are lost.

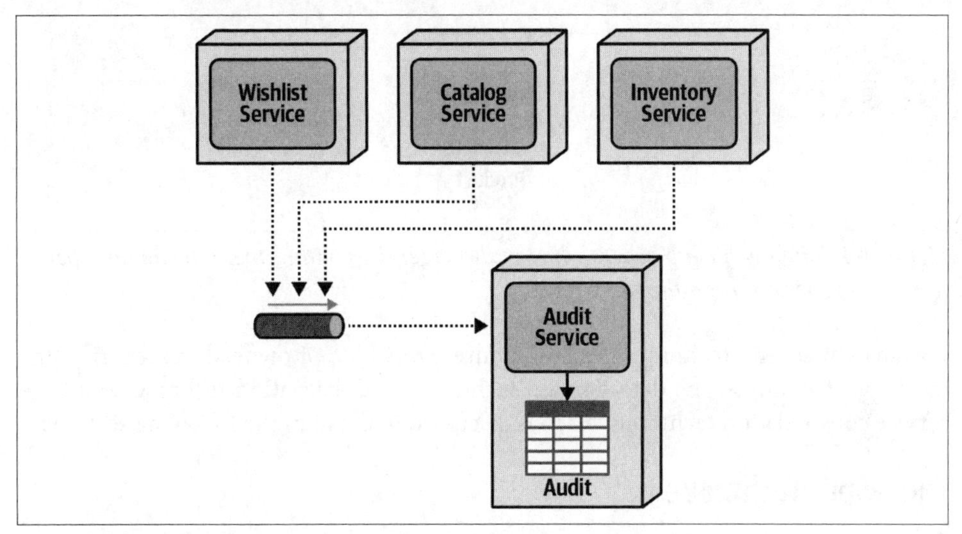

Figure 9-3. Common ownership uses a dedicated service owner

In some cases, it may be necessary for services to read common data they don't own. These read-only access techniques are described in detail in Chapter 10.

Joint Ownership Scenario

One of the more common (and complex) scenarios involving data ownership is *joint ownership*, which occurs when multiple services perform write actions on the same table. This scenario differs from the prior common ownership scenario in that with joint ownership, only a couple of services within the same domain write to the same table, whereas with common ownership, most or all of the services perform write operations on the same table. For example, notice in Figure 9-1 that all services perform write operations on the Audit table (common ownership), whereas only the Catalog and Inventory services perform write operations on the Product table (joint ownership).

Figure 9-4 shows the isolated joint ownership example from Figure 9-1. The Catalog Service inserts new products into the table, removes products no longer offered, and updates static product information as it changes, whereas the Inventory Service is responsible for reading and updating the current inventory for each product as products are queried, sold, or returned.

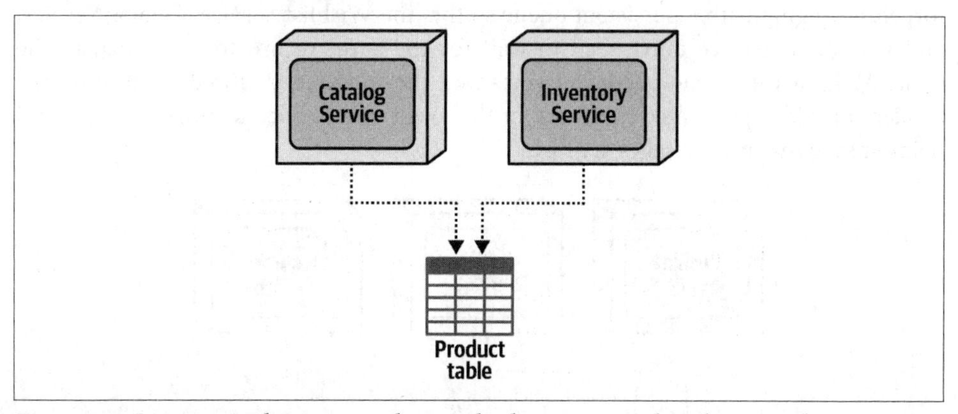

Figure 9-4. Joint ownership occurs when multiple services within the same domain perform write operations on the same table

Fortunately, several techniques exist to address this type of ownership scenario—the table split technique, the data domain technique, the delegation technique, and the service consolidation technique. Each is discussed in detail in the following sections.

Table Split Technique

The *table split technique* breaks a single table into multiple tables so that each service owns a part of the data it's responsible for. This technique is described in detail in the book *Refactoring Databases* and in the companion website (*https://oreil.ly/WJ2kt*).

To illustrate the table split technique, consider the Product table example illustrated in Figure 9-4. In this case, the architect or developer would first create a separate Inventory table containing the product ID (key) and the inventory count (number of items available), pre-populate the Inventory table with data from the existing Product table, then finally remove the inventory count column from the Product table. The source listing in Example 9-1 shows how this technique might be implemented using data definition language (DDL) in a typical relational database.

Example 9-1. DDL source code for splitting up the Product table and moving inventory counts to a new Inventory table

```
CREATE TABLE Inventory
(
product_id VARCHAR(10),
inv_cnt INT
);

INSERT INTO Inventory VALUES (product_id, inv_cnt)
AS SELECT product_id, inv_cnt FROM Product;

COMMIT;

ALTER TABLE Product DROP COLUMN inv_cnt;
```

Splitting the database table moves the joint ownership to a single table ownership scenario: the Catalog Service owns the data in the Product table, and the Inventory Service owns the data in the Inventory table. However, as shown in Figure 9-5, this technique requires communication between the Catalog Service and Inventory Service when products are created or removed to ensure the data remains consistent between the two tables.

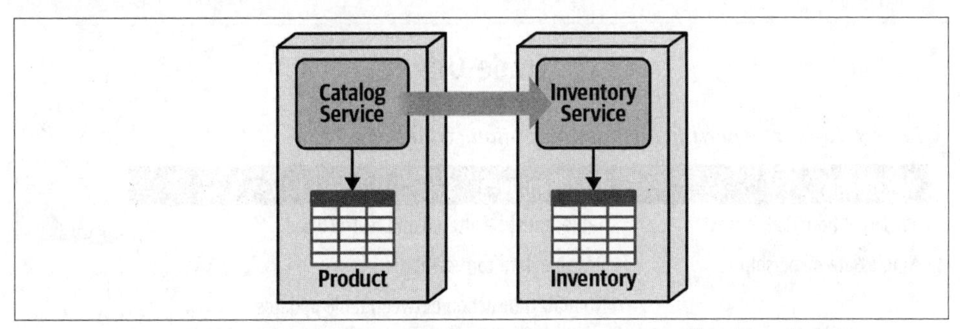

Figure 9-5. Joint ownership can be addressed by breaking apart the shared table

For example, if a new product is added, the Catalog Service generates a product ID and inserts the new product into the Product table. The Catalog Service then must send that new product ID (and potentially the initial inventory counts) to the Inventory Service. If a product is removed, the Catalog Service first removes the product from the Product table, then must notify the Inventory Service to remove the inventory row from the Inventory table.

Synchronizing data between split tables is not a trivial matter. Should communication between the Catalog Service and the Inventory Service be synchronous or asynchronous? What should the Catalog Service do when adding or removing a product and finding that the Inventory Service is not available? These are hard questions to answer, and are usually driven by the traditional *availability* verses *consistency* trade-off commonly found in distributed architectures. Choosing availability means that it's more important that the Catalog Service always be able to add or remove products, even though a corresponding inventory record may not be created in the Inventory table. Choosing consistency means that it's more important that the two tables always remain in sync with each other, which would cause a product creation or removal operation to fail if the Inventory Service is not available. Because network partitioning is necessary in distributed architectures, the CAP theorem (*https://oreil.ly/R1fXW*) states that only one of these choices (consistency or availability) is possible.

The type of communication protocol (synchronous versus asynchronous) also matters when splitting a table. Does the Catalog Service require a confirmation that the corresponding Inventory record is added when creating a new product? If so, then synchronous communication is required, providing better data consistency at the sacrifice of performance. If no confirmation is required, the Catalog Service can use asynchronous fire-and-forget communication, providing better performance at the sacrifice of data consistency. So many trade-offs to consider!

Table 9-1 summarizes the trade-offs associated with the table split technique for joint ownership.

Trade-Offs

Table 9-1. Joint ownership table split technique trade-offs

Advantages	Disadvantages
Preserves bounded context	Tables must be altered and restructured
Single data ownership	Possible data consistency issues
	No ACID transaction between table updates
	Data synchronization is difficult
	Data replication between tables may occur

Data Domain Technique

Another technique for joint ownership is to create a shared *data domain*. This is formed when data ownership is shared between the services, thus creating multiple owners for the table. With this technique, the tables shared by the same services are put into the same schema or database, therefore forming a broader bounded context between the services and the data.

Notice that Figure 9-6 looks close to the original diagram in Figure 9-4 with one noticeable difference—the data domain diagram has the Product table in a separate box outside the context of each owning service. This diagramming technique makes it clear that the table is not owned by or part of the bounded context of either service, but rather shared between them in a broader bounded context.

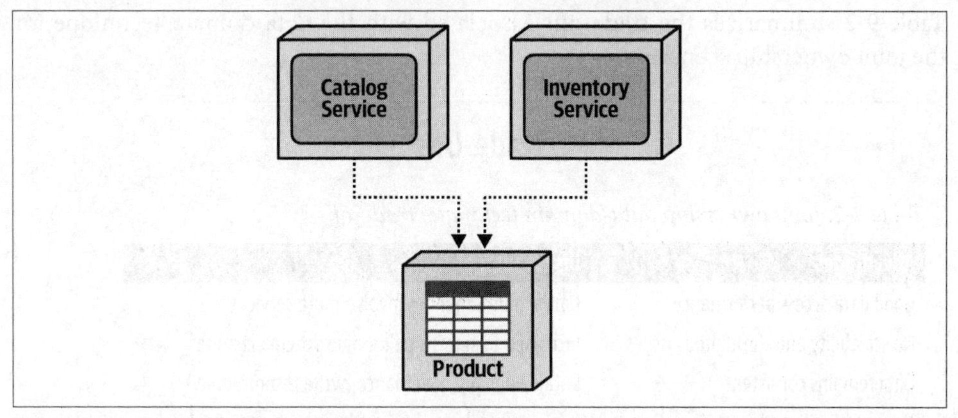

Figure 9-6. With joint ownership, services can share data by using the data domain technique (shared schema)

While data sharing is generally discouraged in distributed architectures (particularly with microservices), it does resolve some of the performance, availability, and data consistency issues found in other joint ownership techniques. Because the services are not dependent on each other, the Catalog Service can create or remove products without needing to coordinate with the Inventory Service, and the Inventory Service can adjust inventory without needing the Catalog Service. Both services become completely independent from each other.

When choosing the data domain technique, always reevaluate why separate services are needed since the data is common to each of the services. Justifications might include scalability differences, fault-tolerance needs, throughput differences, or isolating code volatility (see Chapter 7).

Unfortunately, sharing data in a distributed architecture introduces a number of issues, the first of these being increased effort for changes made to the structure of the data (such as changing the schema of a table). Because a broader bounded context is formed between the services and the data, changes to the shared table structures may require those changes to be coordinated among multiple services. This increases development effort, testing scope, and deployment risk.

Another issue with the data domain technique with regard to data ownership is controlling which services have write responsibility to what data. In some cases, this might not matter, but if it's important to control write operations to certain data, additional effort is required to apply specific governance rules to maintain specific table or column write ownership.

Table 9-2 summarizes the trade-offs associated with the data domain technique for the joint ownership scenario.

Trade-Offs

Table 9-2. Joint ownership data-domain technique trade-offs

Advantages	Disadvantages
Good data access performance	Data schema changes involve more services
No scalability and throughput issues	Increased testing scope for data schema changes
Data remains consistent	Data ownership governance (write responsibility)
No service dependency	Increased deployment risk for data schema changes

Delegate Technique

An alternative method for addressing the joint ownership scenario is the *delegate technique*. With this technique, one service is assigned single ownership of the table and becomes the delegate, and the other service (or services) communicates with the delegate to perform updates on its behalf.

One of the challenges of the delegate technique is knowing which service to assign as the delegate (the sole owner of the table). The first option, called *primary domain priority*, assigns table ownership to the service that most closely represents the primary domain of the data—in other words, the service that does most of the primary entity CRUD operations for the particular entity within that domain. The second option, called *operational characteristics priority*, assigns table ownership to the service needing higher operational architecture characteristics, such as performance, scalability, availability, and throughput.

To illustrate these two options and the corresponding trade-offs associated with each, consider the Catalog Service and Inventory Service joint ownership scenario shown in Figure 9-4. In this example, the Catalog Service is responsible for creating, updating, and removing products, as well as retrieving product information; the Inventory Service is responsible for retrieving and updating product inventory count as well as for knowing when to restock if inventory gets too low.

With the primary domain priority option, the service that performs most of the CRUD operations on the main entity becomes the owner of the table. As illustrated in Figure 9-7, since the Catalog Service performs most of the CRUD operations on product information, the Catalog Service would be assigned as the single owner of the table. This means that the Inventory service must communicate with the Catalog Service to retrieve or update inventory counts since it doesn't own the table.

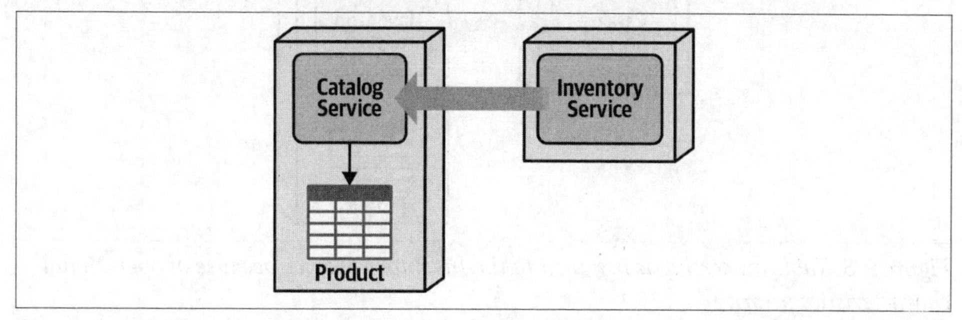

Figure 9-7. Table ownership is assigned to the Catalog service because of domain priority

Like the common ownership scenario described earlier, the delegate technique always forces interservice communication between the other services needing to update the data. Notice in Figure 9-7 that the Inventory Service must send inventory updates through some sort of remote access protocol to the Catalog Service so that it can perform the inventory updates and reads on behalf of the Inventory Service. This communication can either be synchronous or asynchronous. As always in software architecture, more trade-off analysis to consider.

With synchronous communication, the Inventory Service must wait for the inventory to be updated by the Catalog Service, which impacts overall performance but ensures data consistency. Using asynchronous communication to send inventory updates makes the Inventory Service perform much faster, but the data is only eventually consistent. Furthermore, with asynchronous communication, because an error can occur in the Catalog Service while trying to update inventory, the Inventory Service has no guarantee that the inventory was ever updated, impacting data integrity as well.

With the operational characteristics priority option, the ownership roles would be reversed because inventory updates occur at a much faster rate than static product data. In this case, table ownership would be assigned to the Inventory Service, the justification being that updating product inventory is a part of the frequent real-time transactional processing of purchasing products as opposed to the more infrequent administrative task of updating product information or adding and removing products (see Figure 9-8).

With this option, frequent updates to inventory counts can use direct database calls rather than remote access protocols, therefore making inventory operations much faster and more reliable. In addition, the most volatile data (inventory count) is kept highly consistent.

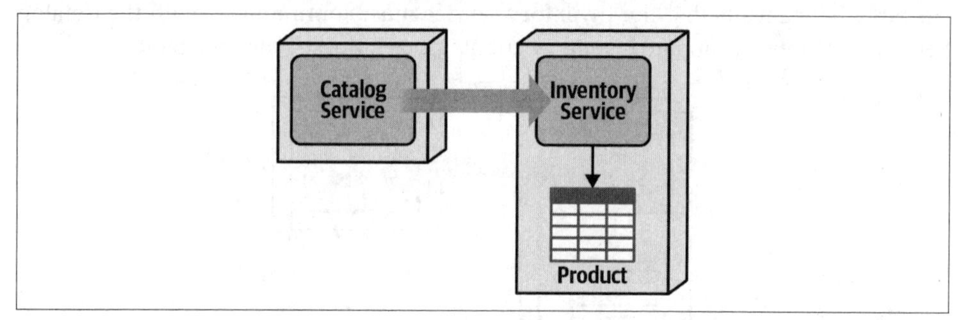

Figure 9-8. Table ownership is assigned to the Inventory Service because of operational characteristics priority

However, one major problem with the diagram illustrated in Figure 9-8 is that of domain management responsibility. The Inventory Service is responsible for managing product inventory, not the database activity (and corresponding error handling) for adding, removing, and updating static product information. For this reason, we usually recommend the domain priority option, and leveraging things like a replicated in-memory cache or a distributed cache to help address performance and fault-tolerance issues.

Regardless of which service is assigned as the delegate (sole table owner), the delegate technique has some disadvantages, the biggest being service coupling and the need for interservice communication. This in turn leads to other issues for nondelegate services, including the lack of an atomic transaction when performing write operations, low performance due to network and processing latency, and low fault tolerance. Because of these issues, the delegate technique is generally better suited for database write scenarios that do not require atomic transactions and that can tolerate eventual consistency through asynchronous communications.

Table 9-3 summarizes the overall trade-offs of the delegate technique.

Trade-Offs

Table 9-3. Joint ownership delegate technique trade-offs

Advantages	Disadvantages
Forms single table ownership	High level of service coupling
Good data schema change control	Low performance for nonowner writes
Abstracts data structures from other services	No atomic transaction for nonowner writes
	Low fault tolerance for nonowner services

Service Consolidation Technique

The delegate approach discussed in the prior section highlights the primary issue associated with joint ownership—service dependency. The *service consolidation technique* resolves service dependency and addresses joint ownership by combining multiple table owners (services) into a single consolidated service, thus moving joint ownership into a single ownership scenario (see Figure 9-9).

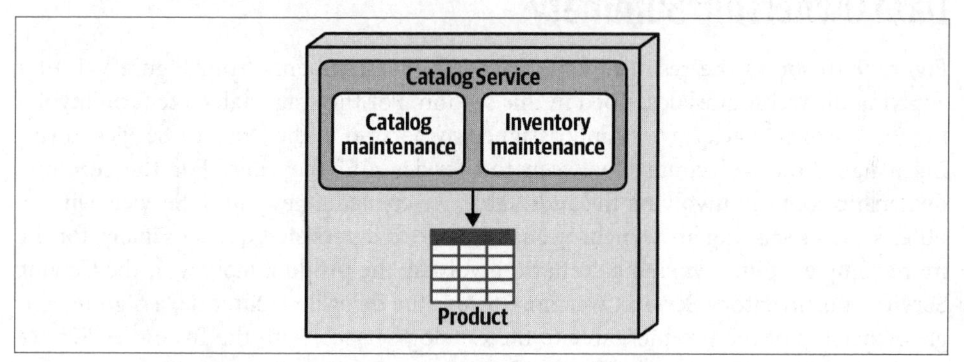

Figure 9-9. Table ownership is resolved by combining services

Like the data domain technique, this technique resolves issues associated with service dependencies and performance, while at the same time addressing the joint ownership problem. However, like the other techniques, it has its share of trade-offs as well.

Combining services creates a more coarse-grained service, thereby increasing the overall testing scope as well as overall deployment risk (the chance of breaking something else in the service when a new feature is added or a bug is fixed). Consolidating services might also impact overall fault tolerance since all parts of the service fail together.

Overall scalability is also impacted when using the service consolidation technique because *all* parts of the service must scale equally, even though some functionality might not need to scale at the same level as other functionality. For example, in Figure 9-9, the catalog maintenance functionality (what used to be in a separate `Catalog` service) must unnecessarily scale to meet the high demands of the inventory retrieval and update functionality.

Table 9-4 summarizes the overall trade-offs of the service consolidation technique.

Trade-Offs

Table 9-4. Joint ownership service consolidation technique trade-offs

Advantages	Disadvantages
Preserves atomic transactions	More coarse-grained scalability
Good overall performance	Less fault tolerance
	Increased deployment risk
	Increased testing scope

Data Ownership Summary

Figure 9-10 shows the resulting table ownership assignments from Figure 9-1 after applying the techniques described in this section. For the single table scenario involving the Wishlist Service, we simply assigned ownership to the Wishlist Service, forming a tight bounded context between the service and the table. For the common ownership scenario involving the audit table, we created a new Audit Service, with all other services sending an asynchronous message to a persisted queue. Finally, for the more complex joint ownership scenario involving the product table with the Catalog Service and Inventory Service, we chose to use the delegate technique, assigning single ownership of the product table to the Catalog Service, with the Inventory Service sending update requests to the Catalog Service.

Once table ownership has been assigned to services, an architect must then validate the table ownership assignments by analyzing business workflows and their corresponding transaction requirements.

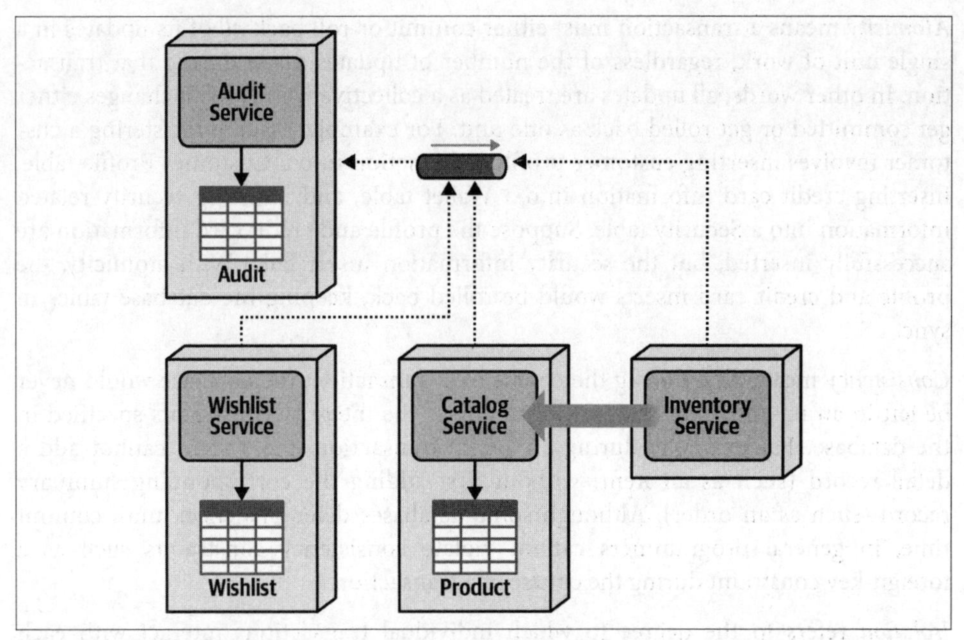

Figure 9-10. Resulting data ownership using delegate technique for joint ownership

Distributed Transactions

When architects and developers think about transactions, they usually think about a single atomic unit of work where multiple database updates are either committed together or all rolled back when an error occurs. This type of atomic transaction is commonly referred to as an *ACID transaction*. As noted in Chapter 6, ACID is an acronym describing the basic properties of an atomic single-unit-of-work database transaction: atomicity, consistency, isolation, and durability.

To understand how distributed transactions work and the trade-offs involved with using a distributed transaction, it's necessary to fully understand the four properties of an ACID transaction. We firmly believe that without an understanding of ACID transactions, an architect cannot perform the necessary trade-off analysis for knowing when (and when not to) use a distributed transaction. Therefore, we will dive into the details of an ACID transaction first, then describe how they differ from distributed transactions.

Atomicity means a transaction must either commit or roll back *all* of its updates in a single unit of work, regardless of the number of updates made during that transaction. In other words, all updates are treated as a collective whole, so all changes either get committed or get rolled back as one unit. For example, assume registering a customer involves inserting customer profile information into a Customer Profile table, inserting credit card information into a Wallet table, and inserting security-related information into a Security table. Suppose the profile and credit card information are successfully inserted, but the security information insert fails. With atomicity, the profile and credit card inserts would be rolled back, keeping the database tables in sync.

Consistency means that during the course of a transaction, the database would never be left in an inconsistent state or violate any of the integrity constraints specified in the database. For example, during an ACID transaction, the system cannot add a detail record (such as an item) without first adding the corresponding summary record (such as an order). Although some databases defer this check until commit time, in general programmers cannot violate consistency constraints such as a foreign-key constraint during the course of a transaction.

Isolation refers to the degree to which individual transactions interact with each other. Isolation protects uncommitted transaction data from being visible to other transactions during the course of the business request. For example, during the course of an ACID transaction, when the customer profile information is inserted into the Customer Profile table, no other services outside of the ACID transaction scope can access the newly inserted information until the entire transaction is committed.

Durability means that once a successful response from a transaction commit occurs, it is guaranteed that *all* data updates are permanent, regardless of further system failures.

To illustrate an ACID transaction, suppose a customer registering for the Sysops Squad application enters all of their profile information, the electronic products they want covered under the support plan, and their billing information on a single user interface screen. This information is then sent to the single Customer Service, as shown in Figure 9-11, which then performs all of the database activity associated with the customer registration business request.

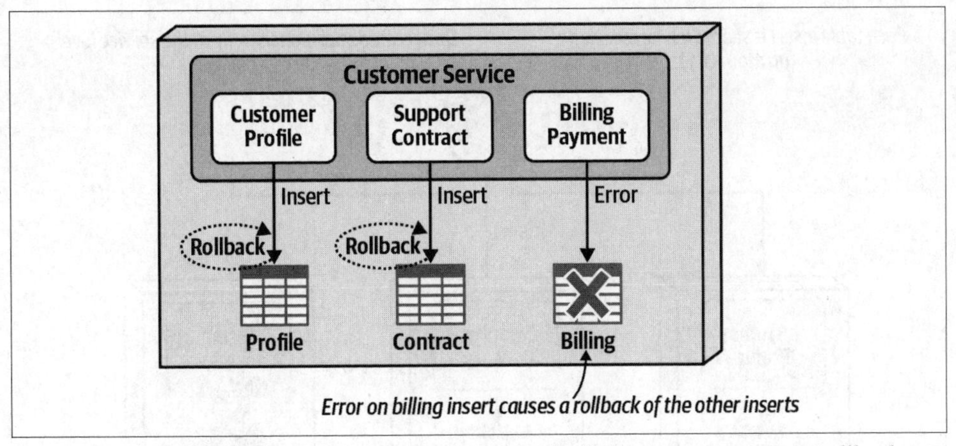

Figure 9-11. With ACID transactions, an error on the billing insert causes a rollback to the other table inserts

First, notice that with an ACID transaction, because an error occurred when trying to insert the billing information, both the profile information and support contract information that were previously inserted are now rolled back (that's the *atomicity* and *consistency* parts of ACID). While not illustrated in the diagram, data inserted into each table during the course of the transaction is not visible to other requests (that's the *isolation* part of ACID).

Note that ACID transactions can exist *within the context of each service* in a distributed architecture, but only if the corresponding database supports ACID properties as well. Each service can perform its own commits and rollbacks to the tables it owns within the scope of the atomic business transaction. However, if the business request spans multiple services, the entire business request itself cannot be an ACID transaction—rather, it becomes a *distributed transaction*.

Distributed transactions occur when an atomic business request containing multiple database updates is performed by separately deployed remote services. Notice in Figure 9-12 that the same request for a new customer registration (denoted by the laptop image representing the customer making the request) is now spread across three separately deployed services—a Customer Profile Service, a Support Contract Service, and a Billing Payment Service.

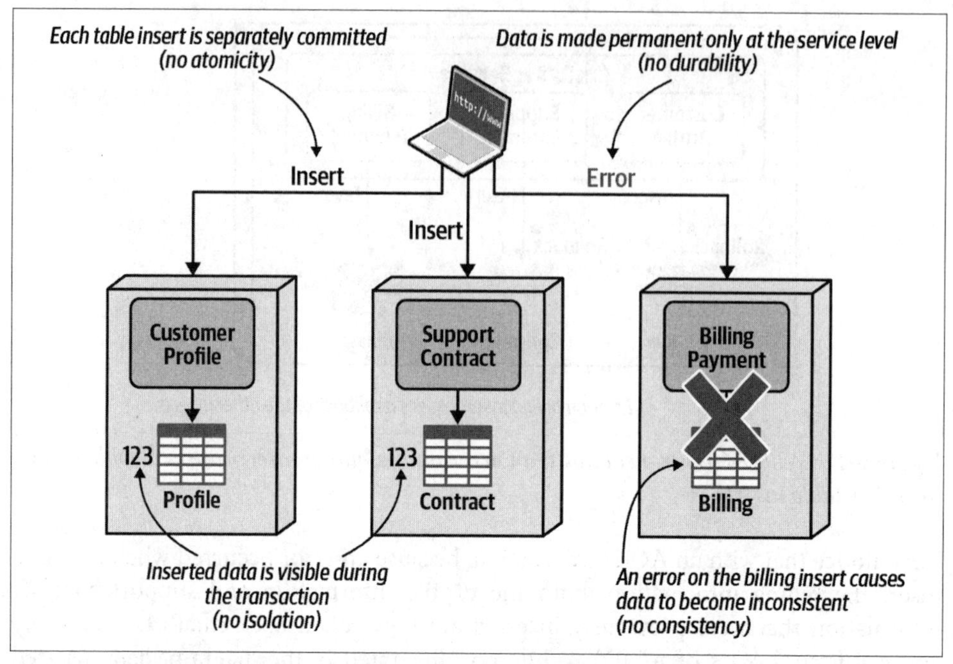

Figure 9-12. Distributed transactions do not support ACID properties

As you can see, distributed transactions do not support ACID properties.

Atomicity is not supported because each separately deployed service commits its own data and performs only one part of the overall atomic business request. In a distributed transaction, atomicity is bound to the *service*, not the *business request* (such as customer registration).

Consistency is not supported because a failure in one service causes the data to be out of sync between the tables responsible for the business request. As shown in Figure 9-12, since the Billing Payment Service insert failed, the Profile table and Contract table are now out of sync with the Billing table (we'll show how to address these issues later in this section). Consistency is also impacted because traditional relational database constraints (such as a foreign key always matching a primary key) cannot be applied during each individual service commit.

Isolation is not supported because once the Customer Profile Service inserts the profile data in the course of a distributed transaction to register a customer, that profile information is available to any other service or request, even though the customer registration process (the current transaction) hasn't completed.

Durability is not supported across the business request—it is supported for only each individual service. In other words, any individual commit of data does not ensure that *all* data within the scope of the entire business transaction is permanent.

Instead of ACID, distributed transactions support something called *BASE*. In chemistry, an *acid* substance and a *base* substance are exactly the opposite. The same is true with atomic and distributed transactions—ACID transactions are opposite of BASE transactions. BASE describes the properties of a distributed transaction: basic availability, soft state, and eventual consistency.

Basic availability (the "BA" part of BASE) means that all of the services or systems in the distributed transaction are expected to be available to participate in the distributed transaction. While asynchronous communication can help decouple services and address availability issues associated with the distributed transaction participants, it unfortunately impacts how long it will take the data to become consistent for the atomic business transaction (see eventual consistency later in this section).

Soft state (the *S* part of BASE) describes the situation where a distributed transaction is in progress and the state of the atomic business request is not yet complete (or in some cases not even known). In the customer registration example shown in Figure 9-12, soft state occurs when the customer profile information is inserted (and committed) in the Profile table, but the support contract and billing information are not. The unknown part of soft state can occur if, using the same example, all three services work in parallel to insert their corresponding data—the exact state of the atomic business request is not known at any point in time until *all three* services report back that the data has been successfully processed. In the case of a workflow using asynchronous communication (see Chapter 11), the in-progress or final state of the distributed transaction is usually difficult to determine.

Eventual consistency (the *E* part of BASE) means that given enough time, all parts of the distributed transaction will complete successfully and all of the data is in sync with one another. The type of eventual consistency pattern used and the way errors are handled dictates how long it will take for all of the data sources involved in the distributed transaction to become consistent.

The next section describes the three types of eventual consistency patterns and the corresponding trade-offs associated with each pattern.

Eventual Consistency Patterns

Distributed architectures rely heavily on eventual consistency as a trade-off for better operational architecture characteristics such as performance, scalability, elasticity, fault tolerance, and availability. While there are numerous ways to achieve eventual consistency between data sources and systems, the three main patterns in use today are the background synchronization pattern, orchestrated request-based pattern, and the event-based pattern.

To better describe each pattern and illustrate how they work, consider again the customer registration process from the Sysops Squad application we discussed earlier in

Figure 9-13. In this example, three separate services are involved in the customer registration process: a Customer Profile Service that maintains basic profile information, a Support Contract Service that maintains products covered under the Sysops Squad repair plan for each customer, and a Billing Payment Service that charges the customer for the support plan. Notice in the figure that customer 123 is a subscriber to the Sysops Squad service, and therefore has data in each of the corresponding tables owned by each service.

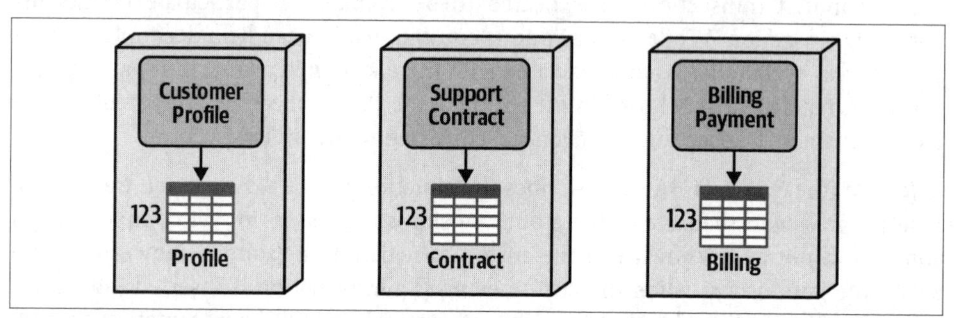

Figure 9-13. Customer 123 is a subscriber in the Sysops Squad application

Customer 123 decides they are no longer interested in the Sysops Squad support plan, so they unsubscribe from the service. As shown in Figure 9-14, the Customer Profile Service receives this request from the user interface, removes the customer from the Profile table, and returns a confirmation to the customer that they are successfully unsubscribed and will no longer be billed. However, data for that customer still exists in the Contract table owned by the Support Contract Service and the Billing table owned by the Billing Payment Service.

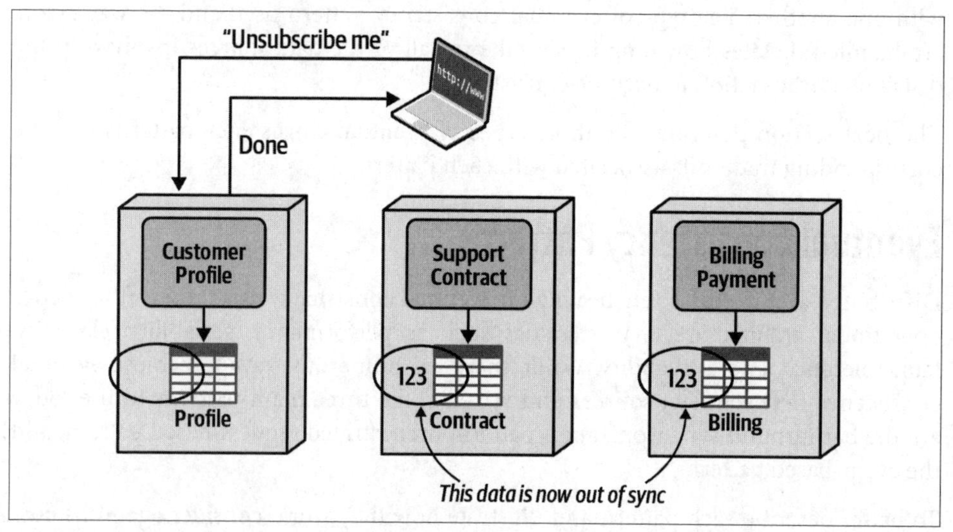

Figure 9-14. Data is out of sync after the customer unsubscribes from the support plan

We will use this scenario to describe each of the eventual consistency patterns for getting all of the data in sync for this atomic business request.

Background Synchronization Pattern

The *background synchronization pattern* uses a separate external service or process to periodically check data sources and keep them in sync with one another. The length of time for data sources to become eventually consistent using this pattern can vary based on whether the background process is implemented as a batch job running sometime in the middle of the night, or a service that wakes up periodically (say, every hour) to check the consistency of the data sources.

Regardless of how the background process is implemented (nightly batch or periodic), this pattern usually has the longest length of time for data sources to become consistent. However, in many cases data sources do not need to be kept in sync immediately. Consider the customer unsubscribe example in Figure 9-14. Once a customer unsubscribes, it really doesn't matter that the support contract and billing information for that customer still exists. In this case, eventual consistency done during the night is a sufficient amount of time to get the data in sync.

One of the challenges of this pattern is that the background process used to keep all the data in sync must know what data has changed. This can be done through an event stream, a database trigger, or reading data from source tables and aligning target tables with the source data. Regardless of the technique used to identify changes, the background process must have knowledge of all the tables and data sources involved in the transaction.

Figure 9-15 illustrates the use of the background synchronization pattern for the Sysops Squad unregister example. Notice that at 11:23:00 the customer issues a request to unsubscribe from the support plan. The Customer Profile Service receives the request, removes the data, and one second later (11:23:01) responds back to the customer that they have been successfully unsubscribed from the system. Then, at 23:00 the background batch synchronization process starts. The background synchronization process detects that customer 123 has been removed either through event streaming or primary table versus secondary table deltas, and deletes the data from the Contract and Billing tables.

This pattern is good for overall responsiveness because the end user doesn't have to wait for the entire business transaction to complete (in this case, unsubscribing from the support plan). But, unfortunately, some serious trade-offs with this eventual consistency pattern.

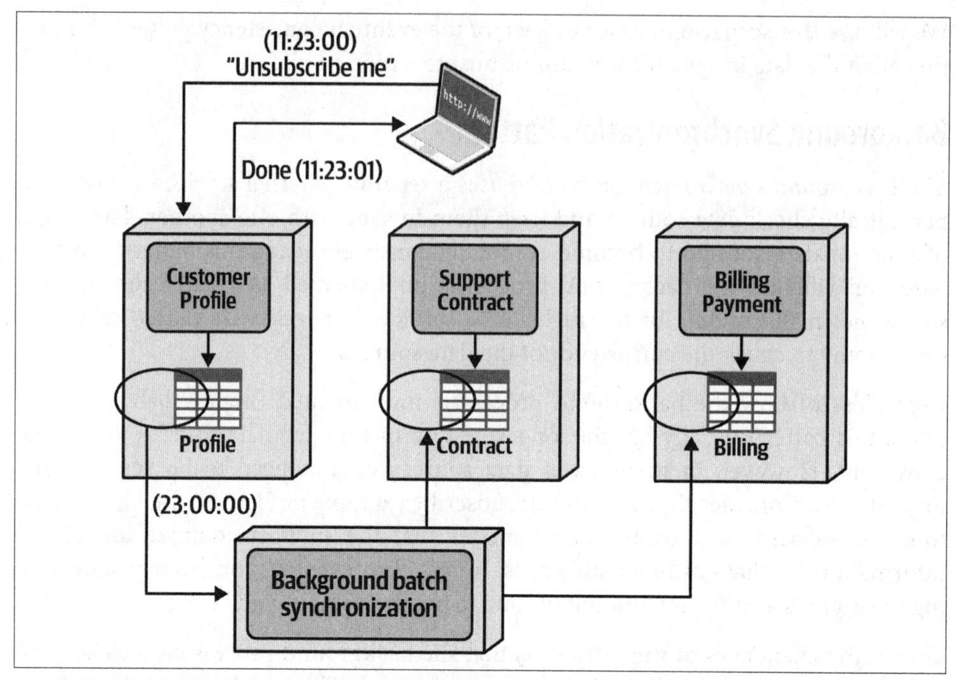

Figure 9-15. The background synchronization pattern uses an external process to ensure data consistency

The biggest disadvantage of the background synchronization pattern is that it couples all of the data sources together, thus breaking every bounded context between the data and the services. Notice in Figure 9-16 that the background batch synchronization process must have write access to each of the tables owned by the corresponding services, meaning that all of the tables effectively have shared ownership between the services and the background synchronization process.

This shared data ownership between the services and the background synchronization process is riddled with issues, and emphasizes the need for tight bounded contexts within a distributed architecture. Structural changes made to the tables owned by each service (changing a column name, dropping a column, and so on) must also be coordinated with an external background process, making changes difficult and time-consuming.

In addition to difficulties with change control, problems occur with regard to duplicated business logic as. In looking at Figure 9-15, it might seem fairly straightforward that the background process would simply perform a `DELETE` operation on all rows in the Contract and Billing tables containing customer 123. However, certain business rules may exist within these services for the particular operation.

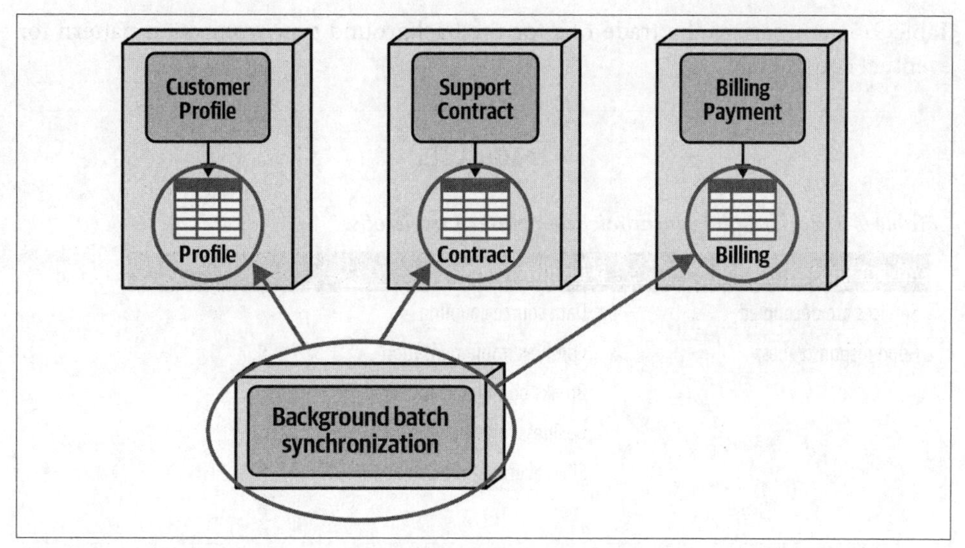

Figure 9-16. The background synchronization pattern is coupled to the data sources, therefore breaking the bounded context and data ownership

For example, when a customer unsubscribes, their existing support contracts and billing history are kept for three months in the event the customer decides to resubscribe to the support plan. Therefore, rather than deleting the rows in those tables, a `remove_date` column is set with a long value representing the date the rows should be removed (a zero value in this column indicates an active customer). Both services check the `remove_date` daily to determine which rows should be removed from their respective tables. The question is, where is that business logic located? The answer, of course, is in the Support Contract and Billing Payment Services—oh, and *also the background batch process!*

The background synchronization eventual consistency pattern is not suitable for distributed architectures requiring tight bounded contexts (such as microservices) where the coupling between data ownership and functionality is a critical part of the architecture. Situations where this pattern is useful are closed (self-contained) heterogeneous systems that don't communicate with each other or share data.

For example, consider a contractor order entry system that accepts orders for building materials, and another separate system (implemented in a different platform) that does contractor invoicing. Once a contractor orders supplies, a background synchronization process moves those orders to the invoicing system to generate invoices. When a contractor changes an order or cancels it, the background synchronization process moves those changes to the invoicing system to update the invoices. This is a good example of systems becoming *eventually consistent*, with the contractor order always in sync between the two systems.

Table 9-5 summarizes the trade-offs for the background synchronization pattern for eventual consistency.

Trade-Offs

Table 9-5. Background synchronization pattern trade-offs

Advantages	Disadvantages
Services are decoupled	Data source coupling
Good responsiveness	Complex implementation
	Breaks bounded contexts
	Business logic may be duplicated
	Slow eventual consistency

Orchestrated Request-Based Pattern

A common approach for managing distributed transactions is to make sure all of the data sources are synchronized during the course of the business request (in other words, while the end user is waiting). This approach is implemented through what is known as the *orchestrated request-based pattern*.

Unlike the previous background synchronization pattern or the event-based pattern described in the next section, the orchestrated request-based pattern attempts to process the entire distributed transaction *during the business request*, and therefore requires some sort of orchestrator to manage the distributed transaction. The orchestrator, which can be a designated existing service or a new separate service, is responsible for managing all of the work needed to process the request, including knowledge of the business process, knowledge of the participants involved, multicasting logic, error handling, and contract ownership.

One way to implement this pattern is to designate one of the primary services (assuming there is one) to manage the distributed transaction. This technique, illustrated in Figure 9-17, designates one of the services to take on the role as orchestrator in addition to its other responsibilities, which in this case is the Customer Profile Service.

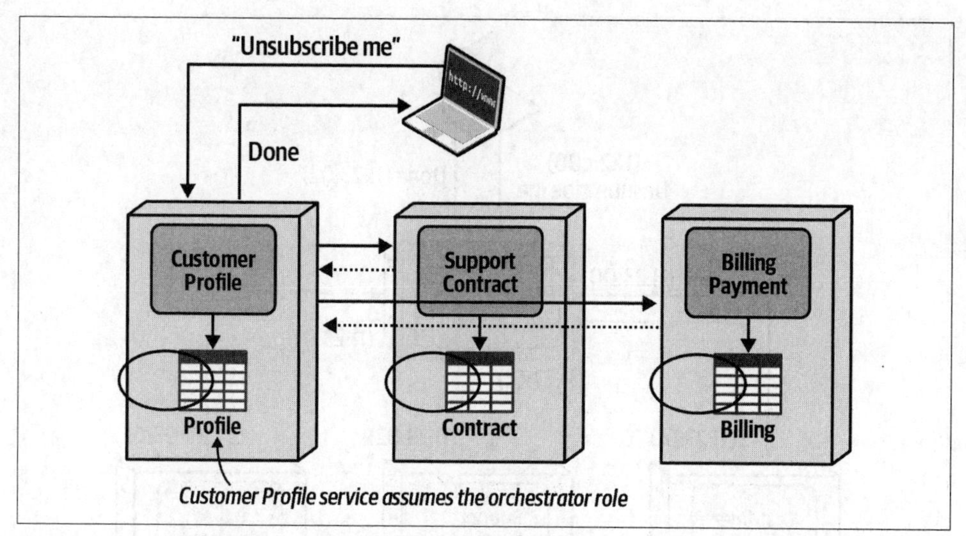

Figure 9-17. The Customer Profile Service takes on the role of an orchestrator for the distributed transaction

Although this approach avoids the need for a separate orchestration service, it tends to overload the responsibilities of the service designated as the distributed transaction orchestrator. In addition to the role of an orchestrator, the designated service managing the distributed transaction must perform its own responsibilities as well. Another drawback to this approach is that it lends itself to tight coupling and synchronous dependencies between services.

The approach we generally prefer when using the orchestrated request-based pattern is to use a dedicated orchestration service for the business request. This approach, illustrated in Figure 9-18, frees up the Customer Profile Service from the responsibility of managing the distributed transaction and places that responsibility on a separate orchestration service.

We will use this separate orchestration service approach to describe how this eventual consistency pattern works and the corresponding trade-offs with this pattern.

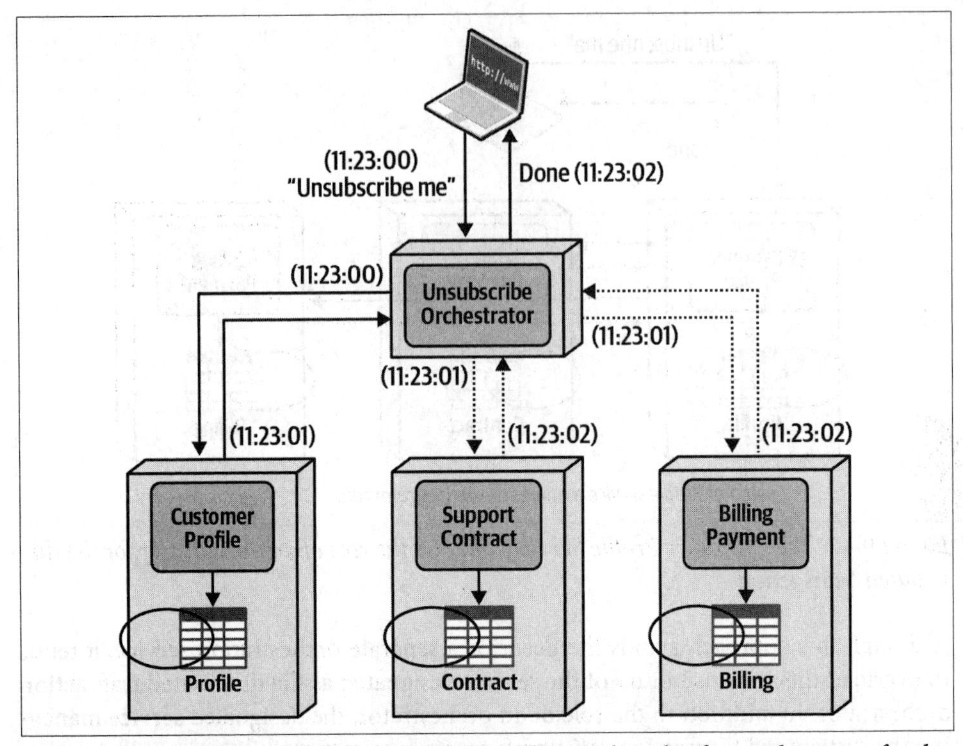

Figure 9-18. A dedicated orchestration service takes on the role of an orchestrator for the distributed transaction

Notice that at 11:23:00 the customer issues a request to unsubscribe from the Sysops Squad support plan. The request is received by the Unsubscribe Orchestrator Service, which then forwards the request synchronously to the Customer Profile Service to remove the customer from the Profile table. One second later, the Customer Profile Service sends back an acknowledgment to the Unsubscribe Orchestrator Service, which then sends parallel requests (either through threads or some sort of asynchronous protocol) to both the Support Contract and Billing Payment Services. Both of these services process the unsubscribe request, and then send an acknowledgment back one second later to the Unsubscribe Orchestrator Service indicating they are done processing the request. Now that all data is in sync, the Unsubscribe Orchestrator Service responds back to the client at 11:23:02 (two seconds after the initial request was made), letting the customer know they were successfully unsubscribed.

The first trade-off to observe is that the orchestration approach generally favors data consistency over responsiveness. Adding a dedicated orchestration service not only adds additional network hops and service calls, but depending on whether the orchestrator executes calls serially or in parallel, additional time is needed for the back-and-forth communication between the orchestrator and the services it's calling.

Response time could be improved in Figure 9-18 by executing the Customer Profile request at the same time as the other services, but we chose to do that operation synchronously for error handling and consistency reasons. For example, if the customer could not be deleted from the Profile table because of an outstanding billing charge, no other action is needed to reverse the operations in the Support Contract and Billing Payment Services. This represents another example of consistency over responsiveness.

Besides responsiveness, the other trade-off with this pattern is complex error handling. While the orchestrated request-based pattern might seem straightforward, consider what happens when the customer is removed from the Profile table and Contract table, but an error occurs when trying to remove the billing information from the Billing table, as illustrated in Figure 9-19. Since the Profile and Support Contract Services individually committed their operations, the Unsubscribe Orchestrator Service must now decide what action to take *while the customer is waiting for the request to be processed*:

1. Should the orchestrator send the request again to the Billing Payment Service for another try?
2. Should the orchestrator perform a compensating transaction and have the Support Contract and Customer Profile Services reverse their update operations?
3. Should the orchestrator respond to the customer that an error occurred and to wait a bit before trying again, while trying to repair the inconsistency?
4. Should the orchestrator ignore the error in hopes that some other process will deal with the issue and respond to the customer that they have been successfully unsubscribed?

This real-world scenario creates a messy situation for the orchestrator. Because this *is* the eventual consistency pattern used, there is no other means to correct the data and get things back in sync (therefore negating options 3 and 4 in the preceding list). In this case, the only real option for the orchestrator is to try to reverse the distributed transaction—in other words, issue a *compensating update* to reinsert the customer in the Profile table and set the `remove_date` column in the Contract table back to zero. This would require the orchestrator to have all of the necessary information to reinsert the customer, and that no side effects occur when creating a new customer (such as initializing the billing information or support contracts).

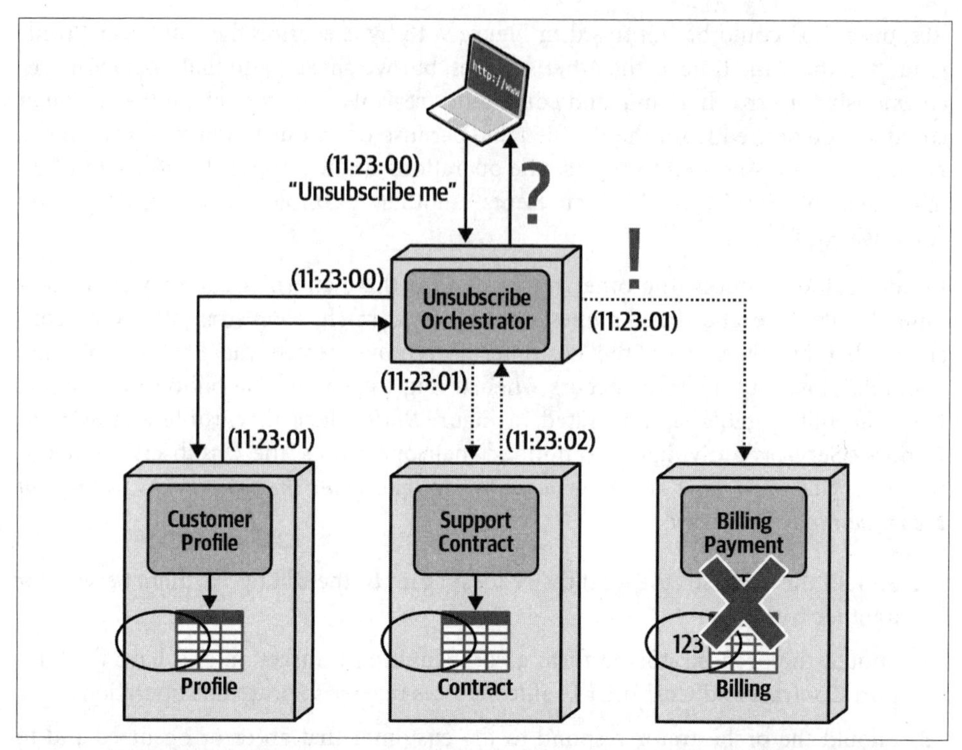

Figure 9-19. Error conditions are very hard to address when using the orchestrated request-based pattern

Another complication with compensating transactions in a distributed architecture is failures that occur during compensation. For example, suppose a compensating transaction was issued to the Customer Profile Service to reinsert the customer, and that operation failed. Now what? Now the data is really out of sync, and there's no other service or process around to repair the problem. Most cases like these typically require human intervention to repair the data sources and get them back in sync. We go into more details about compensating transactions and transactional sagas in "Transactional Saga Patterns" on page 324.

Table 9-6 summarizes the trade-offs for the orchestrated request-based pattern for eventual consistency.

Trade-Offs

Table 9-6. Orchestrated request-based pattern trade-offs

Advantages	Disadvantages
Services are decoupled	Slower responsiveness
Timeliness of data consistency	Complex error handling
Atomic business request	Usually requires compensating transactions

Event-Based Pattern

The *event-based pattern* is one of the most popular and reliable eventual consistency patterns for most modern distributed architectures, including microservices and event-driven architectures. With this pattern, events are used in conjunction with an asynchronous publish-and-subscribe (pub/sub) messaging model to post events (such as `customer unsubscribed`) or command messages (such as `unsubscribe customer`) to a topic or event stream. Services involved in the distributed transaction listen for certain events and respond to those events.

The eventual consistency time is usually short for achieving data consistency because of the parallel and decoupled nature of the asynchronous message processing. Services are highly decoupled from one another with this pattern, and responsiveness is good because the service triggering the eventual consistency event doesn't have to wait for the data synchronization to occur before returning information to the customer.

Figure 9-20 illustrates how the event-based pattern for eventual consistency works. Notice that the customer issues the unsubscribe request to the Customer Profile Service at 11:23:00. The Customer Profile Service receives the request, removes the customer from the Profile table, publishes a message to a message topic or event stream, and returns information one second later letting the customer know they were successfully unsubscribed. At around the same time this happens, both the Support Contract and Billing Payment Services receive the unsubscribe event and perform whatever functionality is needed to unsubscribe the customer, making all the data sources eventually consistent.

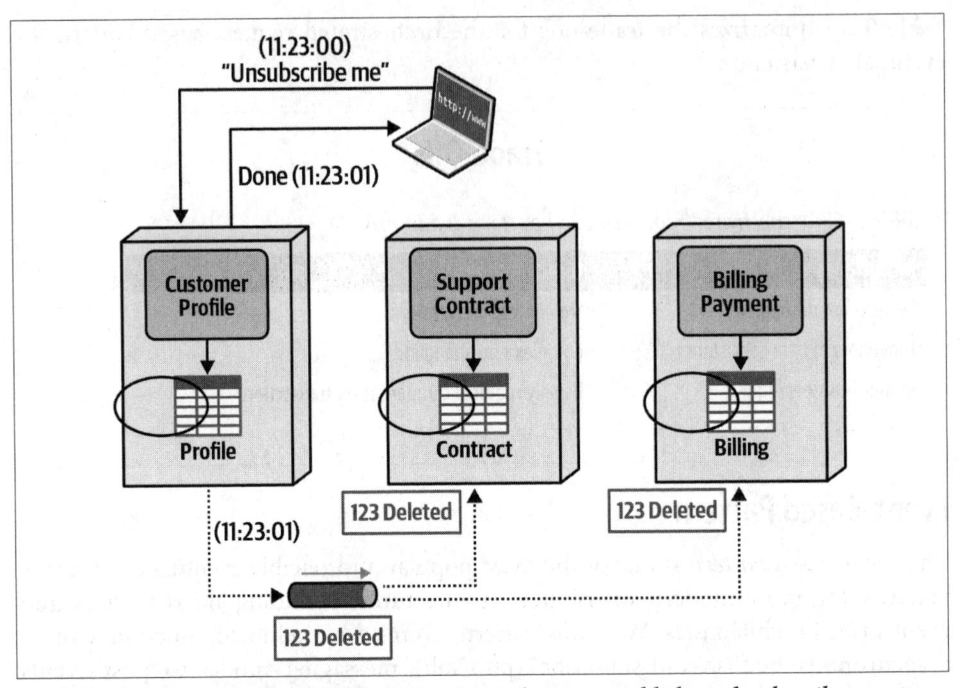

Figure 9-20. The event-based pattern uses asynchronous publish-and-subscribe messaging or event streams to achieve eventual consistency

For implementations using standard topic-based publish-and-subscribe messaging (such as ActiveMQ, RabbitMQ, AmazonMQ, and so on), services responding to the event must be set up as *durable subscribers* to ensure no messages are lost if the message broker or the service receiving the message fails. A durable subscriber is similar in concept to persistent queues in that the subscriber (in this case, the Support Contract Service and Billing Payment Service) does not need to be available at the time the message is published, and subscribers are guaranteed to receive the message once they become available. In the case of event streaming implementations, the message broker (such as Apache Kafka) must always persist the message and make sure it is available in the topic for a reasonable amount of time.

The advantages of the event-based pattern are responsiveness, timeliness of data consistency, and service decoupling. However, similar to all eventual consistency patterns, the main trade-off of this pattern is error handling. If one of the services (for example, the Billing Payment Service illustrated in Figure 9-20) is not available, the fact that it is a durable subscriber means that *eventually* it will receive and process the event when it does become available. However, if the service is processing the event and fails, things get complicated quickly.

Most message brokers will try a certain number of times to deliver a message, and after repeated failures by the receiver, the broker will send the message to a *dead letter queue* (*DLQ*). This is a configurable destination where the event is stored until an automated process reads the message and tries to fix the problem. If it can't be repaired programmatically, the message is then typically sent to a human for manual processing.

Table 9-7 lists the trade-offs for the event-based pattern for eventual consistency.

Trade-Offs

Table 9-7. Event-based pattern trade-offs

Advantages	Disadvantages
Services are decoupled	Complex error handling
Timeliness of data consistency	
Fast responsiveness	

Sysops Squad Saga: Data Ownership for Ticket Processing

`Tuesday, January 18, 09:14`

After talking with Dana and learning about data ownership and distributed transaction management, Sydney and Addison quickly realized that breaking apart data and assigning data ownership to form tight bounded contexts wasn't possible without both teams collaborating on the solution.

"No wonder nothing ever seems to work around here," observed Sydney. "We've always had issues and arguments between us and the database team, and now I see the results of our company treating us as two separate teams."

"Exactly," said Addison. "I'm glad we are working more closely with the data team now. So, from what Dana said, the service that performs write actions on the data table owns the table, regardless of what other services need to access the data in a read-only manner. In that case, looks like the User Maintenance Service needs to own the data."

Sydney agreed, and Addison created a general architecture decision record describing what to do for single-table ownership scenarios:

ADR: Single Table Ownership for Bounded Contexts

Context

When forming bounded contexts between services and data, tables must be assigned ownership to a particular service or group of services.

Decision

When only one service writes to a table, that table will be assigned ownership to that service. Furthermore, services requiring read-only access to a table in another bounded context cannot directly access the database or schema containing that table.

Per the database team, table ownership is defined as the service that performs write operations on a table. Therefore, for single table ownership scenarios, regardless of how many other services need to access the table, only one service is ever assigned an owner, and that owner is the service that maintains the data.

Consequences

Depending on the technique used, services requiring read-only access to a table in another bounded context may incur performance and fault-tolerance issues when accessing data in a different bounded context.

Now that Sydney and Addison better understood table ownership and how to form bounded contexts between the service and the data, they started to work on the survey functionality. The Ticket Completion Service would write the timestamp the ticket was completed and the expert who performed the job to the survey table. The Survey Service would write the timestamp the survey was sent to the customer, and also insert all of the survey results once the survey is received.

"This isn't so hard now that I better understand bounded contexts and table ownership," said Sydney.

"OK, let's move on to the survey functionality," said Addison.

"Oops," said Sydney. "Both the Ticket Completion Service and the Survey Service write to the Survey table."

"That's what Dana called joint-table ownership," said Addison.

"So, what are our options?" asked Sydney.

"Since splitting up the table won't work, it really leaves us with only two options," said Addison. "We can use a common data domain so that both services own the data, or we can use the delegate technique and assign only one service as the owner."

"I like the common data domain. Let both services write to the table and share a common schema," said Sydney.

"Except that won't work in this scenario," said Addison. "The Ticket Completion Service is already talking to the common ticketing data domain. Remember, a service can't connect to multiple schemas."

"Oh, right," said Sydney. "Wait, I know, just add the survey tables to the ticketing data domain schema."

"But now we are starting to combine all the tables back together." said Addison. "Pretty soon we'll be right back to a monolithic database again."

"So what do we do?" asked Sydney.

"Wait, I think I see a good solution here," said Addison. "You know how the Ticket Completion Service has to send a message to the Survey Service anyway to kick off the survey process once a ticket is complete? What if we passed in the necessary data along with that message so that the Survey Service can insert the data when it creates the customer survey?"

"That's brilliant," said Sydney. "That way, the Ticket Completion doesn't need any access to the Survey table."

Addison and Sydney agreed that the Survey Service would own the Survey table, and would use the delegation technique to pass data when the table notifies the Survey Service to kick off the survey process as illustrated in Figure 9-21. Addison wrote an architecture decision record for this decision.

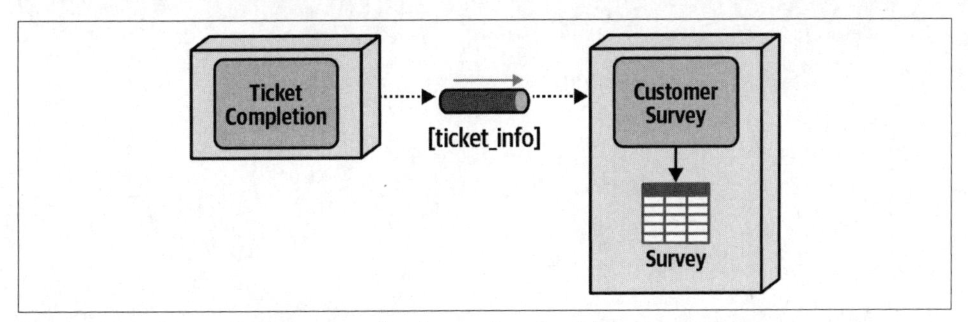

Figure 9-21. Survey Service owns the data using the delegation technique

> *ADR: Survey Service Owns the Survey Table*
>
> *Context*
> Both the Ticket Completion Service and the Survey Service write to the Survey table. Because this is a joint ownership scenario, the alternatives are to use a common shared data domain or use the delegation technique. Table splitting is not an option because of the structure of the Survey table.
>
> *Decision*
> The Survey Service will be the single owner of the Survey table, meaning it is the only service that can perform write operations to that table.
>
> Once a ticket is marked as complete and is accepted by the system, the Ticket Completion Service needs to send a message to the Survey Service to kick off the customer survey processing. Since the Ticket Completion Service is already sending a notification event,

the necessary ticket information can be passed along with that event, thus eliminating the need for the Ticket Completion Service to have any access to the Survey table.

Consequences

All of the necessary data that the Ticket Completion Service needs to insert into the Survey table will need to be sent as part of the payload when triggering the customer survey process.

In the monolithic system, the ticket completion inserted the survey record as part of the completion process. With this decision, the creation of the survey record is a separate activity from the ticket creation process and is now handled by the Survey Service.

CHAPTER 10
Distributed Data Access

`Monday, January 3, 12:43`

"Now that we've assigned ownership of the expert profile table to the User Management Service," said Sydney, "how should the Ticket Assignment Service get to the expert location and skills data? As I said before, with the number of reads it does to the database, it's really not feasible to make a remote call every time it needs to query the table."

"Can you modify the way the assignment algorithm works so that we can reduce the number of queries it needs?" asked Addison.

"Beats me," replied Sydney. "Taylen's the one who usually maintains those algorithms."

Addison and Sydney met with Taylen to discuss the data access issue and to see if Taylen could modify the expert assignment algorithms to reduce the nimber of database calls to the expert profile table.

"Are you kidding me?" asked Taylen. "There's no way I can rewrite the assignment algorithms to do what you are asking. Absolutely no way at all."

"But our only other option is to make remote calls to the User Management Service every time the assignment algorithm needs expert data," said Addison.

"What?" screamed Taylen. "We can't do that!"

"That what I said as well," said Sydney. "That means we are back to square one again. This distributed architecture stuff is hard. I hate to say this, but I am actually starting to miss the monolithic application. Wait, I know. What if we made messaging calls to the User Maintenance Service instead of using REST?"

"That's the same thing," said Taylen. "I still have to wait for the information to come back, whether we use messaging, REST, or any other remote access protocol. That table simply needs to be in the same data domain as the ticketing tables."

"There's got to be another solution to access data we no longer own," said Addison. "Let me check with Logan."

In most monolithic systems using a single database, developers don't give a second thought to reading database tables. SQL table joins are commonplace, and with a simple query all necessary data can be retrieved in a single database call. However, when data is broken into separate databases or schemas owned by distinct services, data access for read operations starts to become hard.

This chapter describes the various ways services can gain read access to data they don't own—in other words, outside the bounded context of the services needing the data. The four patterns of data access we discuss in this chapter include the Interservice Communication pattern, Column Schema Replication pattern, Replicated Cache pattern, and the Data Domain pattern.

Each of these data access patterns has its share of advantages and disadvantages. Yes, once again, *trade-offs*. To better describe each of these patterns, we will return to our Wishlist Service and a Catalog Service example from Chapter 9. The Wishlist Service shown in Figure 10-1 maintains a list of items a customer may want to eventually purchase, and contains the customer ID, item ID, and date the item was added in the corresponding Wishlist table. The Catalog Service is responsible for maintaining all of the items the company sells, and includes the item ID, item description, and static product dimension information, such as the weight, height, length, and so on.

In this example, when a request is made from a customer to display in their wish list, both the item ID *and* and the item description (`item_desc`) are returned to the customer. However, the Wishlist Service does not have the item description in its table; that data is owned by the Catalog Service in a tightly formed bounded context providing change control and data ownership. Therefore, the architect must use one of the data access patterns outlined in this chapter to ensure the Wishlist Service can obtain the product descriptions from the Catalog Service.

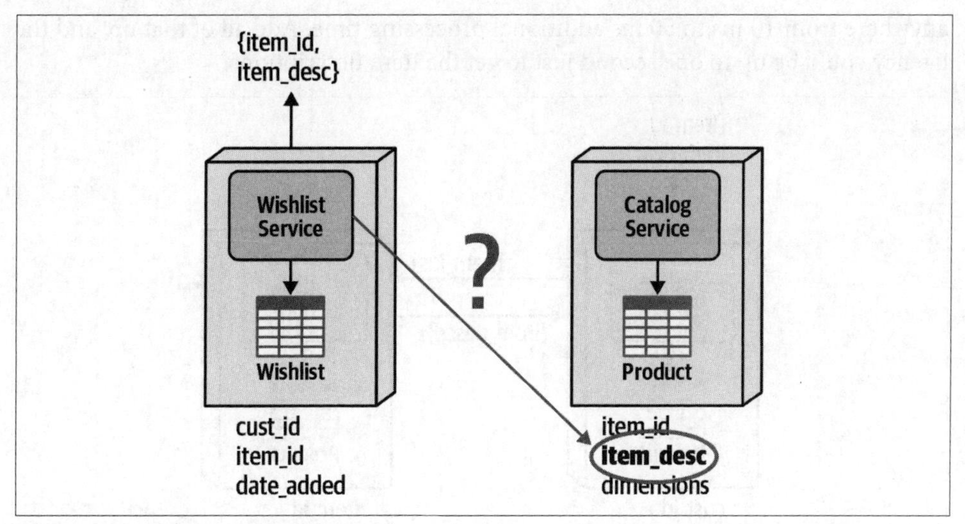

Figure 10-1. Wishlist Service needs item descriptions but doesn't have access to the product table containing the data

Interservice Communication Pattern

The *Interservice Communication pattern* is by far the most common pattern for accessing data in a distributed system. If one service (or system) needs to read data that it cannot access directly, it simply *asks* the owning service or system for it by using some sort of remote access protocol. What can be more simple?

As with most things in software architecture, all is not as it seems. While simple, this common data access technique is unfortunately riddled with disadvantages. Consider Figure 10-2: the Wishlist Service makes a synchronous remote access call to the Catalog Service, passing in a list of item IDs in exchange for a list of corresponding item descriptions.

Notice that for *every* request to get a customer wish list, the Wishlist Service must make a remote call to the Catalog Service to get the item descriptions. The first issue that occurs with this pattern is slower performance due to network latency, security latency, and data latency. *Network latency* is the packet transmission time to and from a service (usually somewhere between 30 ms and 300 ms). Security latency occurs when the endpoint to the target service requires additional authorization to perform the request. *Security latency* can vary greatly depending on the level of security on the endpoint being accessed, but could be anywhere between 20 ms and 400 ms for most systems. *Data latency* describes the situation where multiple database calls need to be made to retrieve the necessary information to pass back to the end user. In this case, rather than a single SQL table join statement, an additional database call must be made by the Catalog Service to retrieve the item description. This might add anywhere from 10 ms to 50 ms additional processing time. Add all of that up, and the latency could be up to one second just to get the item descriptions.

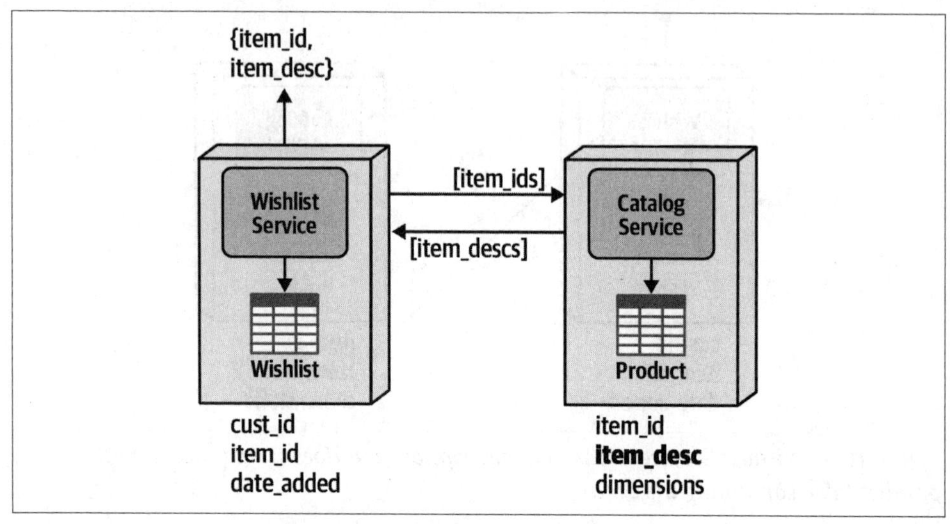

Figure 10-2. Interservice communication data access pattern

Another big disadvantage of this pattern is service coupling. Because the Wishlist must rely on the Catalog Service being available, the services are therefore both semantically and statically coupled, meaning that if the Catalog Service is not available, neither is the Wishlist Service. Furthermore, because of the tight static coupling between the Wishlist Service and the Catalog Service, as the Wishlist Service scales to meet additional demand volume, so must the Catalog Service.

Table 10-1 summarizes the trade-offs associated with the interservice communication data access pattern.

Trade-Offs

Table 10-1. Trade-offs for the Interservice Communication data access pattern

Advantages	Disadvantages
Simplicity	Network, data, and security latency (performance)
No data volume issues	Scalability and throughput issues
	No fault tolerance (availability issues)
	Requires contracts between services

Column Schema Replication Pattern

With the *Column Schema Replication pattern*, columns are replicated across tables, therefore replicating the data and making it available to other bounded contexts. As shown in Figure 10-3, the `item_desc` column is added to the Wishlist table, making that data available to the Wishlist Service without having to ask the Catalog Service for the data.

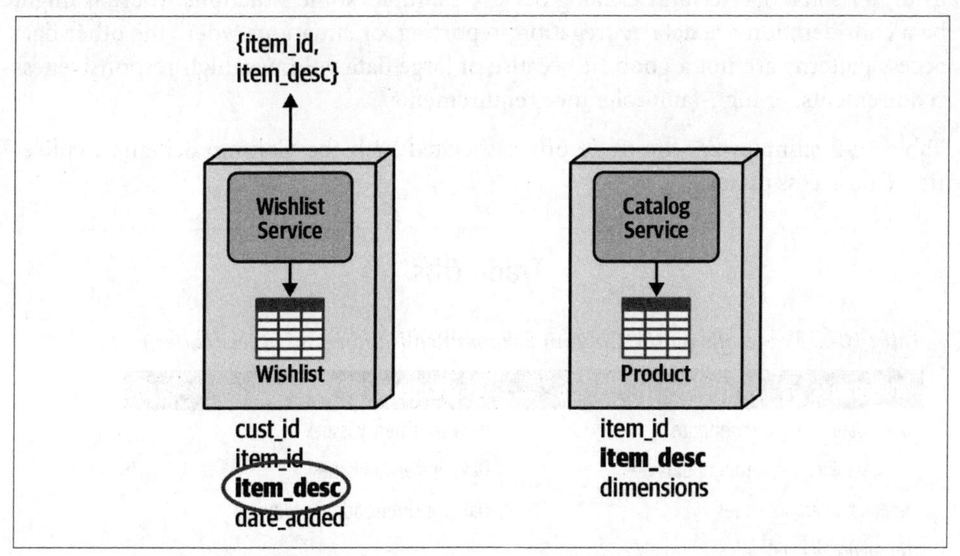

Figure 10-3. With the Column Schema Replication data access pattern, data is replicated to other tables

Data synchronization and data consistency are the two biggest issues associated with the Column Schema Replication data access pattern. Whenever a product is created, removed from the catalog, or a product description changed, the Catalog Service must somehow let the Wishlist Service (and any other services replicating the data) know about the change. This is usually done through asynchronous communications using queues, topics, or event streaming. Unless *immediate* transactional synchronization is required, asynchronous communication is a preferred choice over synchronous communication because it increases responsiveness and reduces the availability dependency between the services.

Another challenge with this pattern is that it is sometimes difficult to govern data ownership. Because the data is replicated in tables belonging to other services, those services can update the data, even though they don't officially *own* the data. This in turn creates even more data consistency issues.

Even though the services are still coupled because of data synchronization, the service requiring read access has immediate access to the data, and can do simple SQL joins or queries to its own table to get the data. This increases performance, fault tolerance, and scalability, all things that were disadvantages with the interservice communication pattern.

While in general we caution against use of this data access pattern for scenarios such as the Wishlist Service and Catalog Service example, some situations where it might be a consideration are data aggregation, reporting, or situations where the other data access patterns are not a good fit because of large data volumes, high responsiveness requirements, or high-fault tolerance requirements.

Table 10-2 summarizes the trade-offs associated with the Column Schema Replication data access pattern.

Trade-Offs

Table 10-2. Trade-offs for the Column Schema Replication data access pattern

Advantages	Disadvantages
Good data access performance	Data consistency issues
No scalability and throughput issues	Data ownership issues
No fault-tolerance issues	Data synchronization is required
No service dependencies	

Replicated Caching Pattern

Most developers and architects think of caching as a technique for increasing overall responsiveness. By storing data within an in-memory cache, retrieving data goes from dozens of milliseconds to only a couple of nanoseconds. However, caching can also be an effective tool for distributed data access and sharing. This pattern leverages *replicated in-memory caching* so that data needed by other services is made available to each service without them having to ask for it. A replicated cache differs from other caching models in that data is held in-memory within each service and is continuously synchronized so that all services have the same exact data at all times.

To better understand the replicated caching model, it's useful to compare it to other caching models to see the differences between them. The *single in-memory* caching model is the simplest form of caching, where each service has its own internal in-memory cache. With this caching model (illustrated in Figure 10-4), in-memory data is not synchronized between the caches, meaning each service has its own unique data specific to that service. While this caching model does help increase responsiveness and scalability within each service, it's not useful for sharing data between services because of the lack of cache synchronization between the services.

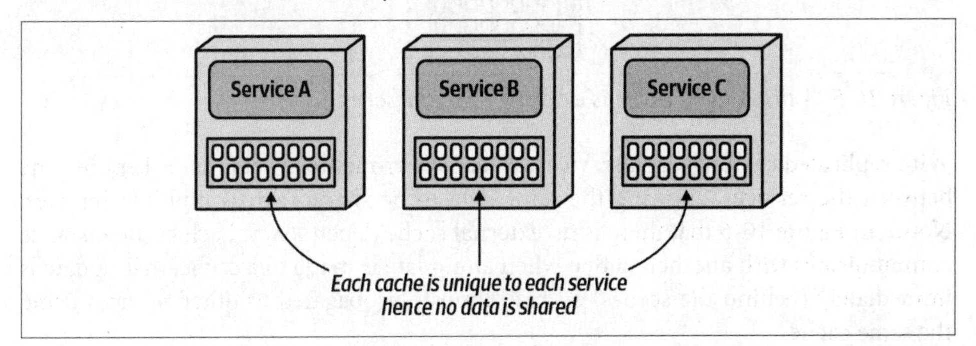

Figure 10-4. With a single in-memory cache, each service contains its own unique data

The other caching model used in distributed architectures is *distributed caching*. As illustrated in Figure 10-5, with this caching model, data is not held in-memory within each service, but rather held externally within a caching server. Services, using a proprietary protocol, make requests to the caching server to retrieve or update shared data. Note that unlike the single in-memory caching model, data can be shared among the services.

The distributed cache model is not an effective caching model to use for the replicated caching data access pattern for several reasons. First, there's no benefit to the fault-tolerance issues found with the Interservice Communication pattern. Rather than depending on a service to retrieve data, the dependency has merely shifted to the caching server.

Because the cache data is centralized and shared, the distributed cache model allows other services to update data, thereby breaking the bounded context regarding data ownership. This can cause data inconsistencies between the cache and the owning database. While this can sometimes be addressed through strict governance, it is nevertheless an issue with this caching model.

Lastly, since access to the centralized distributed cache is through a remote call, network latency adds additional retrieval time for the data, thus impacting overall responsiveness as compared to an in-memory replicated cache.

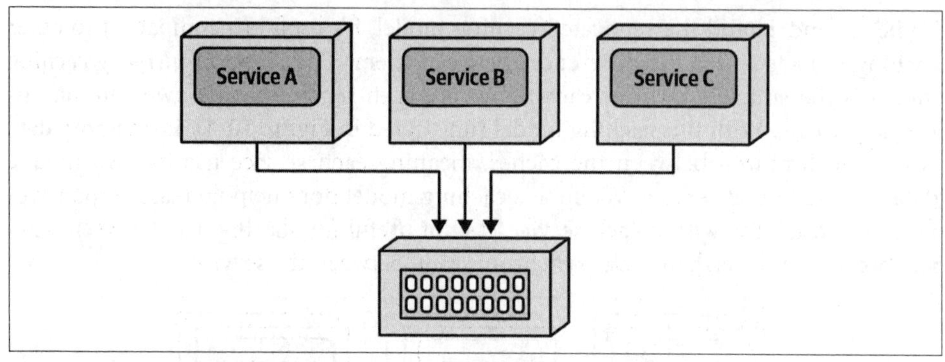

Figure 10-5. A distributed cache is external from the services

With replicated caching, each service has its own in-memory data that is kept in sync between the services, allowing the same data to be shared across multiple services. Notice in Figure 10-6 that there is no external cache dependency. Each cache instance communicates with another so that when an update is made to a cache, that update is immediately (behind the scenes) asynchronously propagated to other services using the same cache.

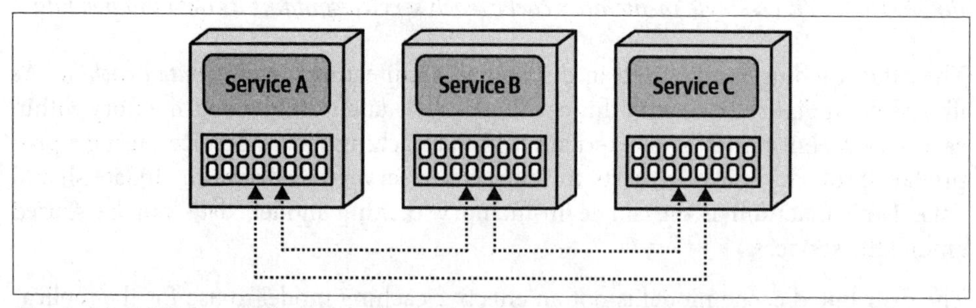

Figure 10-6. With a replicated cache, each service contains the same in-memory data

Not all caching products support replicated caching, so it's important to check with the caching product vendor to ensure support for the replicated caching model. Some of the popular products that do support replicated caching include Hazelcast (*https://hazelcast.com*), Apache Ignite (*https://ignite.apache.org*), and Oracle Coherence (*https://oreil.ly/ISDkz*).

To see how replicated caching can address distributed data access, we'll return to our Wishlist Service and Catalog Service example. In Figure 10-7, the Catalog Service owns an in-memory cache of product descriptions (meaning it is the only service that can modify the cache), and the Wishlist Service contains a read-only in-memory replica of the same cache.

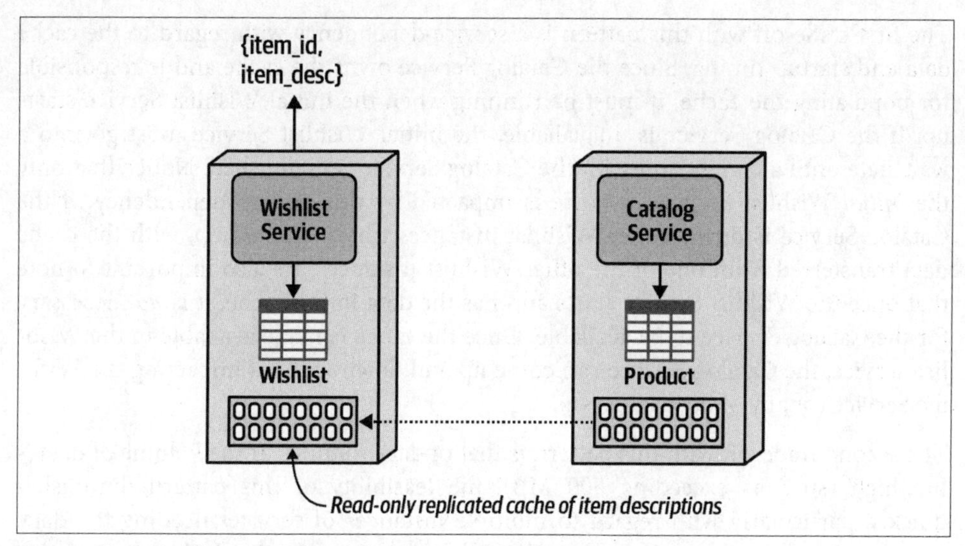

Figure 10-7. Replicated caching data access pattern

With this pattern, the Wishlist Service no longer needs to make calls to the Catalog Service to retrieve product descriptions—they're already in-memory within the Wishlist Service. When updates are made to the product description by the Catalog Service, the caching product will update the cache in the Wishlist Service to make the data consistent.

The clear advantages of the replicated caching pattern are responsiveness, fault tolerance, and scalability. Because no explicit interservice communication is required between the services, data is readily available *in-memory*, providing the fastest possible access to data a service doesn't own. Fault tolerance is also well supported with this pattern. Even if the Catalog Service goes down, the Wishlist Service can continue to operate. Once the Catalog Service comes back up, the caches connect to one another without any disruption to the Wishlist Service. Lastly, with this pattern, the Wishlist Service can scale independently from the Catalog Service.

With all these clear advantages, how could there possibly be a trade-off with this pattern? As the *first law of software architecture* states in our book, *The Fundamentals of Software Architecture* (*https://oreil.ly/J8FPY*), everything in software architecture is a trade-off, and if an architect thinks they have discovered something that *isn't* a trade-off, it means they just haven't *identified* the trade-off yet.

The first trade-off with this pattern is a service dependency with regard to the cache data and startup timing. Since the Catalog Service owns the cache and is responsible for populating the cache, it must be running when the initial Wishlist Service starts up. If the Catalog Service is unavailable, the initial Wishlist Service must go into a wait state until a connection with the Catalog Service is established. Notice that only the *initial* Wishlist Service instance is impacted by this startup dependency; if the Catalog Service is down, other Wishlist instances can be started up, with the cache data transferred from one of the other Wishlist instances. It's also important to note that once the Wishlist Service starts and has the data in the cache, it is *not* necessary for the Catalog Service to be available. Once the cache is made available in the Wishlist Service, the Catalog Service can come up and down without impacting the Wishlist Service (or any of its instances).

The second trade-off with this pattern is that of data volumes. If the volume of data is too high (such as exceeding 500 MB), the feasibility of this pattern diminishes quickly, particularly with regard to multiple instances of services needing the data. Each service instance has its own replicated cache, meaning that if the cache size of 500 MB and 5 instances of a service are required, the total memory used is 2.5 GB. Architects must analyze both the size of the cache *and* the total number of services instances needing the cached data to determine the total memory requirements for the replicated cache.

A third trade-off is that the replicated caching model usually cannot keep the data fully in sync between services if the rate of change of the data (update rate) is too high. This varies based on the size of the data and the replication latency, but in general this pattern is not well suited for highly volatile data (such as product inventory counts). However, for relatively static data (such as a product description), this pattern works well.

The last trade-off associated with this pattern is that of configuration and setup management. Services know about each other in the replicated caching model through TCP/IP broadcasts and lookups. If the TCI/IP broadcast and lookup range is too broad, it can take a long time to establish the socket-level handshake between services. Cloud-based and containerized environments make this particularly challenging because of the lack of control over IP addresses and the dynamic nature of IP addresses associated with these environments.

Table 10-3 lists the trade-offs associated with the replicated cache data access pattern.

Trade-Offs

Table 10-3. Trade-offs associated with the replicated caching data access pattern

Advantages	Disadvantages
Good data access performance	Cloud and containerized configuration can be hard
No scalability and throughput issues	Not good for high data volumes
Good level of fault tolerance	Not good for high update rates
Data remains consistent	Initial service startup dependency
Data ownership is preserved	

Data Domain Pattern

In the previous chapter, we discussed the use of a *data domain* to resolve joint ownership, where multiple services both need to write data to the same table. Tables that are shared between services are put into a single schema that is then shared by both services. That same pattern can be used for data access as well.

Consider the Wishlist Service and Catalog Service problem again, where the Wishlist Service needs access to the product descriptions but does not have access to the table containing those descriptions. Suppose the Interservice Communication pattern is not a feasible solution because of reliability issues with the Catalog Service as well as the performance issues with network latency and the additional data retrieval. Also suppose using the Column Schema Replication pattern is not feasible because of the need for high levels of data consistency. Finally, suppose that the Replicated Cache pattern isn't an option because of the high data volumes. The only other solution is to create a data domain, combining the Wishlist and Product tables in the same shared schema, accessible to both the Wishlist Service and the Catalog Service.

Figure 10-8 illustrates the use of this data access pattern. Notice that the Wishlist and Product tables are no longer owned by either service, but rather shared between them, forming a broader bounded context. With this pattern, gaining access to the product descriptions in the Wishlist Service is a matter of a simple SQL join between the two tables.

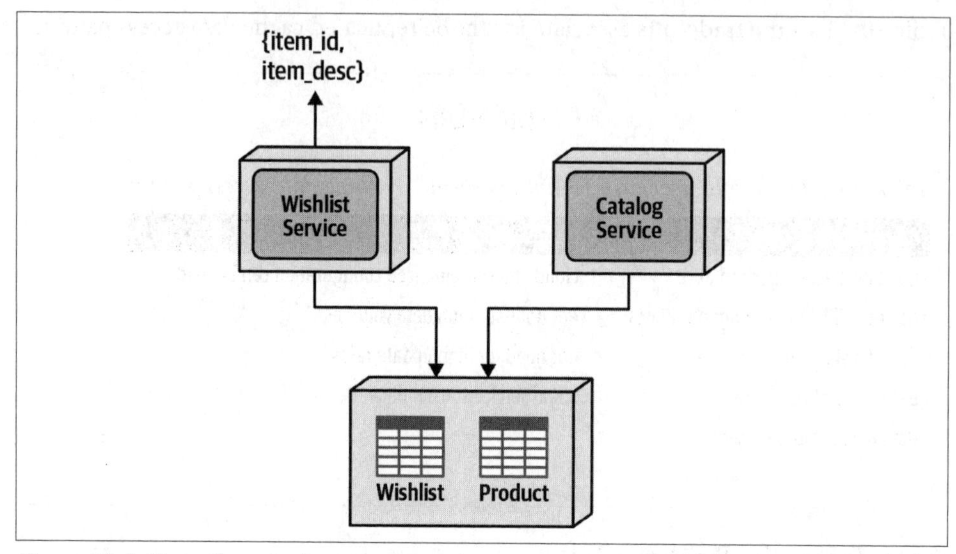

Figure 10-8. Data domain data access pattern

While the sharing of data is generally discouraged in a distributed architecture, this pattern has huge benefits over the other data access patterns. First of all, the services are completely decoupled from each other, thereby resolving any availability dependency, responsiveness, throughput, and scalability issues. Responsiveness is very good with this pattern because the data is available using a normal SQL call, removing the need to do additional data aggregations within the functionality of the service (as is required with the Replicated Cache pattern).

Both data consistency and data integrity rate very high with the Data Domain pattern. Since multiple services access the same data tables, data does not need to be transferred, replicated, or synchronized. Data integrity is preserved in this pattern in the sense that foreign-key constraints can now be enforced between the tables. In addition, other database artifacts, such as views, stored procedures, and triggers, can exist within the data domain. As a matter of fact, the preservation of these integrity constraints and database artifacts is another driver for the use of the Data Domain pattern.

With this pattern, no additional contracts are needed to transfer data between services—the table schema becomes the contract. While this is an advantage for this pattern, it's a trade-off as well. The contracts used with the interservice communication pattern and the Replicated Cache pattern form an abstraction layer over the table schema, allowing changes to the table structures to remain within a tight bounded context and not impact other services. However, this pattern forms a broader bounded context, requiring multiple services to possibly change when the structure to any of the tables in the data domain changes.

Another disadvantage of this pattern is that it can possibly open up security issues associated with data access. For example, in Figure 10-8 the Wishlist Service has complete access to *all* the data within the data domain. While this is OK in the Wishlist and Catalog Service example, there might be times when services accessing the data domain shouldn't have access to certain data. A tighter bounded context with strict service ownership can prevent other services from accessing certain data through the contracts used to pass the data back and forth.

Table 10-4 lists trade-offs associated with the data domain data access pattern.

Trade-Offs

Table 10-4. Trade-offs associated with the data domain data access pattern

Advantages	Disadvantages
Good data access performance	Broader bounded context to manage data changes
No scalability and throughput issues	Data ownership governance
No fault tolerance issues	Data access security
No service dependency	
Data remains consistent	

Sysops Squad Saga: Data Access for Ticket Assignment

Thursday, March 3, 14:59

Logan explained the various methods for data access within a distributed architecture, and also outlined the corresponding trade-offs of each technique. Addison, Sydney, and Taylen then had to come to a decision about which technique to use.

"Unless we start consolidating all of these services, I guess we are stuck with the fact that the Ticket Assignment needs to somehow get to the expert profile data, and fast," said Taylen.

"OK," said Addison. "So service consolidation is out because these services are in entirely different domains, and the shared data domain option is out for the same reasons we talked about before—we cannot have the Ticket Assignment Service connecting to two different databases."

"So, that leaves us with one of two choices." said Sydney. "Either we use interservice communication or replicated caching."

"Wait. Let's explore the replicated caching option for a minute," said Taylen. "How much data are we talking about here?"

"Well," said Sydney, "we have 900 experts in the database. What data does the Ticket Assignment Service need from the expert profile table?"

"It's mostly static information as we get the current expert location feeds from elsewhere. So, it would be the expert's skill, their service location zones, and their standard scheduled availability," said Taylen.

"OK, so that's about 1.3 KB of data per expert. And since we have 900 experts total, that would be... about 1200 KB of data total. And the data is relatively static," said Sydney.

"Hmm, that isn't much data to store in memory," said Taylen.

"Let's not forget that if we used a replicated cache, we would have to take into account how many instances we would have for the User Management Service as well as the Ticket Assignment Service," said Addison. "Just to be on the safe side, we should use the maximum number of instances of each we expect."

"I've got that information," said Taylen. "We expect to have only a maximum of two instances of the User Management Service, and a maximum of four at our highest peak for the Ticket Assignment Service."

"That's not much total in-memory data," observed Sydney.

"No, it's not," said Addison. "OK, let's analyze the trade-offs using the hypothesis-based approach we tried earlier. I suggest that we should go with the in-memory replicated cache option to cache only the data necessary for the Ticket Assignment Service. Any other trade-offs you can think of?"

Both Taylen and Sydney sat there for while trying to think of some negatives for the replicated cache approach.

"What if the User Management Service goes down?" asked Sydney.

"As long as the cache is populated, then the Ticket Assignment Service would be fine," said Addison.

"Wait, you mean to tell me that the data would be in-memory, even if the User Management Service is unavailable?" asked Taylen.

"As long as the User Management Service starts before the Ticket Assignment Service, then yes," said Addison.

"Ah!" said Taylen. "Then there's our first trade-off. Ticket assignment cannot function unless the User Management Service is started. That's not good."

"But," said Addison, "if we made remote calls to the User Management Service and it goes down, the Ticket Assignment Service becomes nonoperational. At least with the replicated cache option, once User Management is up and running, we are no longer dependent on it. So, replicated caching is actually more fault tolerant in this case."

"True," said Taylen. "We just have to be careful about the startup dependency."

"Anything else you can think of as a negative?" asked Addison, knowing another obvious trade-off but wanting the development team to come up with it on their own.

"Um," said Sydney, "yeah. I have one. What caching product are we going to use?"

"Ah," said Addison, "that is in fact another trade-off. Have either of you done replicated caching before? Or anyone on the development team for that matter?"

Both Taylen and Sydney shook their heads.

"Then we have some risk here," said Addison.

"Actually," said Taylen, "I've been hearing a lot about this caching technique for a while and have been dying to try it out. I would volunteer to research some of the products and do some proof-of-concepts on this approach."

"Great," said Addison. "In the meantime, I will research what the licensing cost would be for those products as well, and if there's any technical limitation with respect to our deployment environment. You know, things like availability zone crossovers, firewalls, that sort of stuff."

The team began their research and proof-of-concept work, and found that this is indeed not only a feasible solution cost and effort wise, but would solve the issue of data access to the expert profile table. Addison discussed this approach with Logan, who approved the solution. Addison created an ADR outlining and justifying this decision.

> *ADR: Use of In-Memory Replicated Caching for Expert Profile Data*
>
> *Context*
> The Ticket Assignment Service needs continuous access to the expert profile table, which is owned by the User Management Service in a separate bounded context. Access to the expert profile information can be done through interservice communication, in-memory replicated caching, or a common data domain.
>
> *Decision*
> We will use replicated caching between the User Management Service and the Ticket Assignment Service, with the User Management Service being the sole owner for write operations.
>
> Because the Ticket Assignment Service already connects to the shared ticket data domain schema, it cannot connect to an additional schema. In addition, since the user management functionality and the core ticketing functionality are in two separate domains, we do not want to combine the data tables in a single schema. Therefore, using a common data domain is not an option.
>
> Using an in-memory replicated cache resolves the performance and fault-tolerance issues associated with the interservice communication option.

Consequences

At least one instance of the User Management Service must be running when starting the first instance of the Ticket Assignment Service.

Licensing costs for the caching product would be required for this option.

CHAPTER 11
Managing Distributed Workflows

Tuesday, February 15, 14:34

Austen bolted into Logan's office just after lunch. "I've been looking at the new architecture designs, and I want to help out. Do you need me to write up some ADRs or help with some spikes? I'd be happy to write up the ADR that states that we're only going to use choreography in the new architecture to keep things decoupled."

"Whoa, there, you maniac," said Logan. "Where did you hear that? What gives you that impression?"

"Well, I've been reading a lot about microservices, and everyone's advice seems to be to keep things highly decoupled. When I look at the patterns for communication, it seems that choreography is the most decoupled, so we should always use it, right?"

"*Always* is a tricky term in software architecture. I had a mentor who had a memorable perspective on this, who always said, *Never use absolutes when talking about architecture, except when talking about absolutes.* In other words, *never* say *never*. I can't think of many decisions in architecture where *always* or *never* applies."

"OK," said Austen. "So how *do* architects decide between the different communication patterns?"

As part of our ongoing analysis of the trade-offs associated with modern distributed architectures, we reach the *dynamic* part of quantum coupling, realizing many of the patterns we described and named in Chapter 2. In fact, even our named patterns only touch on the many permutations possible with modern architectures. Thus, an architect should understand the forces at work so that they can make a most objective trade-off analysis.

In Chapter 2, we identified three coupling forces when considering interaction models in distributed architectures: communication, consistency, and coordination, shown in Figure 11-1.

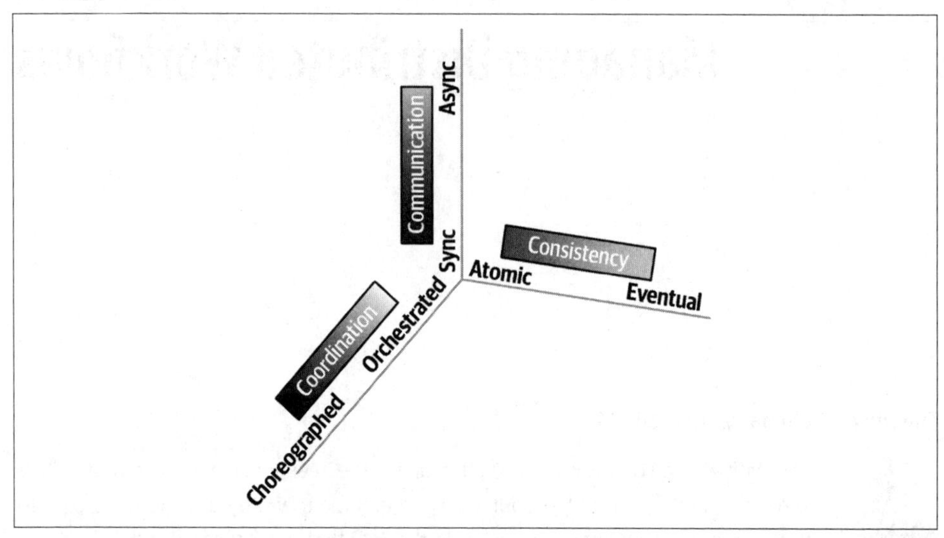

Figure 11-1. The dimensions of dynamic quantum coupling

In this chapter, we discuss *coordination*: combining two or more services in a distributed architecture to form some domain-specific work, along with the many attendant issues.

Two fundamental coordination patterns exist in distributed architectures: orchestration and choreography. The fundamental topological differences between the two styles is illustrated in Figure 11-2.

Orchestration is distinguished by the use of an orchestrator, whereas a choreographed solution does not use one.

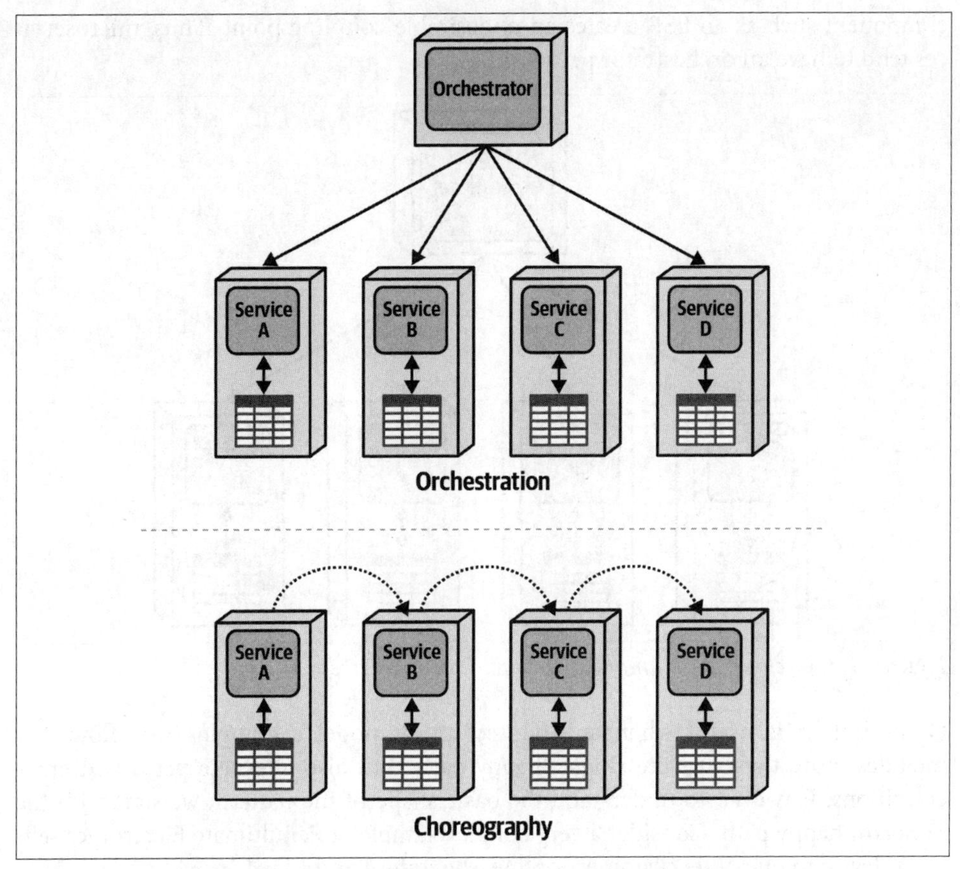

Figure 11-2. Orchestration versus choreography in distributed architectures

Orchestration Communication Style

The *orchestration pattern* uses an *orchestrator* (sometimes called a *mediator*) component to manage workflow state, optional behavior, error handling, notification, and a host of other workflow maintenance. It is named for the distinguishing feature of a musical orchestra, which utilizes a conductor to synchronize the incomplete parts of the overall score to create a unified piece of music. Orchestration is illustrated in the most generic representation in Figure 11-3.

In this example, services A-D are domain services, each responsible for its own bounded context, data, and behavior. The Orchestrator component generally doesn't include any domain behavior outside of the workflow it mediates. Notice that microservices architectures have an orchestrator per workflow, *not* a global orchestrator such as an *enterprise service bus* (ESB) (*https://oreil.ly/KTGrU*). One of the primary goals of the microservices architecture style is decoupling, and using a global component such as an ESB creates an undesirable coupling point. Thus, microservices tend to have an orchestrator per workflow.

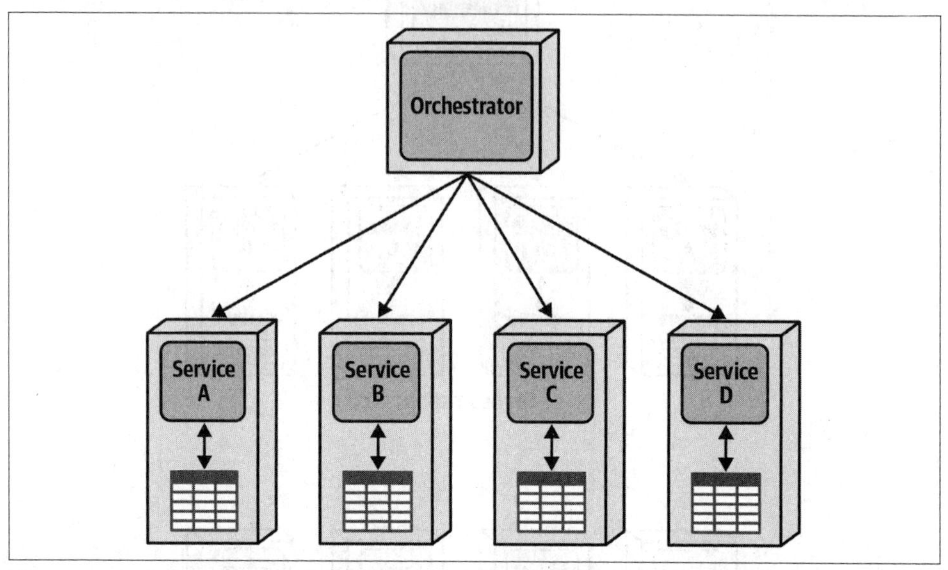

Figure 11-3. Orchestration among distributed microservices

Orchestration is useful when an architect must model a complex workflow that includes more than just the single "happy path," but also alternate paths and error conditions. However, to understand the basic shape of the pattern, we start with the nonerror happy path. Consider a very simple example of Penultimate Electronics selling a device to one of its customers online, shown in Figure 11-4.

This system passes the Place Order request to the Order Placement Orchestrator, which makes a synchronous call to the Order Placement Service, which records the order and returns a status message. Next, the mediator calls the Payment Service, which updates payment information. Next, the orchestrator makes an asynchronous call to the Fulfillment Service to handle the order. The call is asynchronous because no strict timing dependencies exist for order fulfillment, unlike payment verification. For example, if order fulfillment happens only a few times a day, there is no reason for the overhead of a synchronous call. Similarly, the orchestrator then calls the Email Service to notify the user of a successful electronics order.

If the world consisted of only happy paths, software architecture would be easy. However, one of the primary hard parts of software architecture is error conditions and pathways.

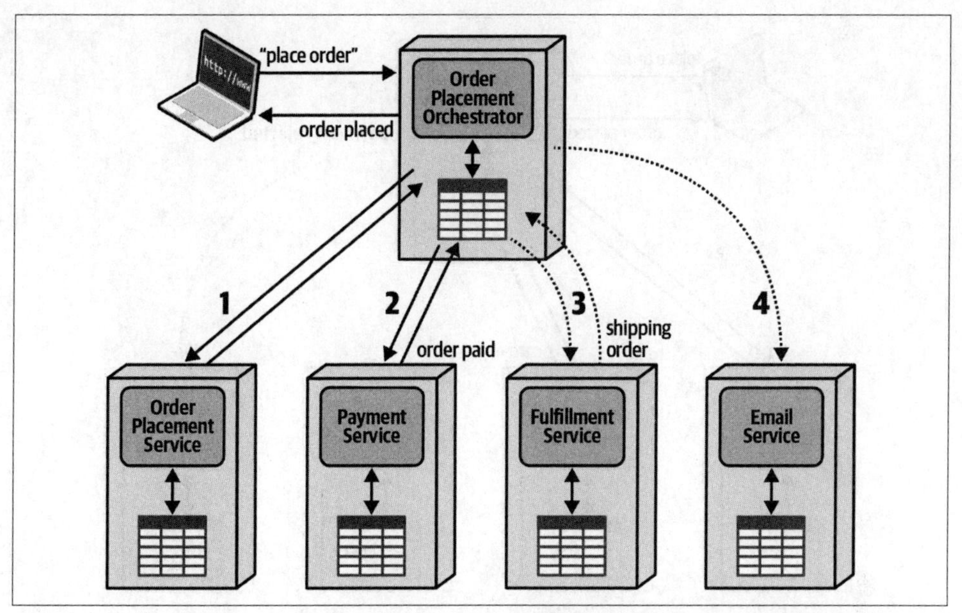

Figure 11-4. A "happy path" workflow using an orchestrator to purchase electronic equipment (note the asynchronous calls denoted by dotted lines for less time-sensitive calls)

Consider two potential error scenarios for electronics purchasing. First, what happens if the customer's payment method is rejected? This error scenario appears in Figure 11-5.

Here, the Order Placement Orchestrator updates the order via the Order Placement Service as before. However, when trying to apply payment, it is rejected by the payment service, perhaps because of an expired credit card number. In that case, the Payment Service notifies the orchestrator, which then places a (typically) asynchronous call to send a message to the Email Service to notify the customer of the failed order. Additionally, the orchestrator updates the state of the Order Placement Service, which still thinks this is an active order.

Notice in this example we're allowing each service to maintain its own transactional state, modeling our "Fairy Tale Saga(seo) Pattern" on page 333. One of the hardest parts of modern architectures is managing transactions, which we cover in Chapter 12.

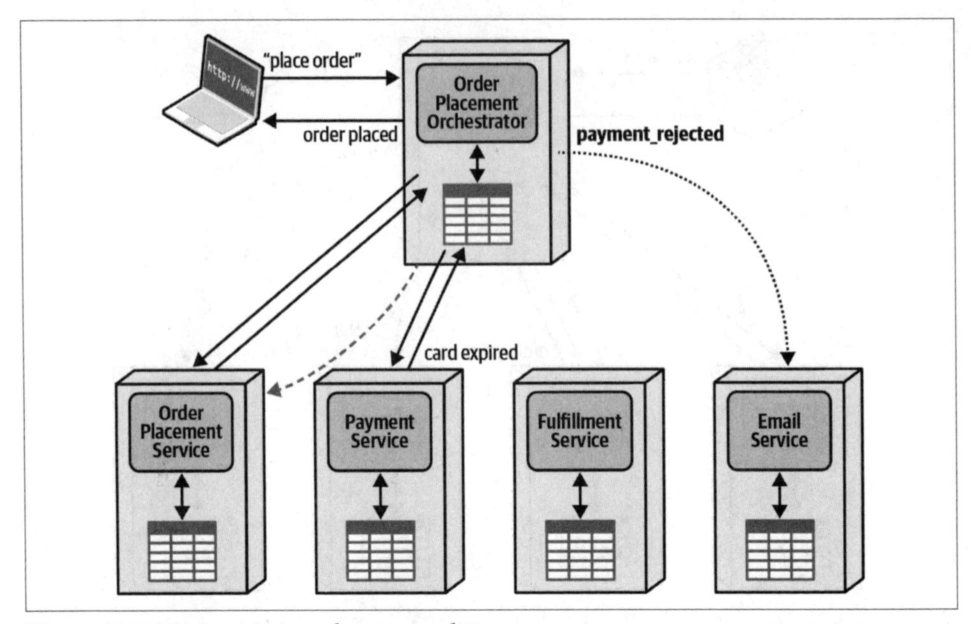

Figure 11-5. Payment rejected error condition

In the second error scenario, the workflow has progressed further along: what happens when the Fulfillment Service reports a back order? This error scenario appears in Figure 11-6.

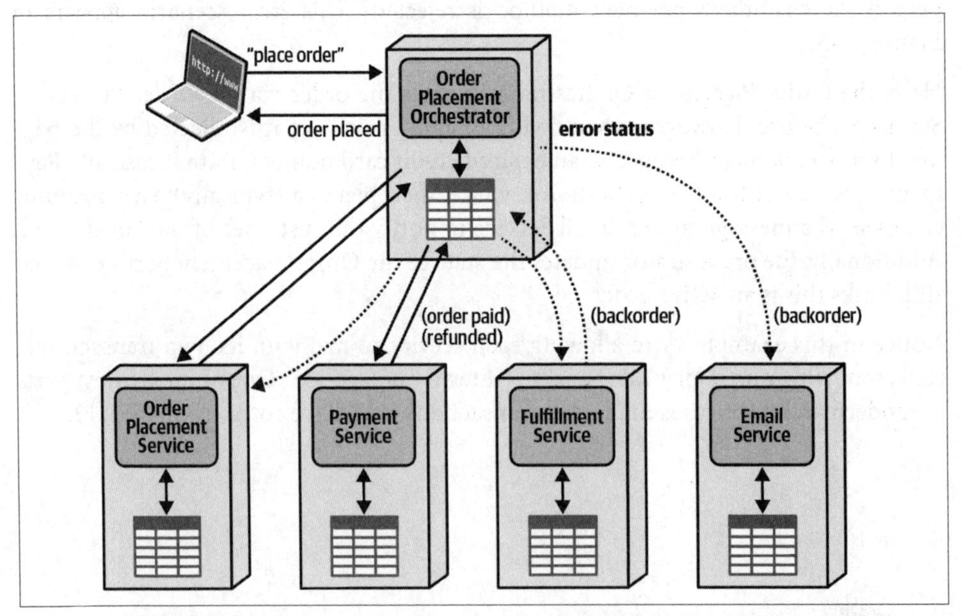

Figure 11-6. When an item is back-ordered, the orchestrator must rectify the state

As you can see, the workflow preceeds as normal until the Fulfillment Service notifies the orchestrator that the current item is out of stock, necessitating a back order. In that case, the orchestrator must refund the payment (this is why many online services don't charge until shipment, not at order time) and update the state of the Order Placement Service.

One interesting characteristic to note in Figure 11-6: even in the most elaborate error scenarios, the architect wasn't required to add additional communication paths that weren't already there to facilitate the normal workflow, which differs from the "Choreography Communication Style" on page 306.

General advantages of the orchestration communication style include the following:

Centralized workflow
: As complexity goes up, having a unified component for state and behavior becomes beneficial.

Error handling
: Error handling is a major part of many domain workflows, assisted by having a state owner for the workflow.

Recoverability
: Because an orchestrator monitors the state of the workflow, an architect may add logic to retry if one or more domain services suffers from a short-term outage.

State management
: Having an orchestrator makes the state of the workflow queriable, providing a place for other workflows and other transient states.

General disadvantages of the orchestration communication style include the following:

Responsiveness
: All communication must go through the mediator, creating a potential throughput bottleneck that can harm responsiveness.

Fault tolerance
: While orchestration enhances recoverability for domain services, it creates a potential single point of failure for the workflow, which can be addressed with redundancy but adds more complexity.

Scalability
: This communication style doesn't scale as well as choreography because it has more coordination points (the orchestrator), which cuts down on potential parallelism. As we discussed in Chapter 2, several dynamic coupling patterns utilize choreography and thus achieve higher scale (notably "Time Travel Saga(sec) Pattern" on page 336 and "Anthology Saga(aec) Pattern" on page 349).

Service coupling
: Having a central orchestrator creates higher coupling between it and domain components, which is sometimes necessary. The orchestration communication style's trade-offs appear in Table 11-1.

Trade-Offs

Table 11-1. Trade-offs for orchestration

Advantage	Disadvantage
Centralized workflow	Responsiveness
Error handling	Fault tolerance
Recoverability	Scalability
State management	Service coupling

Choreography Communication Style

Whereas the Orchestration Communication Style was named for the metaphorical central coordination offered by an orchestrator, the *choreography* pattern visually illustrates intent of the communication style that has no central coordination. Instead, each service participates with the others, similar to dance partners. It isn't an ad hoc performance—the moves were planned beforehand by the choreographer/architect but executed without a central coordinator.

Figure 11-4 described the orchestrated workflow for a customer purchasing electronics from Penultimate Electronics; the same workflow modeled in the choreography communication style appears in Figure 11-7.

In this workflow, the initiating request goes to the first service in the chain of responsibility—in this case, the Order Placement Service. Once it has updated internal records about the order, it sends an asynchronous request that the Payment Service receives. Once payment has been applied, the Payment Service generates a message received by the Fulfillment Service, which plans for delivery and sends a message to the Email Service.

At first glance, the choreography solution seems simpler—fewer services (no orchestrator), and a simple chain of events/commands (messages). However, as with many issues in software architecture, the difficulties lie not with the default paths but rather with boundary and error conditions.

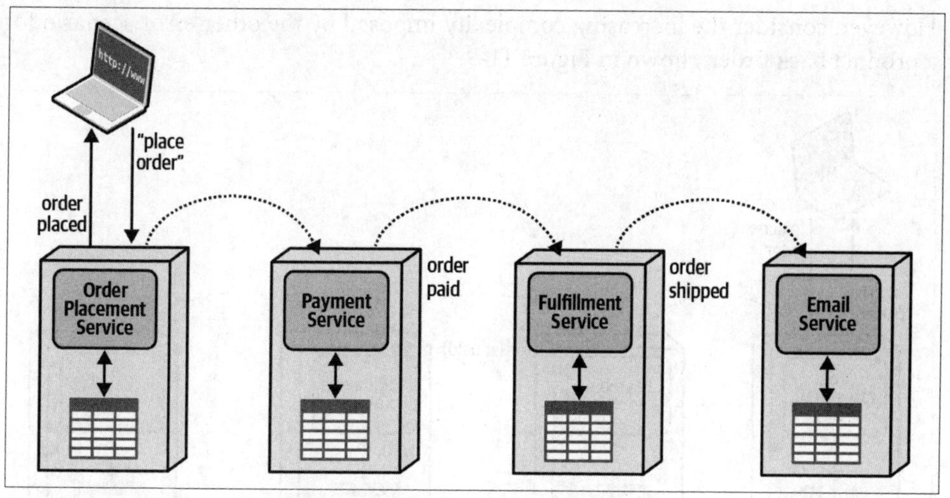

Figure 11-7. Purchasing electronics using choreography

As in the previous section, we cover two potential error scenarios. The first results from failed payment, as illustrated in Figure 11-8.

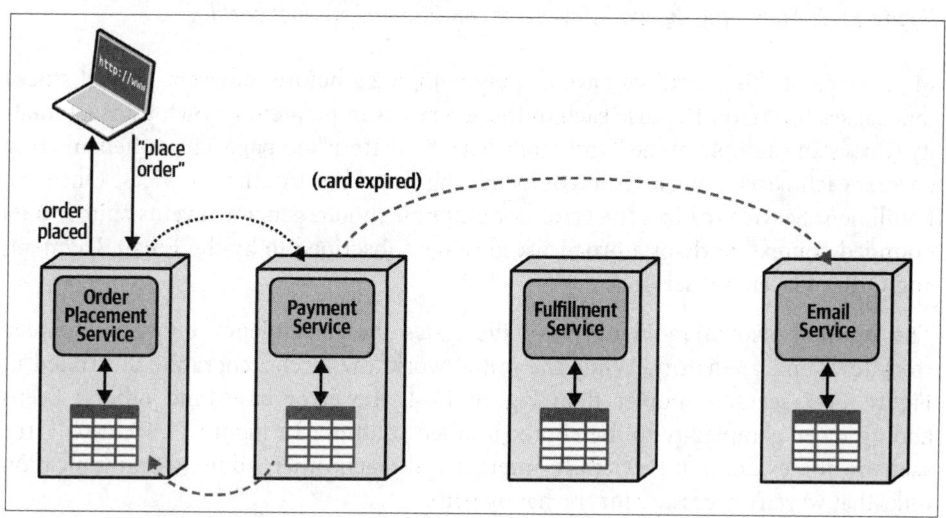

Figure 11-8. Error in payment in choreography

Rather than send a message intended for the Fulfillment Service, the Payment service sends messages indicating failure to the Email Service and back to the Order Placement Service to update the order status. This alternate workflow doesn't appear too complex, with a single new communication link that didn't exist before.

However, consider the increasing complexity imposed by the other error scenario for a product back order, shown in Figure 11-9.

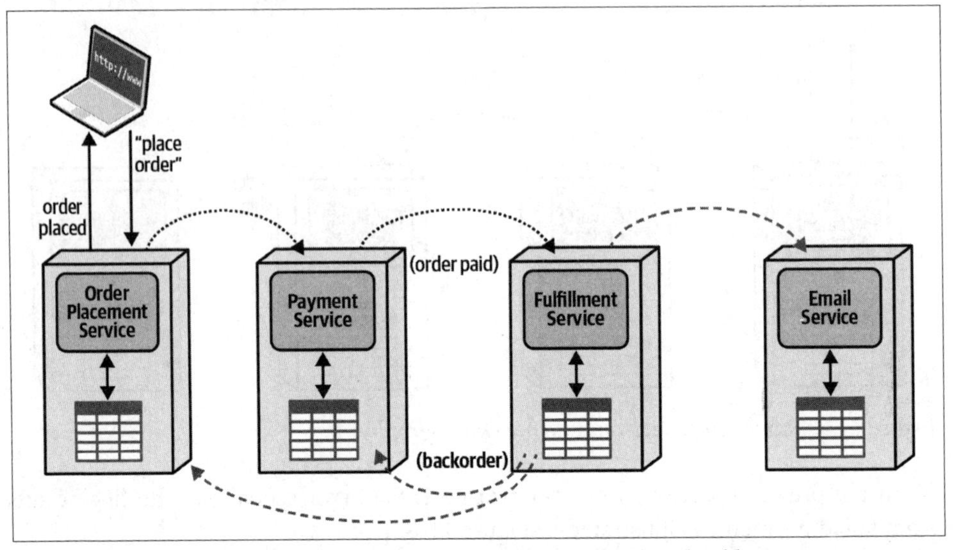

Figure 11-9. Managing the workflow error condition of product backlog

Many steps of this workflow have already completed before the event (out of stock) that causes the error. Because each of these services implement its own transactionality (this is an example of the "Anthology Saga(aec) Pattern" on page 349), when an error occurs, each service must issue compensating messages to other services. Once the Fulfillment Service realizes the error condition, it should generate events suited to its bounded context, perhaps a broadcast message subscribed to by the Email, Payment, and Order Placement services.

The example shown in Figure 11-9 illustrates the dependency between complex workflows and mediators. While the initial workflow in choreography illustrated in Figure 11-7 seemed simpler than Figure 11-4, the error case (and others) keeps adding more complexity to the choreographed solution. In Figure 11-10, each error scenario forces domain services to interact with each other, adding communication links that weren't necessary for the happy path.

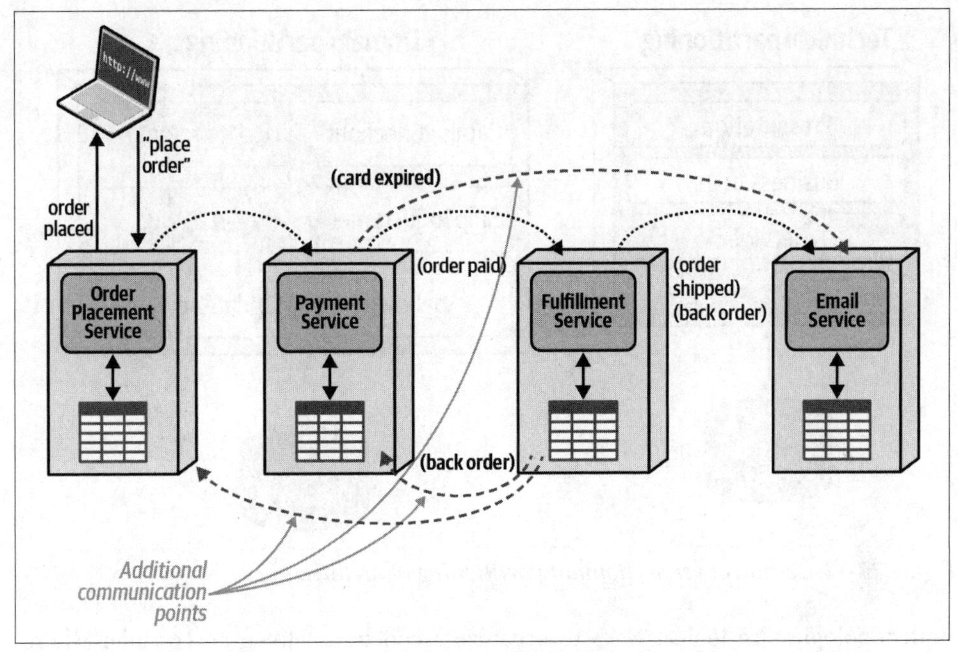

Figure 11-10. Error conditions in choreography typically add communication links

Every workflow that architects need to model in software has a certain amount of *semantic coupling*—the inherent coupling that exists in the problem domain. For example, the process of assigning a ticket to a Sysops Squad member has a certain workflow: a client must request service, skills must be matched to particular specialists, then cross-referenced to schedules and locations. The way an architect models that interaction is the *implementation coupling*.

The semantic coupling of a workflow is mandated by the domain requirements of the solution and must be modeled somehow. However clever an architect is, they cannot reduce the amount of semantic coupling, but their implementation choices may increase it. This doesn't mean that an architect might not push back on impractical or impossible semantics defined by business users—some domain requirements create extraordinarily difficult problems in architecture.

Here is a common example. Consider the standard layered monolithic architecture compared to the more modern style of a modular monolith, shown in Figure 11-11.

The architecture on the left represents the traditional layered architecture, separated by *technical* capabilities such as persistence, business rules, and so on. On the right, the same solution appears, but separated by *domain* concerns such as `Catalog Checkout` and `Update Inventory` rather than technical capabilities.

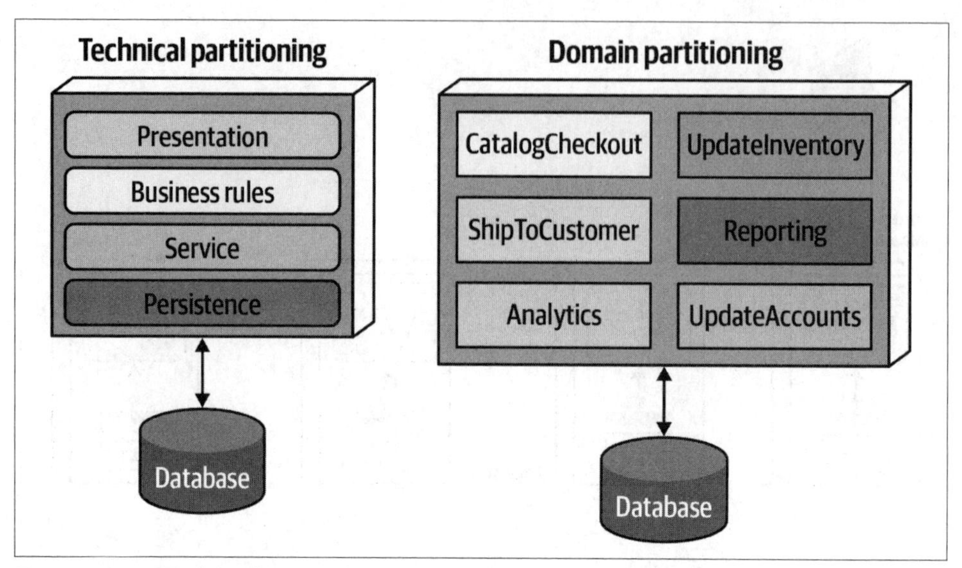

Figure 11-11. Technical versus domain partitioning in architecture

Both topologies are logical ways to organize a codebase. However, consider where domain concepts such as `Catalog Checkout` reside within each architecture, illustrated in Figure 11-12.

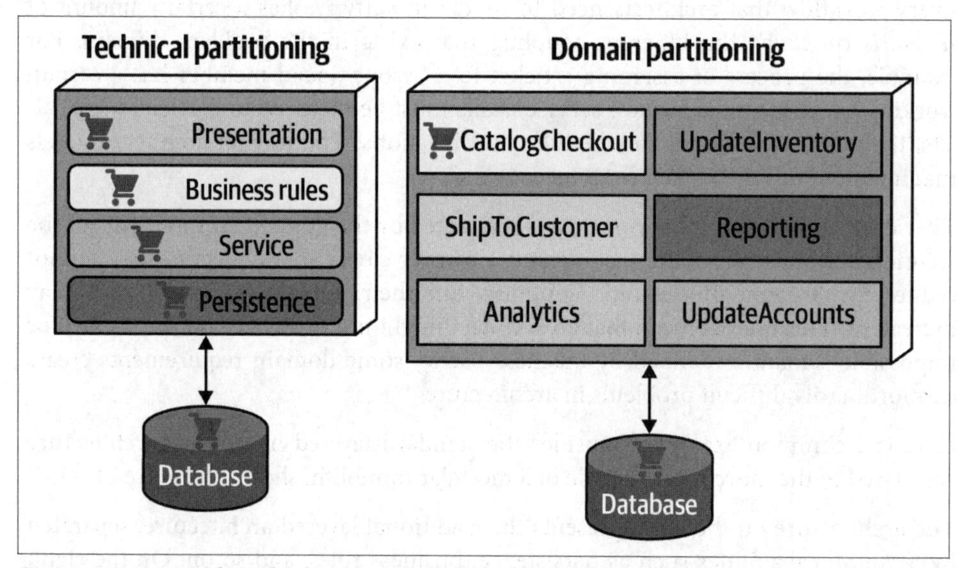

Figure 11-12. Catalog Checkout is smeared across implementation layers in a technically partitioned architecture

`Catalog Checkout` is "smeared" across the layers of the technical architecture, whereas it appears only in the matching domain component and database in the domain partitioned example. Of course, aligning a domain with domain partitioned architecture isn't a revelation—one of the insights of domain-driven design was the primacy of the domain workflows. No matter what, if an architect wants to model a workflow, they must make those moving parts work together. If the architect has organized their architecture the same as the domains, the implementation of the workflow should have similar complexity. However, if the architect has imposed additional layers (as in technical partitioning, shown in Figure 11-12), it increases the overall implementation complexity because now the architect must design for the *semantic* complexity along with the additional *implementation* complexity.

Sometimes the extra complexity is warranted. For example, many layered architectures came from a desire by architects to gain cost savings by consolidating on architecture patterns, such as database connection pooling. In that case, an architect considered the trade-offs of the cost saving associated with technically partitioning database connectivity versus the imposed complexity and cost won in many cases.

The major lesson of the last decade of architecture design is to model the *semantics* of the workflow as closely as possible with the implementation.

An architect can never reduce semantic coupling via implementation, but they can make it worse.

Thus, we can establish a relationship between the semantic coupling and the need for coordination—the more steps required by the workflow, the more potential error and other optional paths appear.

Workflow State Management

Most workflows include transient state about the status of the workflow: what elements have executed, which ones are left, ordering, error conditions, retries, and so on. For orchestrated solutions, the obvious workflow state owner is the orchestrator (although some architectural solutions create stateless orchestrators for higher scale). However, for choreography, no obvious owner for workflow state exists. Many common options exist to manage state in choreography; here are three common ones.

First, the *Front Controller pattern* places the responsibility for state on the first called service in the chain of responsibility, which in this case is Order Placement Service. If that service contains information about both orders and the state of the workflow, some of the domain services must have a communication link to query and update the order state, as illustrated in Figure 11-13.

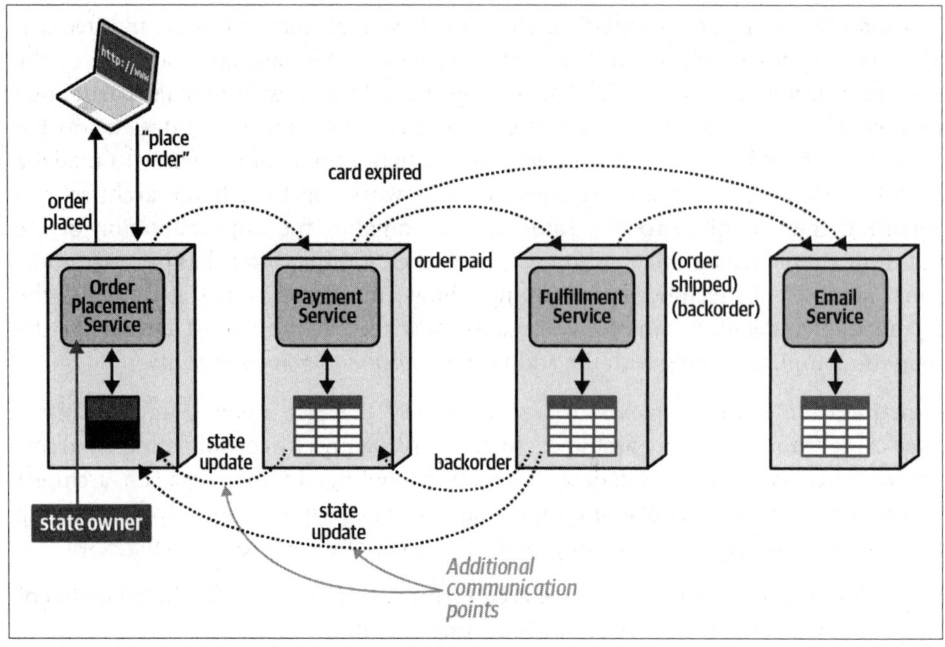

Figure 11-13. In choreography, a Front Controller is a domain service that owns workflow state in addition to domain behavior

In this scenario, some services must communicate back to the Order Placement Service to update the state of the order, as it is the state owner. While this simplifies the workflow, it increases communication overhead and makes the Order Placement Service more complex than one that handled only domain behavior. While the Front Controller pattern has some advantageous characteristics, it also has trade-offs, as shown in Table 11-2.

Trade-Offs

Table 11-2. Trade-offs for the Front Controller pattern

Advantage	Disadvantage
Creates a pseudo-orchestrator within choreography	Adds additional workflow state to a domain service
Makes querying the state of an order trivial	Increases communication overhead
	Detrimental to performance and scale as it increases integration communication chatter

A second way for an architect to manage the transactional state is to keep no transient workflow state at all, relying on querying the individual services to build a real-time snapshot. This is known as *stateless choreography*. While this simplifies the state of the first service, it greatly increases network overhead in terms of chatter between services to build a stateful snapshot. For example, consider a workflow like the simple choreography happy path in Figure 11-7 with no extra state. If a customer wants to know the state of their order, the architect must build a workflow that queries the state of each domain service to determine the most up-to-date order status. While this makes for a highly flexible solution, rebuilding state can be complex and costly in terms of operational architecture characteristics like scalability and performance. Stateless choreography trades high performance for workflow control, as illustrated in Table 11-3.

Trade-Offs

Table 11-3. Stateless choreography trade-offs

Advantage	Disadvantage
Offers high performance and scale	Workflow state must be built on the fly
Extremely decoupled	Complexity rises swiftly with complex workflows

A third solution utilizes *stamp coupling* (described in more detail in "Stamp Coupling for Workflow Management" on page 378), storing extra workflow state in the message contract sent between services. Each domain service updates its part of the overall state and passes that to the next in the chain of responsibility. Thus, any consumer of that contract can check on the status of the workflow without querying each service.

This is a partial solution, as it still does not provide a single place for users to query the state of the ongoing workflow. However, it does provide a way to pass the state between services as part of the workflow, providing each service with additional potentially useful context. As in all features of software architecture, stamp coupling has good and bad characteristics, shown in Table 11-4.

Trade-Offs

Table 11-4. Stamp coupling trade-offs

Advantage	Disadvantage
Allows domain services to pass workflow state without additional queries to a state owner	Contracts must be larger to accommodate workflow state
Eliminates need for a front controller	Doesn't provide just-in-time status queries

In Chapter 13, we discuss how contracts can reduce or increase workflow coupling in choreographed solutions.

Advantages of the choreography communication style include the following:

Responsiveness
: This communication style has fewer single choke points, thus offering more opportunities for parallelism.

Scalability
: Similar to responsiveness, lack of coordination points like orchestrators allows more independent scaling.

Fault tolerance
: The lack of a single orchestrator allows an architect to enhance fault tolerance with the use of multiple instances.

Service decoupling
: No orchestrator means less coupling.

Disadvantages of the choreography communication style include the following:

Distributed workflow
: No workflow owner makes error management and other boundary conditions more difficult.

State management
: No centralized state holder hinders ongoing state management.

Error handling
: Error handling becomes more difficult without an orchestrator because the domain services must have more workflow knowledge.

Recoverability
: Similarly, recoverability becomes more difficult without an orchestrator to attempt retries and other remediation efforts.

Like "Orchestration Communication Style" on page 301, choreography has a number of good and bad trade-offs, often opposites, summarized in Table 11-5.

Trade-Offs

Table 11-5. Trade-offs for the choreography communication style

Advantage	Disadvantage
Responsiveness	Distributed workflow
Scalability	State management
Fault tolerance	Error handling
Service decoupling	Recoverability

Trade-Offs Between Orchestration and Choreography

As with all things in software architecture, neither orchestration nor choreography represent the perfect solution for all possibilities. A number of key trade-offs, including some delineated here, will lead an architect toward one of these two solutions.

State Owner and Coupling

As illustrated in Figure 11-13, state ownership typically resides somewhere, either in a formal mediator acting as an orchestrator, or a front controller in a choreographed solution. In the choreographed solution, removing the mediator forces higher levels of communication between services. This might be a perfectly suitable trade-off. For example, if an architect has a workflow that needs higher scale and typically has few error conditions, it might be worth trading the higher scale of choreography with the complexity of error handling.

However, as workflow complexity goes up, the need for an orchestrator rises proportionally, as illustrated in Figure 11-14.

In addition, the more semantic complexity contained in a workflow, the more utilitarian an orchestrator is. Remember, implementation coupling can't make semantic coupling better, only worse.

Ultimately, the sweet spot for choreography lies with workflows that need responsiveness and scalability, and either don't have complex error scenarios or they are infrequent. This communication style allows for high throughput; it is used by the dynamic coupling patterns "Phone Tag Saga(sac) Pattern" on page 330, "Time Travel Saga(sec) Pattern" on page 336, and "Anthology Saga(aec) Pattern" on page 349. However, it can also lead to extremely difficult implementations when other forces are mixed in, leading to the "Horror Story(aac) Pattern" on page 343.

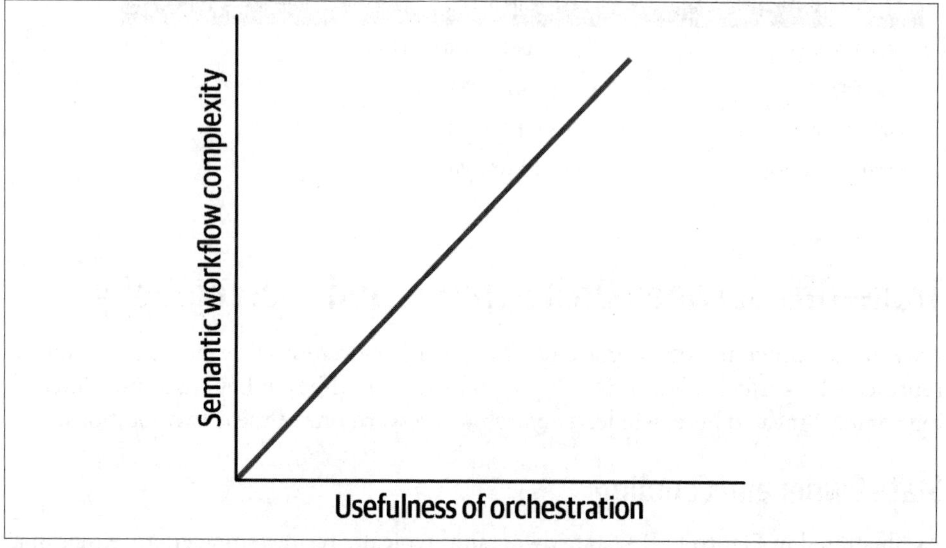

Figure 11-14. As the complexity of the workflow rises, orchestration becomes more useful

On the other hand, *orchestration* is best suited for complex workflows that include boundary and error conditions. While this style doesn't provide as much scale as choreography, it greatly reduces complexity in most cases. This communication style appears in "Epic Saga(sao) Pattern" on page 325, "Fairy Tale Saga(seo) Pattern" on page 333, "Fantasy Fiction Saga(aao) Pattern" on page 340, and "Parallel Saga(aeo) Pattern" on page 346.

Coordination is one of the primary forces that create complexity for architects when determining how to best communicate between microservices. Next, we investigate how this force intersects with another primary force, *consistency*.

Sysops Squad Saga: Managing Workflows

`Thursday, March 15, 11:00`

Addison and Austen arrived at Logan's office right on time, armed with a presentation and ritual coffee urn from the kitchen.

"Are you ready for us?" asked Addison.

"Sure," said Logan. "Good timing—just got off a conference call. Are y'all ready to talk about workflow options for the primary ticket flow?"

"Yes!" said Austen. "I think we should use choreography, but Addison thinks orchestration, and we can't decide."

"Give me an overview of the workflow we're looking at."

"It's the primary ticket workflow," said Addison. "It involves four services; here are the steps."

Customer-facing operations

1. Customer submits a trouble ticket through the Ticket Management Service and receives a ticket number.

Background operations

1. The Ticket Assignment Service finds the right Sysops expert for the trouble ticket.
2. The Ticket Assignment Service routes the trouble ticket to the systems expert's mobile device.
3. The customer is notified via the Notification Service that the Sysops expert is on their way to fix the problem.
4. The expert fixes the problem and marks the ticket as complete, which is sent to the Ticket Management Service.
5. The Ticket Management Service communicates with the Survey Service to tell the customer to fill out the survey.

"Have you modeled both solutions?" asked Logan.

"Yes. The drawing for choreography is in Figure 11-15."

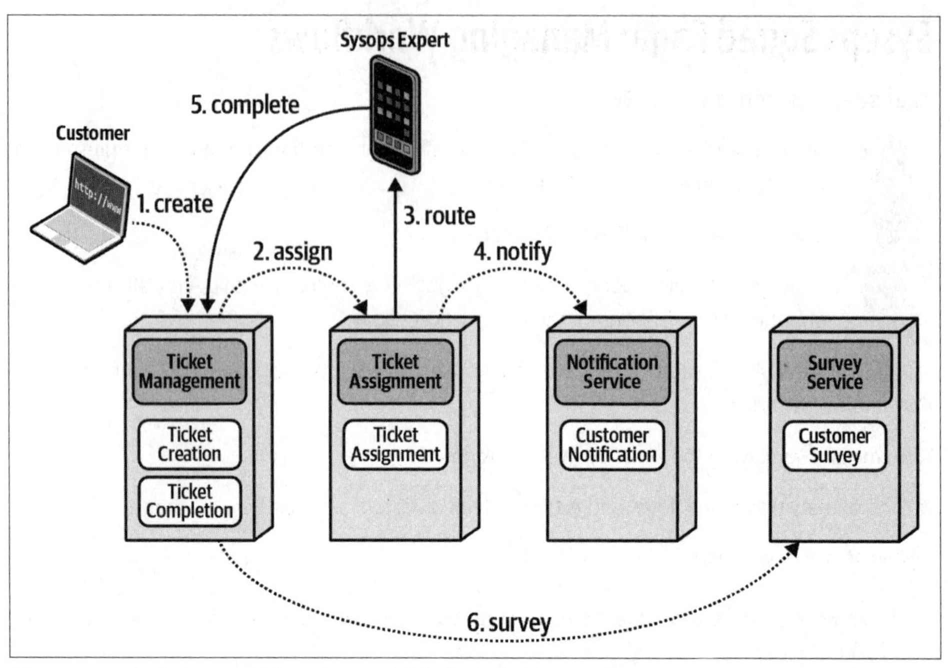

Figure 11-15. Primary ticket flow modeled as choreography

"...and the model for orchestration is in Figure 11-16."

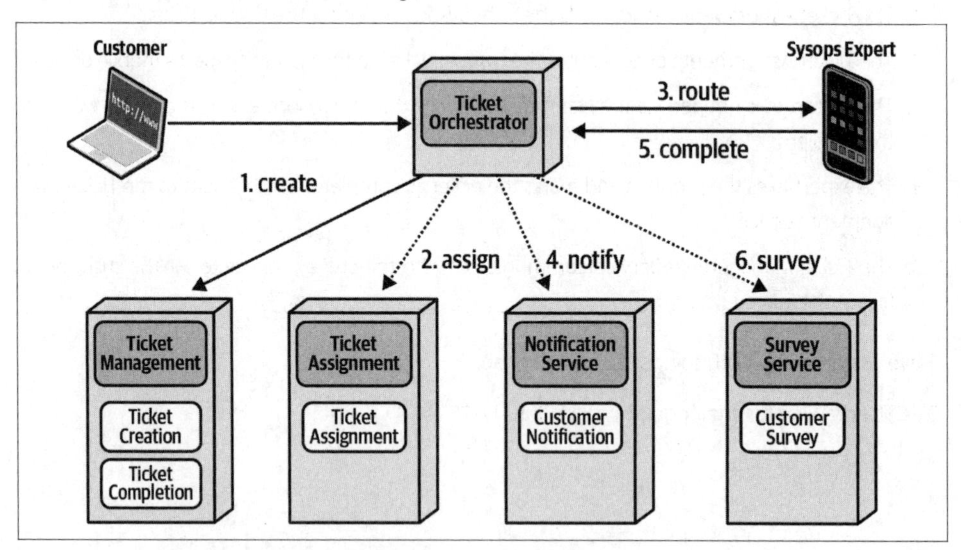

Figure 11-16. Primary ticket workflow modeled as orchestration

Logan pondered the diagrams for a moment, then pronounced, "Well, there doesn't seem to be an obvious winner here. You know what that means."

Austen piped up, "Trade-offs!"

"Of course," laughed Logan. " Let's think about the likely scenarios and see how each solution reacts to them. What are the primary issues you are concerned with?"

"The first is lost or misrouted tickets. The business has been complaining about it, and it has become a priority," said Addison.

"OK, which handles that problem better—orchestration or choreography?"

"Easier control of the workflow sounds like the orchestrator version is better—we can handle all the workflow issues there," volunteered Austen.

"OK, let's build a table of issues and preferred solutions in Table 11-6."

Trade-Offs

Table 11-6. Trade-off between orchestration and choreography for ticket workflow

Orchestration	Choreography
Workflow control	

"What's the next issue we should model?" Addison asked.

"We need to know the status of a trouble ticket at any given moment—the business has requested this feature, and it makes it easier to track several metrics. That implies we need an orchestrator so that we can query the state of the workflow."

"But you don't have to have an orchestrator for that—we can query any given service to see if it has handled a particular part of the workflow, or use stamp coupling," said Addison.

"That's right—this isn't a zero-sum game," said Logan. "It's possible that both or neither work just as well. We'll give both solutions credit in our updated table in Table 11-7."

Trade-Offs

Table 11-7. Updated trade-offs between orchestration and choreography for ticket workflow

Orchestration	Choreography
Workflow control	
State query	State query

"OK, what else?"

"Just one more that I can think of," Addison said. "Tickets can get canceled by the customer, and tickets can get reassigned because of expert availability, lost connections to the expert's mobile device, or expert delays at a customer site. Therefore, proper error handling is important. That means orchestration?"

"Yes, generally. Complex workflows must go somewhere, either in an orchestrator or scattered through services. It's nice to have a single place to consolidate error handling. And choreography definitely does not score well here, so we'll update our table in Table 11-8."

Trade-Offs

Table 11-8. Final trade-offs between orchestration and choreography for ticket workflow

Orchestration	Choreography
Workflow control	
State query	State query
Error handling	

"That looks pretty good. Any more?"

"Nothing that's not obvious," said Addison. "We'll write this up in an ADR; in case we think of any other issues, we can add them there."

ADR: Use Orchestration for Primary Ticket Workflow

Context

For the primary ticket workflow, the architecture must support easy tracking of lost or mis-tracked messages, excellent error handling, and the ability to track ticket status. Either an orchestration solution illustrated in Figure 11-16 or a choreography solution illustrated in Figure 11-15 will work.

Decision

We will use orchestration for the primary ticketing workflow.

We modeled orchestration and choreography and arrived at the trade-offs in Table 11-8.

Consequences

Ticketing workflow might have scalability issues around a single orchestrator, which should be reconsidered if current scalability requirements change.

CHAPTER 12

Transactional Sagas

`Thursday, March 31, 16:55`

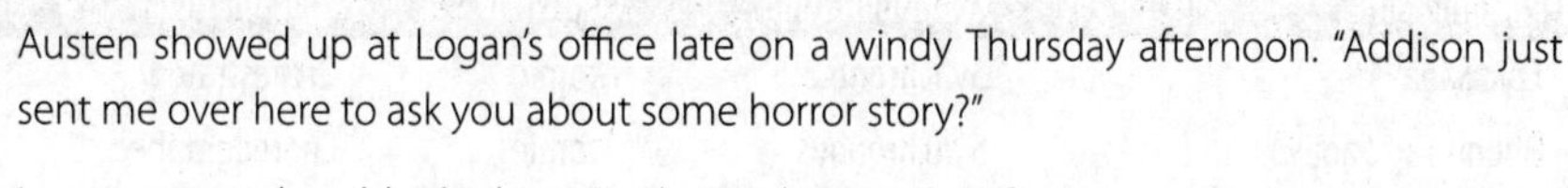

Austen showed up at Logan's office late on a windy Thursday afternoon. "Addison just sent me over here to ask you about some horror story?"

Logan stopped and looked up. "Is that a description of whatever crazy extreme sport you're doing this weekend? What is it this time?"

"It's late spring, so a bunch of us are going ice skating on the thawing lake. We're wearing body suits, so it's really a combination of skating and swimming. But that's not what Addison meant at all. When I showed Addison my design for the Ticketing workflow, I was immediately instructed to come to you and tell you I've created a horror story."

Logan laughed. "Oh, I see what's going on—you stumbled into the Horror Story saga communication pattern. You designed a workflow with asynchronous communication, atomic transactionality, and choreography, right?"

"How did you know?"

"That's the Horror Story saga pattern, or really, anti-pattern. There are eight generic saga patterns we start from, so it's good to know what they are, because each has a different balance of trade-offs."

The concept of a *saga* in architecture predates microservices, originally concerned with limiting the scope of database locks in early distributed architectures—the paper largely assumed to have coined the concept is from the Proceedings of the 1987 ACM conference. In his book *Microservices Patterns* (Manning Publications) and also outlined in the "Saga Pattern" (*https://oreil.ly/drXJa*) section of his website, Chris Richardson describes the *saga pattern* for microservices as a sequence of local transactions where each update publishes an event, thus triggering the next update in the sequence. If any of those updates fail, the saga issues a series of compensating updates to undo the prior changes made during the saga.

However, recall from Chapter 2 that this is only one of eight possible types of sagas. In this section, we dive much deeper and look at the inner workings of transactional sagas and how to manage them, particularly when errors occur. After all, since distributed transactions lack atomicity (see "Distributed Transactions" on page 263), what makes them interesting is when problems occur.

Transactional Saga Patterns

In Chapter 2, we introduced a matrix that juxtaposed each of the intersecting dimensions when architects must choose how to implement a transactional saga, reproduced in Table 12-1.

Table 12-1. The matrix of dimensional intersections for distributed architectures

Pattern name	Communication	Consistency	Coordination
Epic Saga$^{(sao)}$	Synchronous	Atomic	Orchestrated
Phone Tag Saga$^{(sac)}$	Synchronous	Atomic	Choreographed
Fairy Tale Saga$^{(seo)}$	Synchronous	Eventual	Orchestrated
Time Travel Saga$^{(sec)}$	Synchronous	Eventual	Choreographed
Fantasy Fiction Saga$^{(aao)}$	Asynchronous	Atomic	Orchestrated
Horror Story$^{(aac)}$	Asynchronous	Atomic	Choreographed
Parallel Saga$^{(aeo)}$	Asynchronous	Eventual	Orchestrated
Anthology Saga$^{(aec)}$	Asynchronous	Eventual	Choreographed

We provide whimsical names for each combination, all derived from types of sagas. However, the pattern names exist to help differentiate the possibilities, and we don't want to provide a memorization test to associate a pattern name to a set of characteristics, so we have added a superscript to each saga type indicating the values of the three dimensions listed in alphabetical order (as in Table 12-1). For example, the *Epic Saga$^{(sao)}$ pattern* indicates the values of *synchronous, atomic*, and *orchestrated* for communication, consistency, and coordination. The superscripts help you associate names to character sets more easily.

While architects will utilize some of the patterns more than others, they all have legitimate uses and differing sets of trade-offs.

We illustrate each possible communication combination with both a three-dimensional representation of the intersection of the three forces in space along with an example workflow using generic distributed services, which we refer to as *isomorphic diagrams*. These diagrams show interactions between services in the most generic way, toward our goal of showing architect concepts in the simplest form. In each of these diagrams, we use the set of generic symbols shown in Figure 12-1.

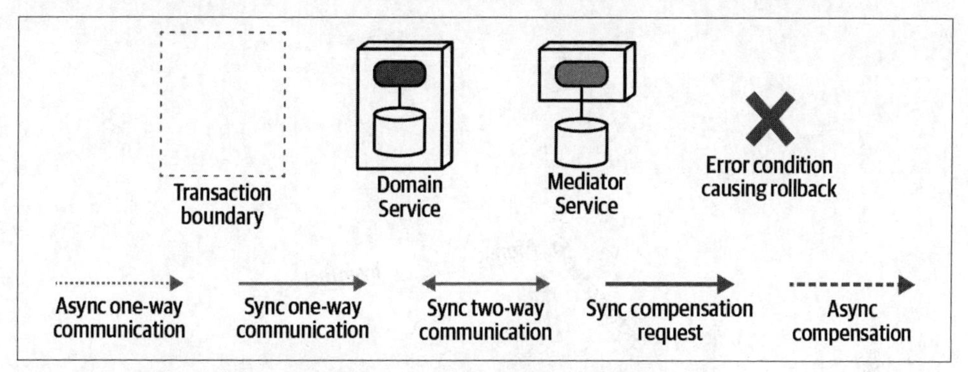

Figure 12-1. Legend for ISO architecture interaction diagrams

For each of the architecture patterns, we do not show every possible interaction, which would become repetitive. Instead, we identify and illustrate the differentiating features of the pattern—what makes its behavior unique among the patterns.

Epic Saga(sao) Pattern

This type of communication is the "traditional" saga pattern as many architects understand it, also called an *Orchestrated Saga* because of its coordination type. Its dimensional relationships appear in Figure 12-2.

This pattern utilizes *synchronous* communication, *atomic* consistency, and *orchestrated* coordination. The architect's goal when choosing this pattern mimics the behavior of monolithic systems—in fact, if a monolithic system were added to this diagram in Figure 12-2, it would be the origin (0, 0, 0), lacking distribution entirely. Thus, this communication style is most familiar with architects and developers of traditional transactional systems.

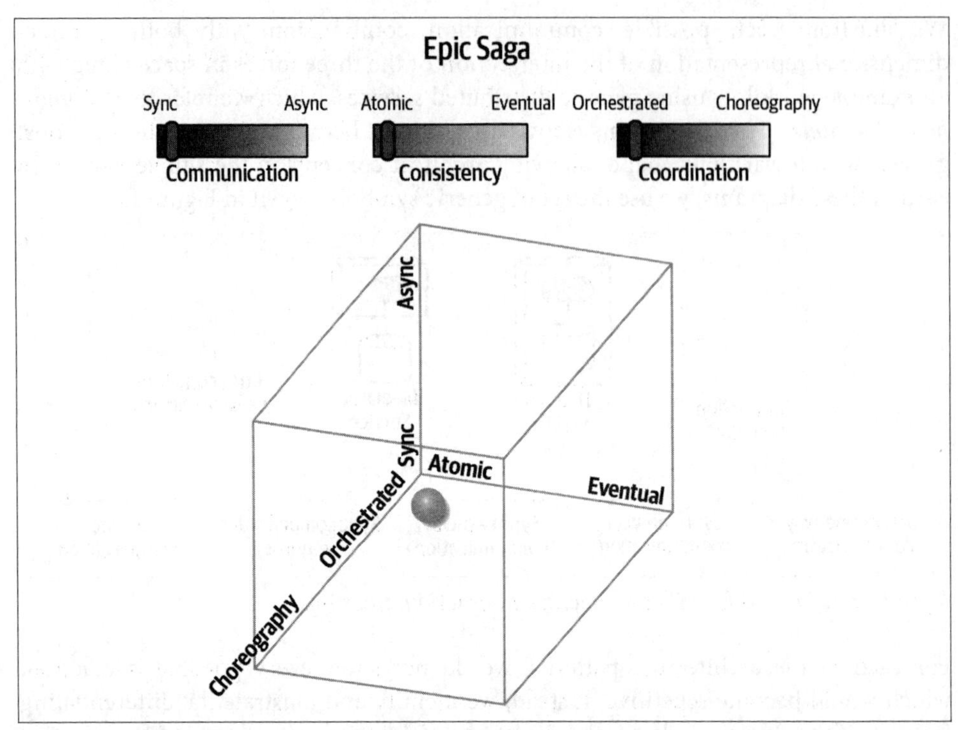

Figure 12-2. The Epic Saga(sao) pattern's dynamic coupling (communication, consistency, coordination) relationships

The isomorphic representation of the Epic Saga(sao) pattern appears in Figure 12-3.

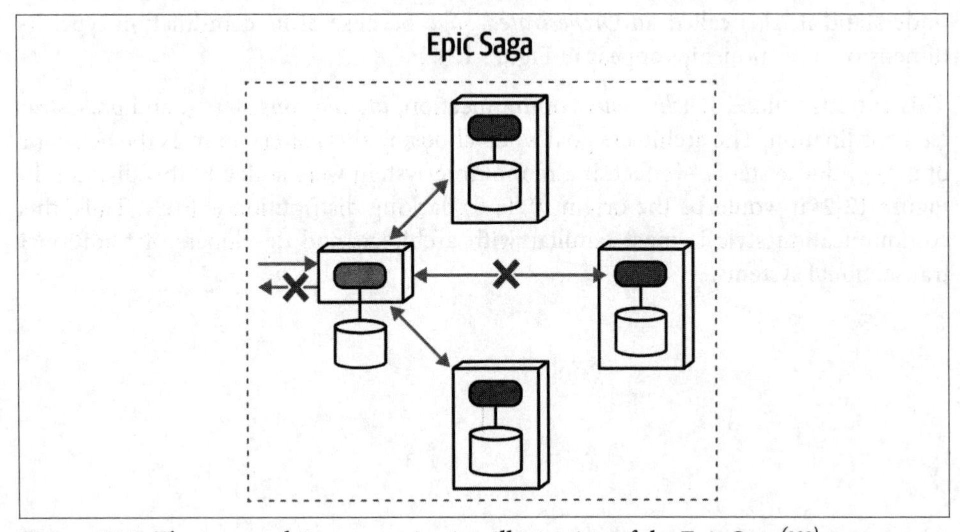

Figure 12-3. The isomorphic communication illustration of the Epic Saga(sao) pattern

Here, an *orchestrator* service orchestrates a workflow that includes updates for three services, expected to occur transactionally—either all three calls succeed or none do. If one of the calls fails, they all fail and return to the previous state. An architect can solve this coordination problem in a variety of ways, all complex in distributed architectures. However, such transactions limit the choice of databases and have legendary failure modes.

Many nascent or naive architects trust that, because a pattern exists for a problem, it represents a clean solution. However, the pattern is recognition of only commonality, not solvability. Distributed transactions provide an excellent example of this phenomenon—architects accustomed to modeling transactions in nondistributed systems sometimes believe that moving that capability to the distributed world is an incremental change. However, transactions in distributed architectures present a number of challenges, which become proportionally worse depending on the complexity of the semantic coupling of the problem.

Consider a common implementation of the Epic Saga[(sao)] pattern, utilizing compensating transactions. A *compensating update* is one that reverses a data write action performed by another service (such as reversing an update, reinserting a previously deleted row, or deleting a previously inserted row) during the course of the distributed transaction scope. While compensating updates attempt to reverse changes in order to bring distributed data sources back to their original state prior to the start of the distributed transaction, they are riddled with complex issues, challenges, and trade-offs.

A compensating transaction pattern assigns a service to monitor the transactional completeness of a request, as shown in Figure 12-4.

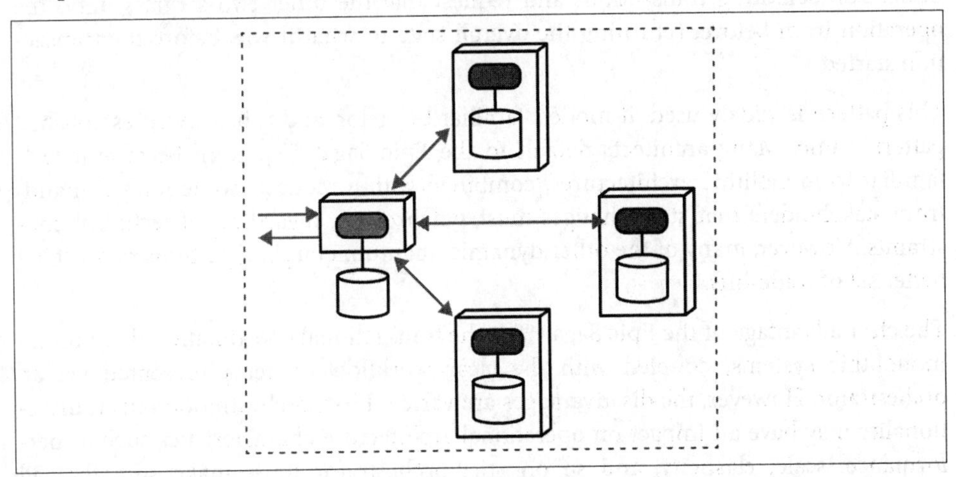

Figure 12-4. A successful orchestrated transactional Epic Saga using a compensating transaction

However, as with many things in architecture, the error conditions cause the difficulties. In a compensating transaction framework, the mediator monitors the success of calls, and issues compensating calls to other services if one or more of the requests fail, as shown in Figure 12-5.

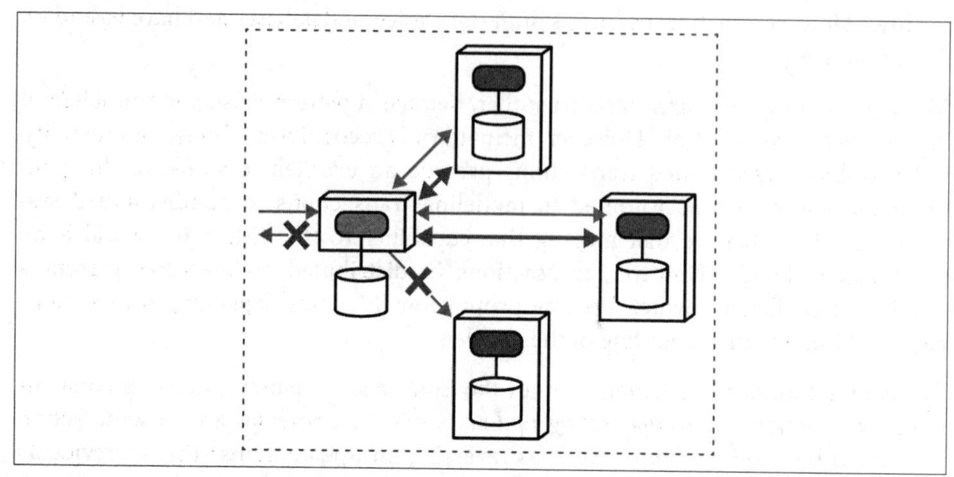

Figure 12-5. When an error occurs, a mediator must send compensating requests to other services

A mediator both accepts requests and mediates the workflow, and synchronous calls to the first two services succeed. However, when trying to make the call to the last service, it fails (from a possibly a wide variety of both domain and operational reasons). Because the goal of the Epic Saga(sao) is atomic consistency, the mediator must utilize compensating transactions and request that the other two services undo the operation from before, returning the overall state to what it was before the transaction started.

This pattern is widely used: it models familiar behavior, and it has a well-established pattern name. Many architects default to the Epic Saga(sao) pattern because it feels familiar to monolithic architectures, combined with a request (sometimes demand) from stakeholders that state changes must synchronize, regardless of technical constraints. However, many of the other dynamic quantum coupling patterns may offer a better set of trade-offs.

The clear advantage of the Epic Saga(sao) is the transactional coordination that mimics monolithic systems, coupled with the clear workflow owner represented via an orchestrator. However, the disadvantages are varied. First, orchestration plus transactionality may have an impact on operational architecture characteristics such as performance, scale, elasticity, and so on—the orchestrator must make sure that all participants in the transaction have succeeded or failed, creating timing bottlenecks. Second, the various patterns used to implement distributed transactionality (such as compensating transactions) succumb to a wide variety of failure modes and boundary conditions, along with adding inherent complexity via undo operations. Distributed transactions present a host of difficulties and thus are best avoided if possible.

The Epic Saga$^{(sao)}$ pattern features the following characteristics:

Coupling level
: This pattern exhibits extremely high levels of coupling across all possible dimensions: synchronous communication, atomic consistency, and orchestrated coordination—it is in fact the most highly coupled pattern in the list. This isn't surprising, as it models the behavior of highly coupled monolithic system communication, but creates a number of issues in distributed architectures.

Complexity level
: Error conditions and other intensive coordination added to the requirement of atomicity add complexity to this architecture. The synchronous calls this architecture uses mitigate some of the complexity, as architects don't have to worry about race conditions and deadlocks during calls.

Responsiveness/availability
: Orchestration creates a bottleneck, especially when it must also coordinate transactional atomicity, which reduces responsiveness. This pattern uses synchronous calls, further impacting performance and responsiveness. If any of the services are not available or an unrecoverable error occurs, this pattern will fail.

Scale/elasticity
: Similar to *responsiveness*, the bottleneck and coordination required to implement this pattern make scale and other operational concerns difficult.

While the Epic Saga$^{(sao)}$ is popular because of familiarity, it creates a number of challenges, both from a design and operational characteristics standpoint, as shown in Table 12-2.

Table 12-2. Ratings for the Epic Saga$^{(sao)}$

Epic Saga$^{(sao)}$ pattern	Ratings
Communication	Synchronous
Consistency	Atomic
Coordination	Orchestrated
Coupling	Very high
Complexity	Low
Responsiveness/availability	Low
Scale/elasticity	Very low

Fortunately, architects need not default to patterns that, while seemingly familiar, create accidental complexity—a variety of other patterns exist with differing sets of trade-offs. Refer to the "Sysops Squad Saga: Atomic Transactions and Compensating Updates" on page 358 for a concrete example of the Epic Saga(sao) and some of the complex challenges it presents (and how to address those challenges).

Phone Tag Saga(sac) Pattern

The *Phone Tag Saga(sac) pattern* changes one of the dimensions of the Epic Saga(sao), changing *coordination* from *orchestrated* to *choreographed*; this change is illustrated in Figure 12-6.

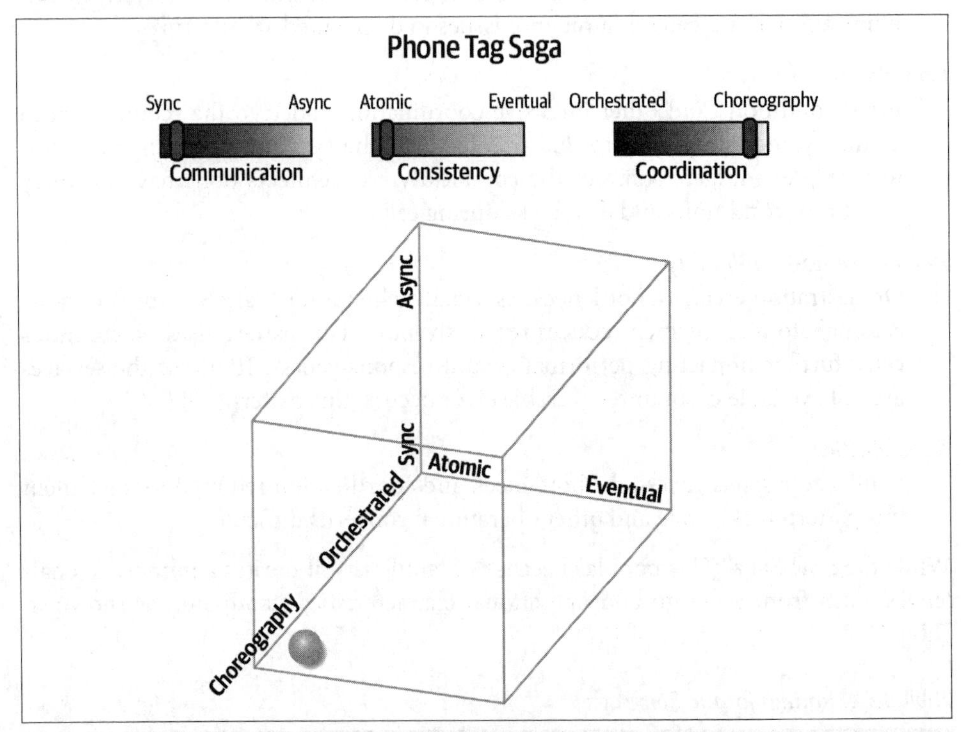

Figure 12-6. The Phone Tag pattern utilizes loosely coupled communication

The pattern name is *Phone Tag* because it resembles a well-known children's game known as *Telephone* in North America: children form a circle, and one person whispers a secret to the next person, who passes it along to the next, until the final version is spoken by the last person. In Figure 12-6, choreography is favored over orchestration, creating the corresponding change in the structural communication shown in Figure 12-7.

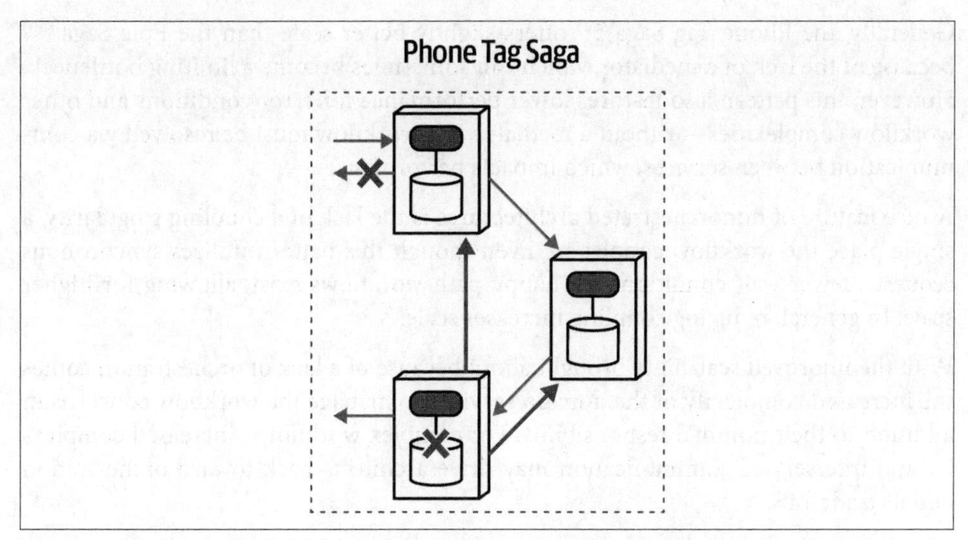

Figure 12-7. Because of a lack of orchestration, each participant must coordinate status

The Phone Tag Saga(sac) pattern features atomicity but also choreography, meaning that the architect designates no formal orchestrator. Yet atomicity requires some degree of coordination. In Figure 12-7, the initially called service becomes the coordination point (sometimes called the *front controller*). Once it has finished its work, it passes a request on to the next service in the workflow, which continues until the workflow succeeds. However, if an error condition occurs, each service must have built-in logic to send compensating requests back along the chain.

Because the architectural goal is transactional atomicity, logic to coordinate that atomicity must reside somewhere. Thus, domain services must contain more logic about the workflow context they participate within, including error handling and routing. For complex workflows, the *front controller* in this pattern will become as complex as most mediators, reducing the appeal and applicability of this pattern. Thus, this pattern is commonly used for simple workflows that need higher scale, but with a potential performance impact.

How does choreography versus orchestration improve operational architecture characteristics like scale? Using choreography even with synchronous communication cuts down on bottlenecks—in nonerror conditions, the last service in the workflow can return the result, allowing for higher throughput and fewer choke points. Performance for happy path workflows can be faster than in an Epic Saga(sao) because of lack of coordination. However, error conditions will be much slower without a mediator—each service must unwind the call chain, which also increases coupling between services.

Generally, the Phone Tag Saga(sac) offers slightly better scale than the Epic Saga(sao) because of the lack of a mediator, which can sometimes become a limiting bottleneck. However, this pattern also features lower performance for error conditions and other workflow complexities—without a mediator, the workflow must be resolved via communication between services, which impacts performance.

A nice feature of nonorchestrated architectures is the lack of a coupling singularity, a single place the workflow couples to. Even though this pattern utilizes synchronous requests, fewer wait conditions for happy path workflows exist, allowing for higher scale. In general, reducing coupling increases scale.

With the improved scalability brought about because of a lack of orchestration comes the increased complexity of the domain services to manage the workflow concerns in addition to their nominal responsibility. For complex workflows, increased complexity and interservice communication may drive architects back toward orchestration and its trade-offs.

The Phone Tag Saga(sac) has a fairly rare combination of features—generally, if an architect chooses *choreography*, they also choose *asynchronicity*. However, in some cases where an architect might choose this combination instead: synchronous calls ensure that each domain service completes its part of the workflow before invoking the next, eliminating race conditions. If error conditions are easy to resolve, or domain services can utilize idempotence and retries, then architects can build higher parallel scale using this pattern compared to an Epic Saga(sao).

The Phone Tag Saga(sac) pattern has the following characteristics:

Coupling level
: This pattern relaxes one of the coupling dimensions of the Epic Saga(sao) pattern, utilizing a choreographed rather than orchestrated workflow. Thus, this pattern is slightly less coupled, but with the same transactional requirement, meaning that the complexity of the workflow must be distributed between the domain services.

Complexity level
: This pattern is significantly more complex than the Epic Saga(sao); complexity in this pattern rises linearly proportionally to the semantic complexity of the workflow: the more complex the workflow, the more logic must appear in each service to compensate for lack of orchestrator. Alternatively, an architect might add workflow information to the messages themselves as a form of *stamp coupling* (see "Stamp Coupling for Workflow Management" on page 378) to maintain state but adding to the overhead context required by each service.

Responsiveness/availability
: Less orchestration generally leads to better responsiveness, but error conditions in this pattern become more difficult to model without an orchestrator, requiring more coordination via callbacks and other time-consuming activities.

Scale/elasticity
: Lack of orchestration translates to fewer bottlenecks, generally increasing scalability, but only slightly. This pattern still utilizes tight coupling around two of the three dimensions, so scalability isn't a highlight, especially if error conditions are common.

The ratings for the Phone Tag Saga(sac) appear in Table 12-3.

Table 12-3. Ratings for the Phone Tag Saga(sac)

Phone Tag Saga(sac)	Ratings
Communication	Synchronous
Consistency	Atomic
Coordination	Choreographed
Coupling	High
Complexity	High
Responsiveness/availability	Low
Scale/elasticity	Low

The Phone Tag Saga(sac) pattern is better for simple workflows that don't have many common error conditions. While it offers a few better characteristics than the Epic Saga(sao), the complexity introduced by lack of an orchestrator offsets many of the advantages.

Fairy Tale Saga(seo) Pattern

Typical fairy tales provide happy stories with easy-to-follow plots, thus the name *Fairy Tale Saga(seo)*, which utilizes synchronous communication, eventual consistency, and orchestration, as shown in Figure 12-8.

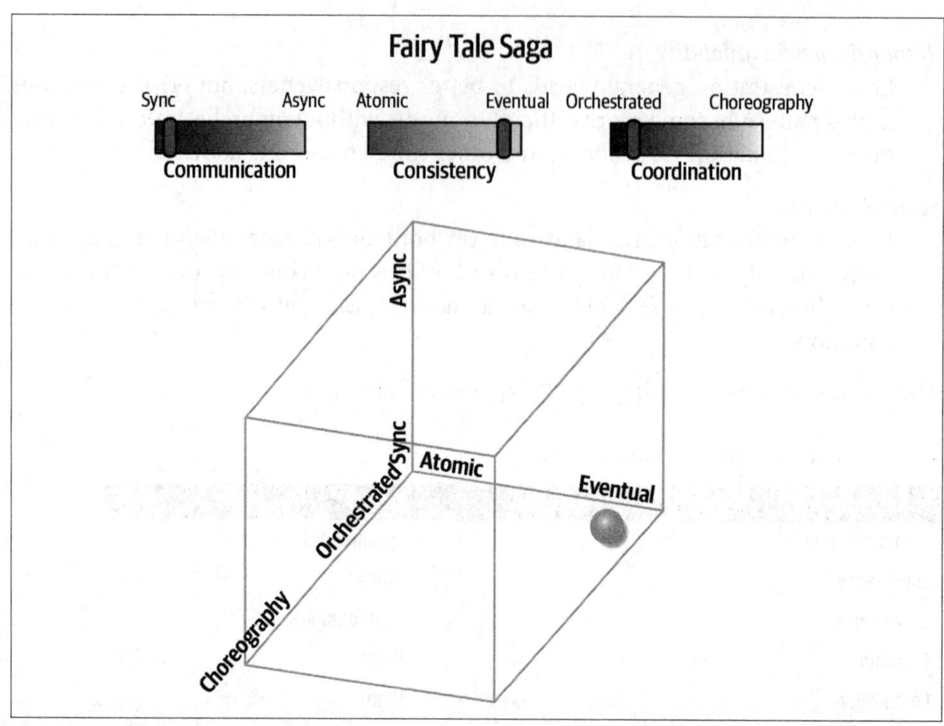

Figure 12-8. The Fairy Tale Saga(seo) illustrates eventual consistency

This communication pattern relaxes the difficult atomic requirement, providing many more options for architects to design systems. For example, if a service is down temporarily, eventual consistency allows for caching a change until the service restores. The communication structure for the Fairy Tale Saga(seo) is illustrated in Figure 12-9.

In this pattern, an orchestrator exists to coordinate request, response, and error handling. However, the orchestrator isn't responsible for managing transactions, which each domain service retains responsibility for (for examples of common workflows, see Chapter 11). Thus the orchestrator can manage compensating calls, but without the requirement of occurring within an active transaction.

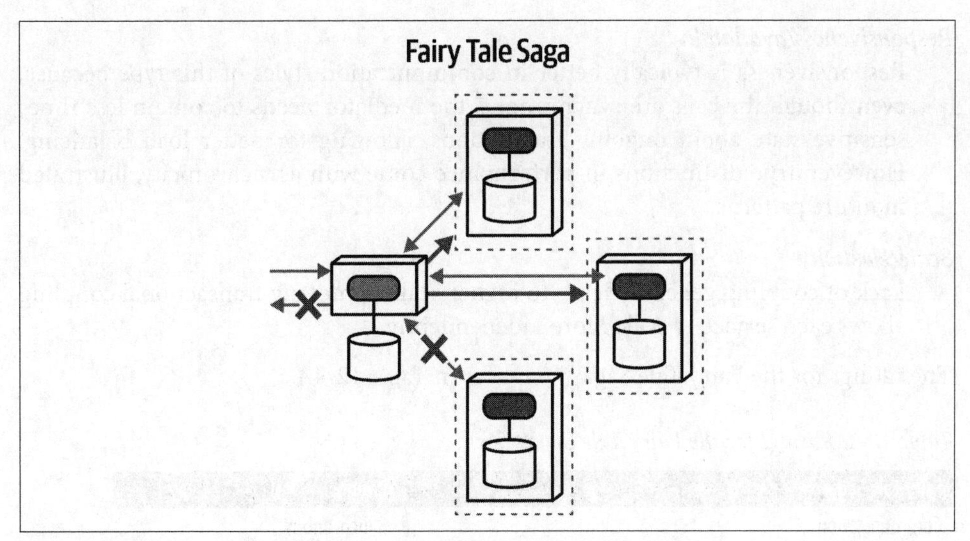

Figure 12-9. Isomorphic illustration of a Fairy Tale interaction

This is a much more attractive pattern and appears commonly in many microservices architectures. Having a mediator makes managing workflows easier, synchronous communication is the easier of the two choices, and eventual consistency removes the most difficult coordination challenge, especially for error handling.

The biggest appealing advantage of the Fairy Tale Saga(seo) is the lack of holistic transactions. Each domain service manages its own transactional behavior, relying on eventual consistency for the overall workflow.

Compared to many other patterns, this pattern generally exhibits a good balance of trade-offs:

Coupling level
: The Fairy Tale Saga(seo) features high coupling, with two of the three coupling drivers maximized in this pattern (synchronous communication and orchestrated coordination). However, the worse driver of coupling complexity—transactionality—disappears in this pattern in favor of eventual consistency. The orchestrator must still manage complex workflows, but without the stricture of doing so within a transaction.

Complexity level
: Complexity for the Fairy Tale Saga(seo) is quite low; it includes the most convenient options (orchestrated, synchronicity) with the loosest restriction (eventual consistency). Thus the name Fairy Tale Saga(seo)—a simple story with a happy ending.

Responsiveness/availability

Responsiveness is typically better in communication styles of this type because, even though the calls are synchronous, the mediator needs to contain less time-sensitive state about ongoing transactions, allowing for better load balancing. However, true distinctions in performance come with asynchronicity, illustrated in future patterns.

Scale/elasticity

Lack of coupling generally leads to higher scale; removing transactional coupling allows each service to scale more independently.

The ratings for the Fairy Tale Saga(seo) appear in Table 12-4.

Table 12-4. Ratings for the Fairy Tale Saga(seo)

Fairy Tale Saga(seo)	Ratings
Communication	Synchronous
Consistency	Eventual
Coordination	Orchestrated
Coupling	High
Complexity	Very low
Responsiveness/availability	Medium
Scale/elasticity	High

If an architect can take advantage of eventual consistency, this pattern is quite attractive, combining the easy moving parts with the fewest scary restrictions, making it a popular choice among architects.

Time Travel Saga(sec) Pattern

The *Time Travel Saga(sec)* pattern features synchronous communication, and eventual consistency, but choreographed workflow. In other words, this pattern avoids a central mediator, placing the workflow responsibilities entirely on the participating domain services, as illustrated in Figure 12-10.

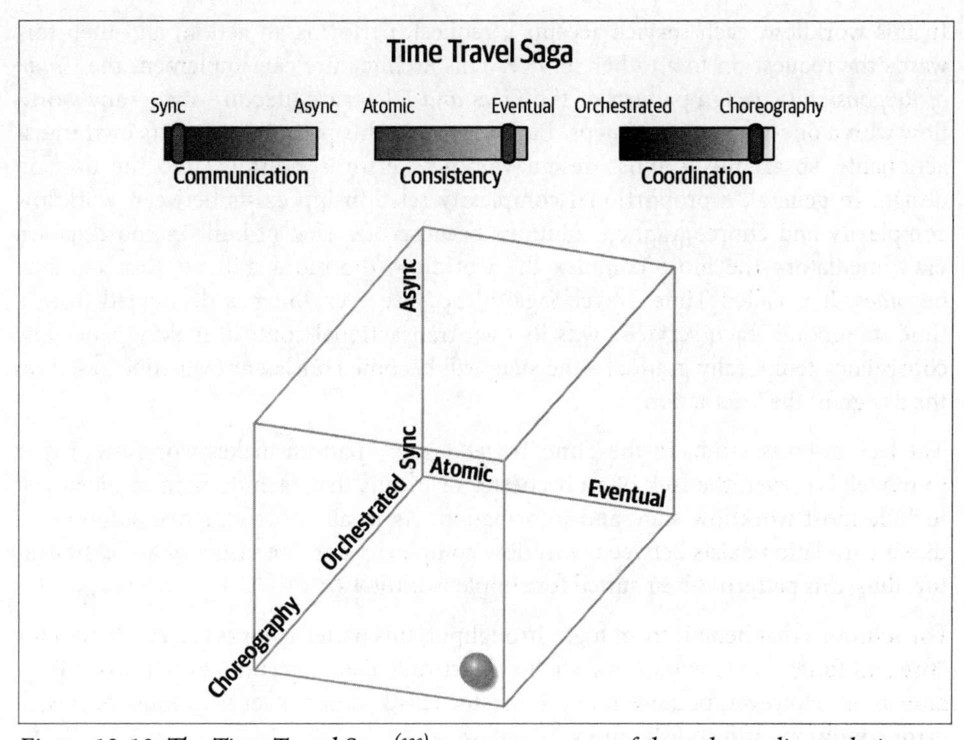

Figure 12-10. The Time Travel Saga(sec) pattern uses two of three decoupling techniques

The structural topology illustrates the lack of orchestration, shown in Figure 12-11.

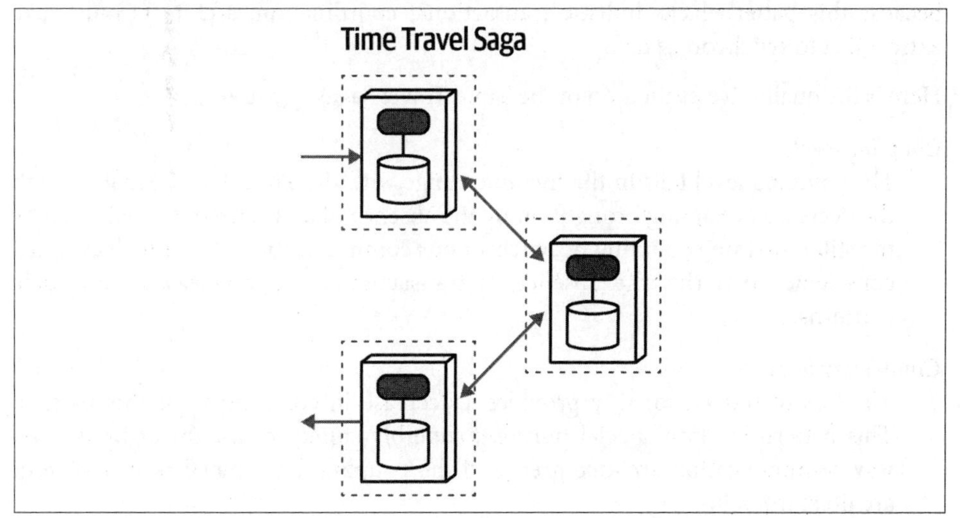

Figure 12-11. Complex workflows become difficult to manage without orchestration

In this workflow, each service accepts a request, performs an action, and then forwards the request on to another service. This architecture can implement the *Chain of Responsibility* design pattern or the *Pipes and Filters* architecture style—any workflow with a one-way series of steps. Each service in this pattern "owns" its own transactionality, so architects must design workflow error conditions into the domain design. In general, a proportional complexity relationship exists between workflow complexity and choreographed solutions because of a lack of built-in coordination via a mediator—the more complex the workflow, the more difficult choreography becomes. It is called Time Travel Saga(sec) because everything is decoupled from a time standpoint: each service owns its own transactional context, making workflow consistency temporally gradual—the state will become consistent over time based on the design of the interaction.

The lack of transactions in the Time Travel Saga(sec) pattern makes workflows easier to model; however, the lack of an orchestrator means that each domain service must include most workflow state and information. As in all choreographed solutions, a direct correlation exists between workflow complexity and the utility of an orchestrator; thus, this pattern is best suited for simple workflows.

For solutions that benefit from high throughput, this pattern works extremely well for "fire and forget" style workflows, such as electronic data ingestion, bulk transactions, and so on. However, because no orchestrator exists, domain services must deal with error conditions and coordination.

Lack of coupling increases scalability with this pattern; only adding asynchronicity would make it more scalable (as in the Anthology Saga(aec) pattern). However, because this pattern lacks holistic transactional coordination, architects must take extra effort to synchronize data.

Here is the qualitative evaluation of the Time Travel Saga(sec) pattern:

Coupling level

The coupling level falls in the medium range with the Time Travel Saga(sec), with the decreased coupling brought on by the absence of an orchestrator balanced by the still remaining coupling of synchronous communication. As with all eventual consistency patterns, the absence of transactional coupling eases many data concerns.

Complexity level

The loss of transactionality provides a decrease in complexity for this pattern. This pattern is quasi-special-purpose, superbly suited to fast throughput, one-way communication architectures, and the coupling level matches that style of architecture well.

Responsiveness/availability

Responsiveness scores a *medium* with this architectural pattern: it is quite high for built-to-purpose systems, as described previously, and quite low for complex error handling. Because no orchestrator exists in this pattern, each domain service must handle the scenario to restore eventual consistency in the case of an error condition, which will cause a lot of overhead with synchronous calls, impacting responsiveness and performance.

Scale/elasticity

This architecture pattern offers extremely good scale and elasticity; it could only be made better with asynchronicity (see the Anthology Saga(aec) pattern).

The ratings for the Time Travel Saga(sec) pattern appear in Table 12-5.

Table 12-5. Ratings for the Time Travel Saga(sec)

Time Travel Saga(sec)	Ratings
Communication	Synchronous
Consistency	Eventual
Coordination	Choreographed
Coupling	Medium
Complexity	Low
Responsiveness/availability	Medium
Scale/elasticity	High

The Time Travel Saga(sec) pattern provides an on-ramp to the more complex but ultimately scalable Anthology Saga(aec) pattern. Architects and developers find dealing with synchronous communication easier to reason about, implement, and debug; if this pattern provides adequate scalability, teams don't have to embrace the more complex but more scalable alternatives.

Fantasy Fiction Saga(aao) Pattern

The *Fantasy Fiction Saga(aao)* uses *atomic* consistency, *asynchronous* communication, and *orchestrated* coordination, as shown in Figure 12-12.

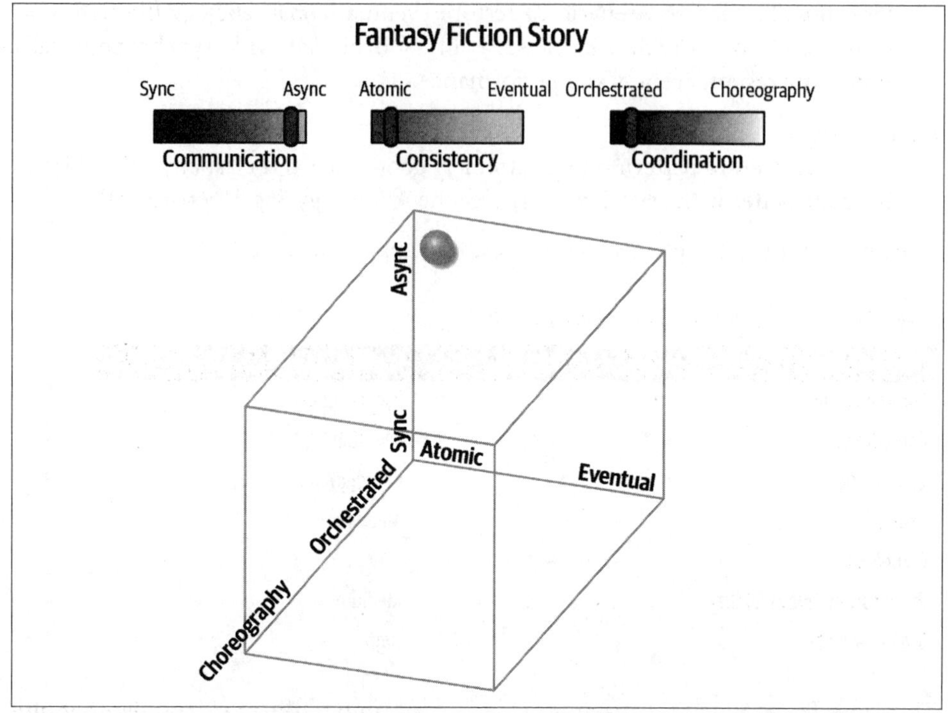

Figure 12-12. Asynchronous communication makes transactionality difficult in this pattern

The structure representation shown in Figure 12-13 starts to show some of the difficulties with this pattern.

Just because a combination of architectural forces exists doesn't mean it forms an attractive pattern, yet this relatively implausible combination has uses. This pattern resembles the Epic Saga(sao) in all aspects except for *communication*—this pattern uses *asynchronous* rather than *synchronous* communication. Traditionally, one way that architects increase the responsiveness of distributed systems is by using asynchronicity, allowing operations to occur in parallel rather than serially. This may seem like a good way to increase the perceived performance over an Epic Saga(sao).

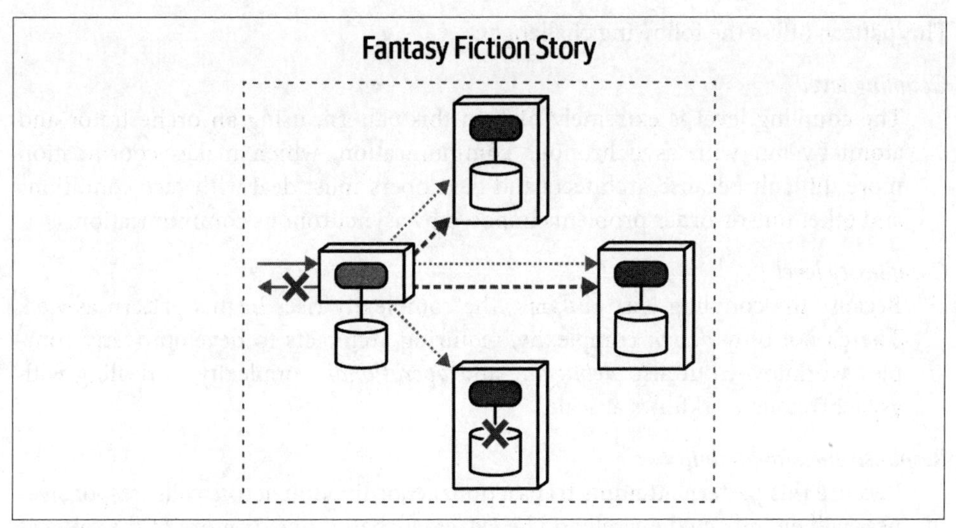

Figure 12-13. The Fantasy Fiction Saga(aao) pattern is far-fetched because transaction coordination for asynchronous communication presents difficulties

However, asynchronicity isn't a simple change—it adds many layers of complexity to architecture, especially around coordination, requiring much more complexity in the mediator. For example, suppose a transactional workflow *Alpha* begins. Because everything is asynchronous, while Alpha is pending, transactional workflow Beta begins. Now, the mediator must keep track of the state of all ongoing transactions in pending state.

It gets worse. Suppose that workflow *Gamma* begins, but the first call to the domain service depends on the still pending outcome of Alpha—how can an architect model this behavior? While possible, the complexity grows and grows.

Adding asynchronicity to orchestrated workflows adds asynchronous transactional state to the equation, removing serial assumptions about ordering and adding the possibilities of deadlocks, race conditions, and a host of other parallel system challenges.

This pattern offers the following challenges:

Coupling level
: The coupling level is extremely high in this pattern, using an orchestrator and atomicity but with asynchronous communication, which makes coordination more difficult because architects and developers must deal with race conditions and other out-of-order problems imposed by asynchronous communication.

Complexity level
: Because the coupling is so difficult, the complexity rises in this pattern as well. There's not only *design* complexity, requiring architects to develop overly complex workflows, but also *debugging* and *operational* complexity of dealing with asynchronous workflows at scale.

Responsiveness/availability
: Because this pattern attempts transactional coordination across calls, responsiveness will be impacted overall and be extremely bad if one or more of the services isn't available.

Scale/elasticity
: High scale is virtually impossible in transaction systems, even with asynchronicity. Scale is much better in the similar pattern Parallel Saga(aeo), which switches *atomic* to *eventual consistency*.

The ratings for the Fantasy Fiction Saga(aao) pattern appear in Table 12-6.

Table 12-6. Ratings for the Fantasy Fiction Saga(aao)

Fantasy Fiction	Ratings
Communication	Asynchronous
Consistency	Atomic
Coordination	Orchestrated
Coupling	High
Complexity	High
Responsiveness/availability	Low
Scale/elasticity	Low

This pattern is unfortunately more popular than it should be, mostly from the misguided attempt to improve the performance of Epic Saga(sao) while maintaining transactionality; a better option is usually Parallel Saga(aeo).

Horror Story(aac) Pattern

One of the patterns must be the worst possible combination; it is the aptly named *Horror Story(aac) pattern*, characterized by *asynchronous* communication, *atomic* consistency, and *choreographed* coordination, illustrated in Figure 12-14.

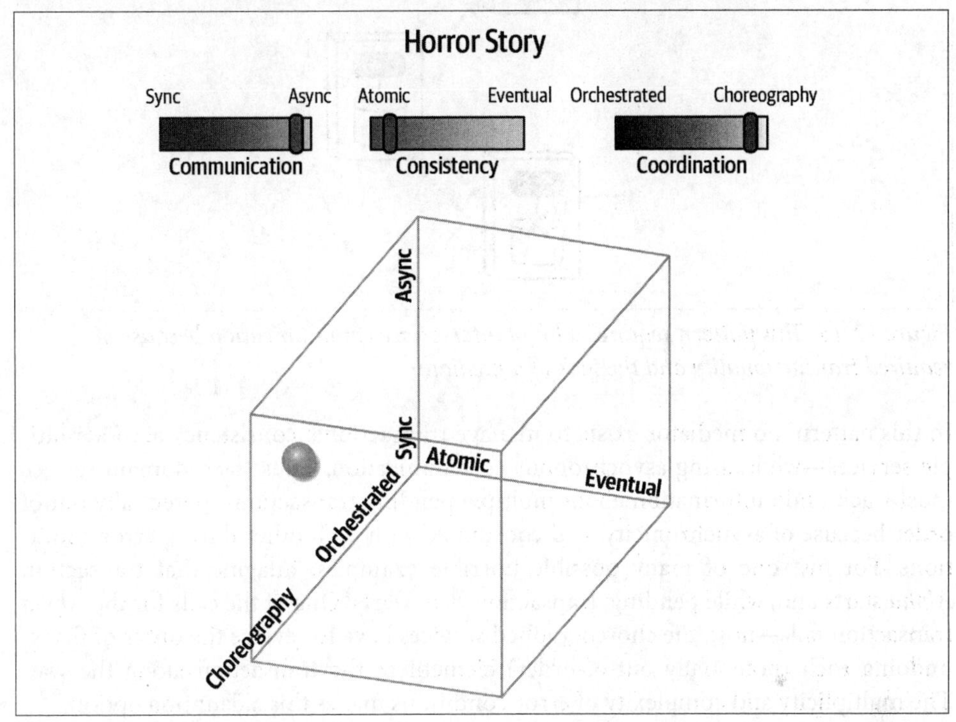

Figure 12-14. The most difficult combination: achieving transactionality while asynchronous and choreographed

Why is this combination so horrible? It combines the most stringent coupling around consistency (*atomic*) with the two loosest coupling styles, *asynchronous* and *choreography*. The structural communication for this pattern appears in Figure 12-15.

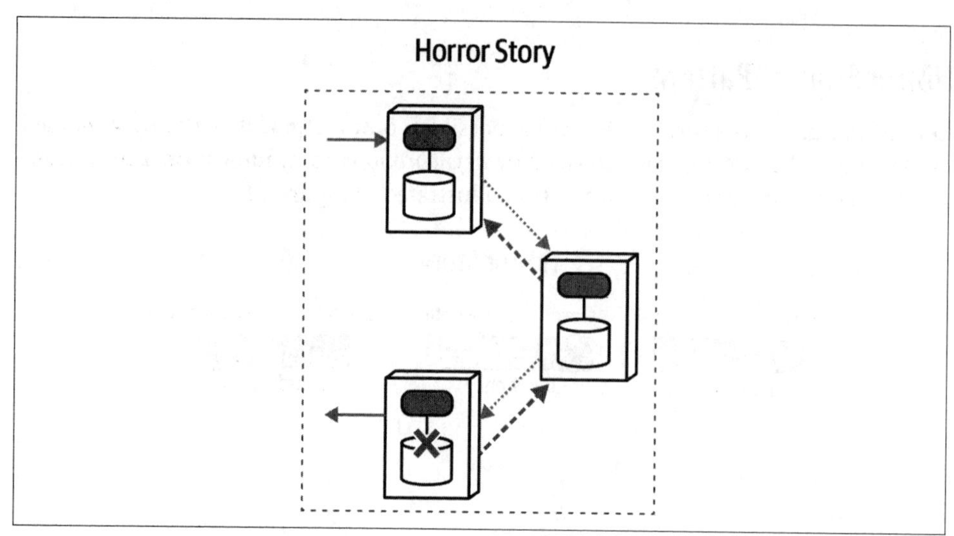

Figure 12-15. This pattern requires a lot of interservice communication because of required transactionality and the lack of a mediator

In this pattern, no mediator exists to manage transactional consistency across multiple services—while using asynchronous communication. Thus, each domain service must track undo information about multiple pending transactions, potentially out of order because of asynchronicity, and coordinate with each other during error conditions. For just one of many possible horrible examples, imagine that transaction *Alpha* starts and, while pending, transaction *Beta* starts. One of the calls for the *Alpha* transaction fails—now, the choreographed services have to reverse the order of firing, undoing each (potentially out-of-order) element of the transaction along the way. The multiplicity and complexity of error conditions makes this a daunting option.

Why might an architect choose this option? Asynchronicity is appealing as a performance boost, yet the architect may still try to maintain transactional integrity, which has many myriad failure modes. Instead, an architect would be better off choosing the Anthology Saga(aec) pattern, which removes holistic transactionality.

The qualitative evaluations for the Horror Story(aac) pattern are as follows:

Coupling level

Surprisingly, the coupling level for this pattern isn't the worst (that "honor" goes to the Epic Saga(sao) pattern). While this pattern does attempt the worst kind of single coupling (transactionality), it relieves the other two, lacking both a mediator and the coupling—increasing synchronous communication.

Complexity level

Just as the name implies, the complexity of this pattern is truly horrific, the worst of any because it requires the most stringent requirement (transactionality) with the most difficult combination of other factors to achieve that (asynchronicity and choreography).

Scale/elasticity

This pattern does scale better than ones with a mediator, and asynchronicity also adds the ability to perform more work in parallel.

Responsiveness/availability

Responsiveness is *low* for this pattern, similar to the other patterns that require holistic transactions: coordination for the workflow requires a large amount of interservice "chatter," hurting performance and responsiveness.

The trade-offs for the Horror Story(aac) pattern appear in Table 12-7.

Table 12-7. Ratings for the Horror Story(aac)

Horror Story(aac)	Ratings
Communication	Asynchronous
Consistency	Atomic
Coordination	Choreographed
Coupling	Medium
Complexity	Very high
Responsiveness/availability	Low
Scale/elasticity	Medium

The aptly named Horror Story(aac) pattern is often the result of a well-meaning architect starting with an Epic Saga(sao) pattern, noticing slow performance because of complex workflows, and realizing that techniques to improve performance include asynchronous communication and choreography. However, this thinking provides an excellent example of not considering all the entangled dimensions of a problem space. In isolation, asynchronous communication improves performance. However, as architects, we cannot consider it in isolation when it is entangled with other architecture dimensions, such as consistency and coordination.

Parallel Saga$^{(aeo)}$ Pattern

The *Parallel Saga$^{(aeo)}$* pattern is named after the "traditional" Epic Saga$^{(sao)}$ pattern with two key differences that ease restrictions and therefore make it an easier pattern to implement: asynchronous communication and eventual consistency. The dimensional diagram of the Parallel Saga$^{(aeo)}$ pattern appears in Figure 12-16.

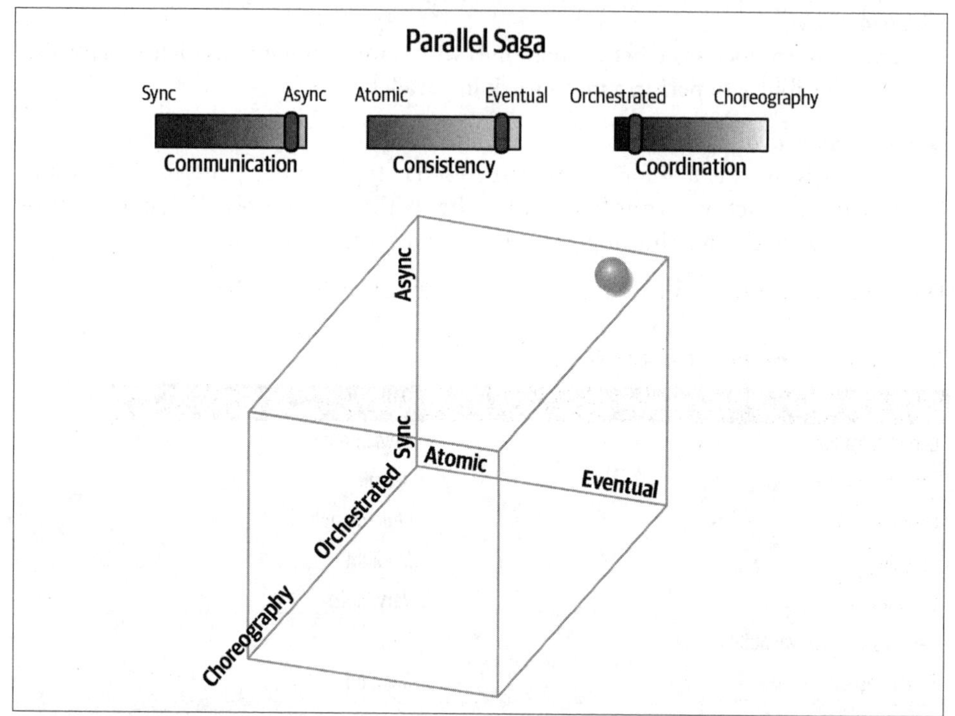

Figure 12-16. Parallel Saga$^{(aeo)}$ offers performance improvements over traditional sagas

The most difficult goals in the Epic Saga$^{(sao)}$ pattern revolve around transactions and synchronous communication, both of which cause bottlenecks and performance degradation. As shown in Figure 12-16, the pattern loosens both restraints.

The isomorphic representation of Parallel Saga$^{(aeo)}$ appears in Figure 12-17.

This pattern uses a mediator, making it suitable for complex workflows. However, it uses asynchronous communication, allowing for better responsiveness and parallel execution. Consistency in the pattern lies with the domain services, which may require some synchronization of shared data, either in the background or driven via the mediator. As in other architectural problems that require coordination, a mediator becomes quite useful.

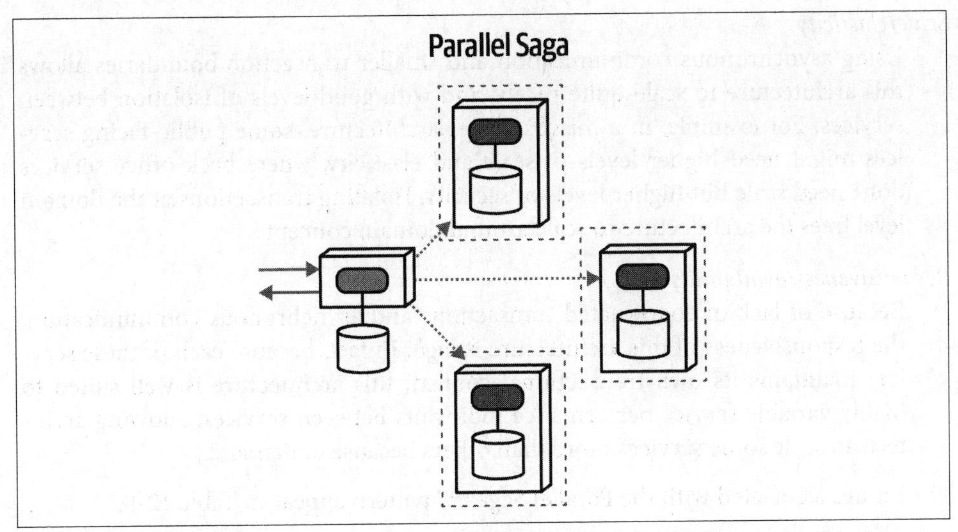

Figure 12-17. Each service owns its own transactionality; the mediator coordinates request and response

For example, if an error occurs during the execution of a workflow, the mediator can send asynchronous messages to each involved domain service to compensate for the failed change, which may entail retries, data synchronization, or a host of other remediations.

Of course, the loosening of constraints implies that some benefits will be traded off, which is the nature of software architecture. Lack of transactionality imposes more burden on the mediator to resolve error and other workflow issues. Asynchronous communication, while offering better responsiveness, makes resolving timing and synchronization issues difficult—race conditions, deadlocks, queue reliability, and a host of other distributed architecture headaches reside in this space.

The Parallel Saga(aeo) pattern exhibits the following qualitative scores:

Coupling level
: This pattern has a low coupling level, isolating the coupling-intensifying force of transactions to the scope of the individual domain services. It also utilizes asynchronous communication, further decoupling services from wait states, allowing for more parallel processing but adding a time element to an architect's coupling analysis.

Complexity level
: The complexity of the Parallel Saga(aeo) is also low, reflecting the lessening of coupling stated previously. This pattern is fairly easy for architects to understand, and orchestration allows for simpler workflow and error-handling designs.

Scale/elasticity

Using asynchronous communication and smaller transaction boundaries allows this architecture to scale quite nicely, and with good levels of isolation between services. For example, in a microservices architecture, some public-facing services might need higher levels of scale and elasticity, where back office services don't need scale but higher levels of security. Isolating transactions at the domain level frees the architecture to scale around domain concepts.

Responsiveness/availability

Because of lack of coordinated transactions and asynchronous communication, the responsiveness of this architecture is *high*. In fact, because each of these services maintains its own transactional context, this architecture is well suited to highly variable service performance footprints between services, allowing architects to scale some services more than others because of demand.

The ratings associated with the Parallel Saga(aeo) pattern appear in Table 12-8.

Table 12-8. Ratings for the Parallel Saga(aeo)

Parallel Saga(aeo)	Ratings
Communication	Asynchronous
Consistency	Eventual
Coordination	Orchestrated
Coupling	Low
Complexity	Low
Responsiveness/availability	High
Scale/elasticity	High

Overall, the Parallel Saga(aeo) pattern offers an attractive set of trade-offs for many scenarios, especially with complex workflows that need high scale.

Anthology Saga(aec) Pattern

The *Anthology Saga(aec) pattern* provides the exact opposite set of characteristics to the traditional Epic Saga(sao) pattern: it utilizes *asynchronous* communication, *eventual* consistency, and *choreographed* coordination, providing the least coupled exemplar among all these patterns. The dimensional view of the Anthology Saga(aec) pattern appears in Figure 12-18.

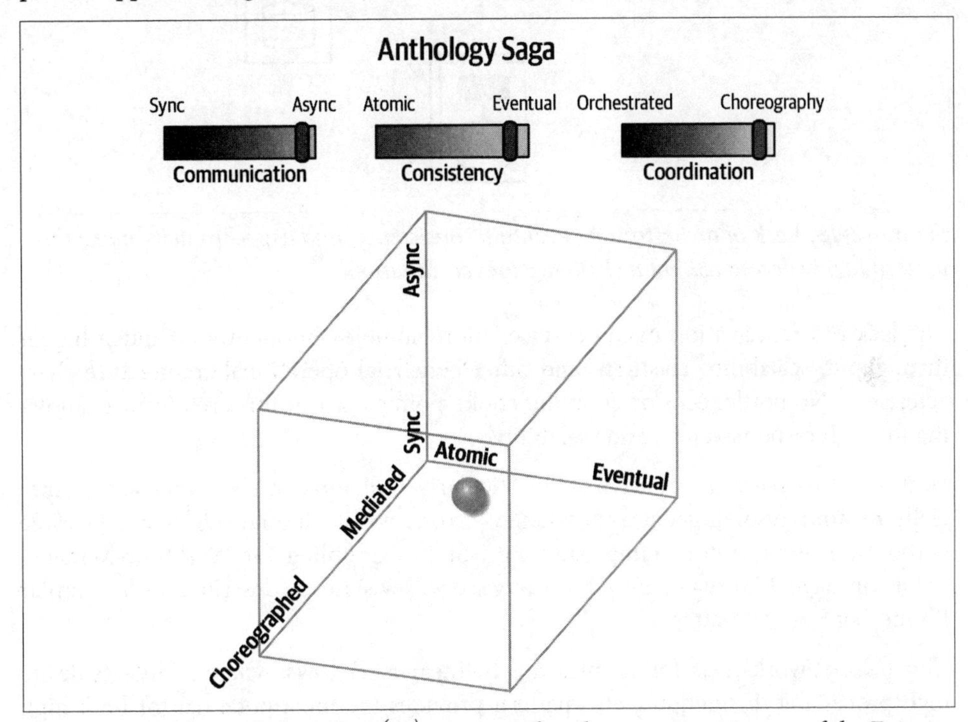

Figure 12-18. The Anthology Saga(aec) pattern offers the opposite extremes of the Epic Saga, and is therefore the least coupled pattern

The anthology pattern uses message queues to send asynchronous messages to other domain services without orchestration, as illustrated in Figure 12-19.

As you can see, each service maintains its own transactional integrity, and no orchestrator exists, forcing each domain service to include more context about the workflows they participate in, including error handling and other coordination strategies.

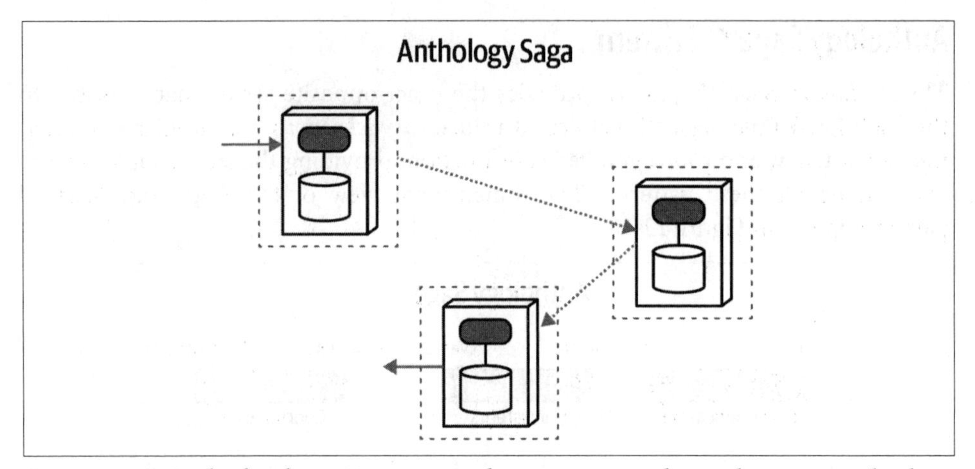

Figure 12-19. Lack of orchestration, eventual consistency, and asynchronicity make this pattern highly decoupled but a challenge for coordination

The lack of orchestration makes services more complex but allows for much higher throughput, scalability, elasticity, and other beneficial operational architecture characteristics. No bottlenecks or coupling choke points exist in this architecture, allowing for high responsiveness and scalability.

However, this pattern doesn't work particularly well for complex workflows, especially around resolving data consistency errors. While it may not seem possible without an orchestrator, stamp coupling ("Stamp Coupling for Workflow Management" on page 378) may be used to carry workflow state, as described in the similar Phone Tag Saga[(sac)] pattern.

This pattern works best for simple, mostly linear workflows, where architects desire high processing throughput. This pattern provides the most potential for both high performance and scale, making it an attractive choice when those are key drivers for the system. However, the degree of decoupling makes coordination difficult, prohibitively so for complex or critical workflows.

The short-story-inspired Anthology Saga[(aec)] pattern has the following characteristics:

Coupling level
: Coupling for this pattern is the lowest for any other combination of forces, creating a highly decoupled architecture well suited for high scale and elasticity.

Complexity level
: While the coupling is extremely low, complexity is correspondingly high, especially for complex workflows where an orchestrator (lacking here) is convenient.

Scale/elasticity
: This pattern scores the highest in the scale and elasticity category, correlating with the overall lack of coupling found in this pattern.

Responsiveness
: Responsiveness is high in this architecture because of a lack of speed governors (transactional consistency, synchronous communication) and use of responsiveness accelerators (choreographed coordination).

The ratings table for the Anthology Saga[(aec)] pattern appear in Table 12-9.

Table 12-9. Ratings for the Anthology Saga[(aec)]

Anthology Saga[(aec)]	Ratings
Communication	Asynchronous
Consistency	Eventual
Coordination	Choreographed
Coupling	Very low
Complexity	High
Responsiveness/availability	High
Scale/elasticity	Very high

The Anthology Saga[(aec)] pattern is well suited to extremely high throughput communication with simple or infrequent error conditions. For example, a *Pipes and Filters* architecture would fit this pattern exactly.

Architects can implement the patterns described in this section in a variety of ways. For example, architects can manage transactional sagas through atomic transactions by using compensating updates or by managing transactional state with eventual consistency. This section showed the advantages and disadvantages of each approach, which will help an architect decide which transactional saga pattern to use.

State Management and Eventual Consistency

State management and eventual consistency leverage *finite state machines* (see "Saga State Machines" on page 352) to always know the current state of the transactional saga, and to also eventually correct the error condition through retries or some sort of automated or manual corrective action. To illustrate this approach, consider the Fairy Tale Saga[(seo)] implementation of the ticket completion example illustrated in Figure 12-20.

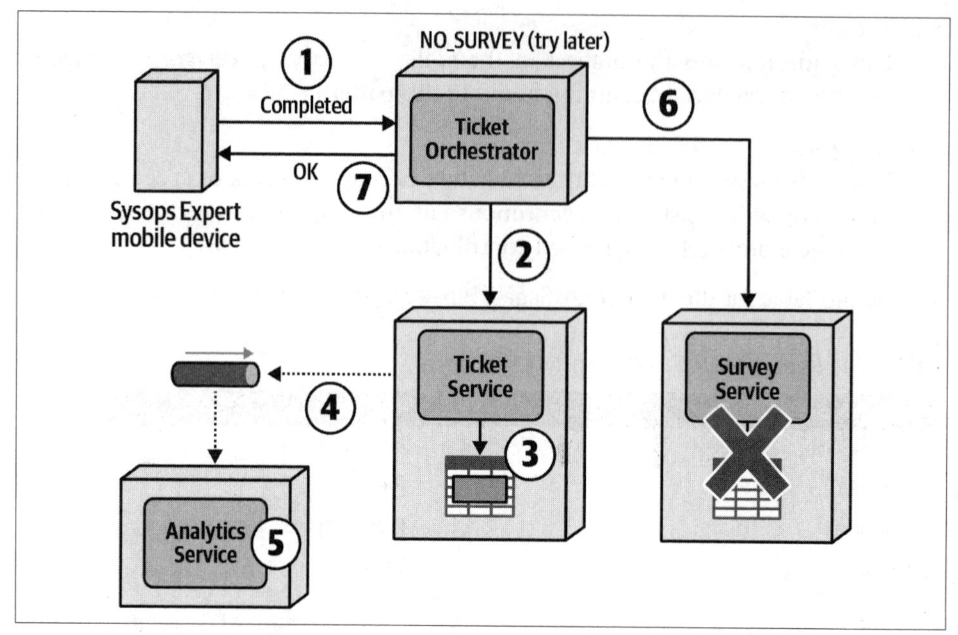

Figure 12-20. The Fairy Tale Saga leads to better responsiveness, but leaves data sources out of sync with one another until they can be corrected

Notice that the Survey Service is not available during the scope of the distributed transaction. However, with this type of saga, rather than issue a compensating update, the state of the saga is changed to `NO_SURVEY` and a successful response is sent to the Sysops Expert (step 7 in the diagram). The Ticket Orchestrator Service then works asynchronously (behind the scenes) to resolve the error programmatically by retries and error analysis. If it cannot resolve the error, the Ticket Orchestrator Service sends the error to an administrator or supervisor for manual repair and processing.

By managing the *state* of the saga rather than issuing compensating updates, the end user (in this case, the Sysops Squad expert) doesn't need to be concerned that the survey was not sent to the customer—that responsibility is for the Ticket Orchestrator Service to worry about. Responsiveness is good from the end user's perpective, and the user can work on other tasks while the errors are handled by the system.

Saga State Machines

A *state machine* is a pattern that describes all of the possible paths that can exist within a distributed architecture. A state machine always starts with a beginning state that launches the transactional saga, then contains transition states and corresponding action that should occur when the transition state happens.

To illustrate how a saga state machine works, consider the following workflow of a new problem ticket created by a customer in the Sysops Squad system:

1. The customer enters a new problem ticket into the system.
2. The ticket is assigned to the next available Sysops Squad expert.
3. The ticket is then routed to the expert's mobile device.
4. The expert receives the ticket and works on the issue.
5. The expert finishes the repair and marks the ticket as complete.
6. A survey is sent to the customer.

The various states that can exist within this transactional saga, as well as the corresponding transition actions, are illustrated in Figure 12-21. Notice that the transactional saga begins with the `START` node indicating the saga entry point, and terminates with the `CLOSED` node indicating the saga exit point.

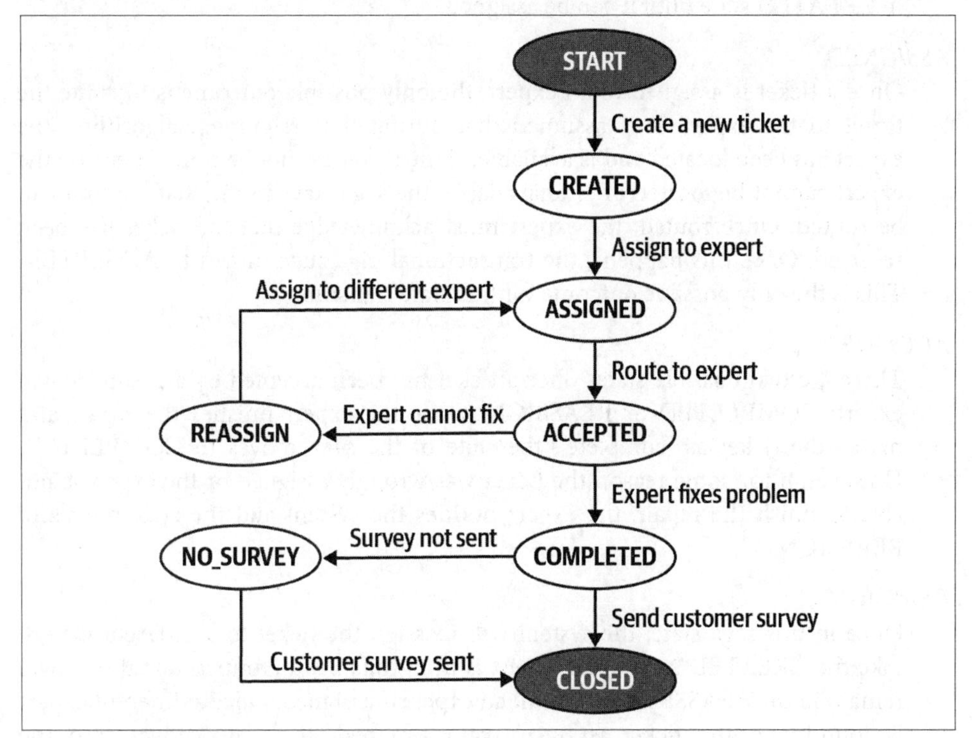

Figure 12-21. State diagram for creating a new problem ticket

The following items describe in more detail this transactional saga and the corresponding states and transition actions that happen within each state:

START
: The transactional saga starts with a customer entering a new problem ticket into the system. The customer's support plan is verified, and the ticket data is validated. Once the ticket is inserted into the ticket table in the database, the transactional saga state moves to CREATED and the customer is notified that the ticket has been successfully created. This is the only possible outcome for this state transition—any errors within this state prevent the saga from starting.

CREATED
: Once the ticket is successfully created, it is assigned to a Sysops Squad expert. If no expert is available to service the ticket, it is held in a wait state until an expert is available. Once an expert is assigned, the saga state moves to the ASSIGNED state. This is the only outcome for this state transition, meaning the ticket is held in CREATED state until it can be assigned.

ASSIGNED
: Once a ticket is assigned to an expert, the only possible outcome is to route the ticket to the expert. It is assumed that during the assignment algorithm, the expert has been located and is available. If the ticket cannot be routed because the expert cannot be located or is unavailable, the saga stays in this state until it can be routed. Once routed, the expert must acknowledge that the ticket has been received. Once this happens, the transactional saga state moves to ACCEPTED. This is the only possible outcome for this state transition.

ACCEPTED
: There are two possible states once a ticket has been accepted by a Sysops Squad expert: COMPLETED or REASSIGN. Once the expert finishes the repair and marks the ticket as "complete," the state of the saga moves to COMPLETED. However, if for some reason the ticket was wrongly assigned or the expert is not able to finish the repair, the expert notifies the system and the state moves to REASSIGN.

REASSIGN
: Once in this saga state, the system will reassign the ticket to a different expert. Like the CREATED state, if an expert is not available, the transactional saga will remain in the REASSIGN state until an expert is assigned. Once a different expert is found and the ticket is once again assigned, the state moves into the ASSIGNED state, waiting to be accepted by the other expert. This is the only possible outcome for this state transition, and the saga remains in this state until an expert is assigned to the ticket.

COMPLETED

The two possible states once an expert completes a ticket are CLOSED or NO_SURVEY. When the ticket is in this state, a survey is sent to the customer to rate the expert and the service, and the saga state is moved to CLOSED, thus ending the transaction saga. However, if the Survey Service is unavailable or an error occurs while sending the survey, the state moves to NO_SURVEY, indicating that the issue was fixed but no survey was sent to the customer.

NO_SURVEY

In this error condition state, the system continues to try sending the survey to the customer. Once successfully sent, the state moves to CLOSED, marking the end of the transactional saga. This is the only possible outcome of this state transaction.

In many cases, it's useful to put the list of all possible state transitions and the corresponding transition action in some sort of table. Developers can then use this table to implement the state transition triggers and possible error conditions in an orchestration service (or respective services if using choreography). An example of this practice is shown in Table 12-10, which lists all the possible states and actions that are triggered when the state transition occurs.

Table 12-10. Saga state machine for a new problem ticket in the Sysops Squad system

Initiating state	Transition state	Transaction action
START	CREATED	Assign ticket to expert
CREATED	ASSIGNED	Route ticket to assigned expert
ASSIGNED	ACCEPTED	Expert fixes problem
ACCEPTED	COMPLETED	Send customer survey
ACCEPTED	REASSIGN	Reassign to a different expert
REASSIGN	ASSIGNED	Route ticket to assigned expert
COMPLETED	CLOSED	Ticket saga done
COMPLETED	NO_SURVEY	Send customer survey
NO_SURVEY	CLOSED	Ticket saga done

The choice between using compensating updates or state management for distributed transaction workflows depends on the situation as well as trade-off analysis between responsiveness and consistency. Regardless of the technique used to manage errors within a distributed transaction, the state of the distributed transaction should be known and also managed.

Table 12-11 summarizes the trade-offs associated with using state management rather than atomic distributed transactions with compensating updates.

Trade-Offs

Table 12-11. Trade-offs associated with state management rather than atomic distributed transactions with compensating updates

Advantages	Disadvantages
Good responsiveness	Data may be out of sync when errors occur
Less impact to end user for errors	Eventual consistency may take some time

Techniques for Managing Sagas

Distributed transactions are not something that can be simply "dropped into" a system. They cannot be downloaded or purchased using some sort of framework or product like ACID transaction managers—they must be designed, coded, and maintained by developers and architects.

One of the techniques we like to use to help manage distributed transactions is to leverage annotations (Java) or custom attributes (C#), or other similar artifacts in other languages. While these language artifacts themselves don't contain any actual functionality, they do provide a programmatic way of capturing and documenting the transactional sagas in the system, as well as provide a means for associating services with transactional sagas.

The source listings in Example 12-1 (Java) and Example 12-2 (C#) show an example of implementing these annotations and custom attributes. Notice that in both implementations, the transactional sagas (`NEW_TICKET`, `CANCEL_TICKET`, and so on) are contained within the `Transaction` enum, providing a single place within the source code for listing and documenting the various sagas that exist within an application context.

Example 12-1. Source code defining a transactional saga annotation (Java)

```
@Retention(RetentionPolicy.RUNTIME)
@Target(ElementType.TYPE)
public @interface Saga {
   public Transaction[] value();

   public enum Transaction {
      NEW_TICKET,
      CANCEL_TICKET,
      NEW_CUSTOMER,
      UNSUBSCRIBE,
      NEW_SUPPORT_CONTRACT
   }
}
```

Example 12-2. Source code defining a transactional saga attribute (C#)

```
[AttributeUsage(AttributeTargets.Class)]
class Saga : System.Attribute {
   public Transaction[] transaction;

   public enum Transaction {
      NEW_TICKET,
      CANCEL_TICKET,
      NEW_CUSTOMER,
      UNSUBSCRIBE,
      NEW_SUPPORT_CONTRACT
   };
}
```

Once defined, these annotations or attributes can be used to identify services that are involved in the transactional saga. For example, the source code listing in Example 12-3 shows that the Survey Service (identified by the `SurveyServiceAPI` class as the service entry point) is involved in the `NEW_TICKET` saga, whereas the Ticket Service (identified by the `TicketServiceAPI` class as the service entry point) is involved in two sagas: the `NEW_TICKET` and the `CANCEL_TICKET`.

Example 12-3. Source code showing the use of the transactional saga annotation (Java)

```
@ServiceEntrypoint
@Saga(Transaction.NEW_TICKET)
public class SurveyServiceAPI {
   ...
}

@ServiceEntrypoint
@Saga({Transaction.NEW_TICKET,)
      Transaction.CANCEL_TICKET})
public class TicketServiceAPI {
   ...
}
```

Notice how the `NEW_TICKET` saga includes the Survey Service and the Ticket Service. This is valuable information to a developer because it helps them define the testing scope when making changes to a particular workflow or saga, and also lets them know what other services might be impacted by a change to one of the services within the transactional saga.

Using these annotations and custom attributes, architects and developers can write simple command-line interface (CLI) tools to walk through a codebase or source code repository to provide saga information in real time. For example, using a simple custom code-walk tool, a developer, architect, or even a business analyst can query what services are involved for the `NEW_TICKET` saga:

```
$ ./sagatool.sh NEW_TICKET -services

-> Ticket Service
-> Assignment Service
-> Routing Service
-> Survey Service

$
```

A custom code-walking tool can look at each class file in the application context containing the `@ServiceEntrypoint` custom annotation (or attribute) and check the `@Saga` custom annotation for the presence of the particular saga (in this case, `Transaction.NEW_TICKET`). This sort of custom tool is not complicated to write, and can help provide valuable information when managing transactional sagas.

Sysops Squad Saga: Atomic Transactions and Compensating Updates

Tuesday, April 5, 09:44

Addison and Austen met first thing with Logan to hash out the issues around transactionality in the new microservices architecture in the longish conference room.

Logan began, "I know that not everyone is on the same page about how what you've read applies to what we're doing here. So, I've prepared some workflows and diagrams to help everyone get on the same page. Today, we're discussing marking a ticket complete in the system. For this workflow, the Sysops Squad expert completes a job and marks the ticket as "complete" using the mobile application on the expert's mobile device. I want to talk about the Epic Saga pattern and the issues around compensating updates. I've created a diagram to illustrate this workflow in Figure 12-22 Can everyone see it?"

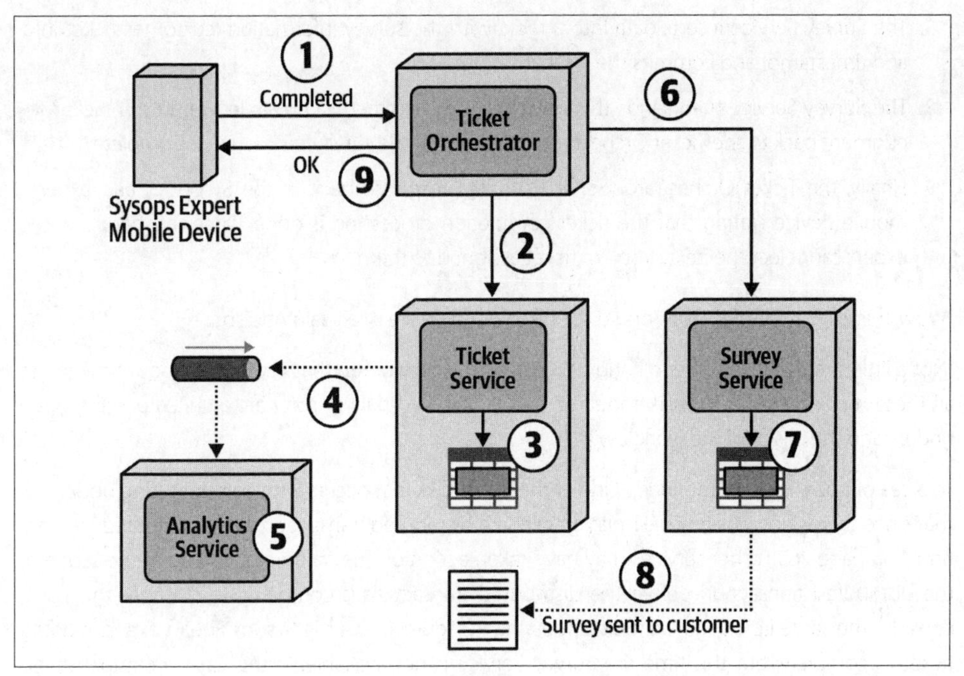

Figure 12-22. The epic saga requires the ticket status to be updated and survey to be sent in one synchronous atomic operation

Logan continued, "I've also created a list that describes each step. The circled numbers on the diagram match up with the workflow."

1. The Sysops Squad expert marks the ticket as complete using an app on their mobile device, which is synchronously received by the Ticket Orchestrator Service.
2. The Ticket Orchestrator Service sends a synchronous request to the Ticket Service to change the state of the ticket from "in-progress" to "complete."
3. The Ticket Service updates the ticket number to "complete" in the database table and commits the update.
4. As part of the ticket completion process, the Ticket Service asynchronously sends ticketing information (such as ticket repair time, ticket wait time, duration, and so on) to a queue to be picked up by the Analytics Service. Once sent, the Ticket Service sends an acknowledgment to the Ticket Orchestrator Service that the update is complete.
5. At about the same time, the Analytics Service asynchronously receives the updated ticket analytics and starts to process the ticket information.
6. The Ticket Orchestrator Service then sends a synchronous request to the Survey Service to prepare and send the customer survey to the customer.

7. The Survey Service inserts data into a table with the survey information (customer, ticket info, and timestamp) and commits the insert.
8. The Survey Service then sends the survey to the customer via email and returns an acknowledgment back to the Ticket Orchestrator Service that the survey processing is complete.
9. Finally, the Ticket Orchestrator Service sends a response back to the Sysops Squad expert's mobile device stating that the ticket completion processing is done. Once this happens, the expert can select the next problem ticket assigned to them.

"Wow, this is really helpful. How long did it take you to create this?" said Addison.

"Not a little time, but it's come in handy. You aren't the only group that's confused about how to get all these moving pieces to work together. This is the hard part of software architecture. Everyone understand the basics of the workflow?"

To a sea of nods, Logan continued, "One of the first issues that occurs with compensating updates is that since there's no transactional isolation within a distributed transaction (see "Distributed Transactions" on page 263), other services may have taken action on the data updated within the scope of the distributed transaction before the distributed transaction is complete. To illustrate this issue, consider the same Epic Saga example appearing in Figure 12-23: the Sysops Squad expert marks a ticket as complete, but this time the Survey Service is not available. In this case, a compensating update (step 7 in the diagram) is sent to the Ticket Service to reverse the update, changing the ticket state from *completed* back to *in-progress* (step 8 in the diagram)."

"Notice also in Figure 12-23 that since this is an atomic distributed transaction, an error is then sent back to the Sysops Squad expert indicating that the action was not successful and to try again. Now, a question for you: why should the Sysops Squad expert have to worry that the survey is not sent?"

Austen pondered a moment. "But wasn't that part of the workflow in the monolith? All that stuff happened within a transaction, if I remember correctly."

"Yeah, but I always thought that was weird, just never said anything," said Addison. "I don't see why the expert should worry about the survey. The expert just wants to get on to the next ticket assigned to them."

"Right," Logan said. "This is the issue with atomic distributed transactions—the end user is unnecessarily semantically coupled to the business process. But notice that Figure 12-23 also illustrates the issue with the lack of transaction isolation within a distributed transaction. Notice that as part of the original update to mark the ticket as *complete,* the Ticket Service asynchronously sent the ticket information to a queue (step 4 in the diagram) to be processed by the Analytics Service (step 5). However, when the compensating update is issued to the Ticket Service (step 7), the ticket information has already been processed by the Analytics Service in step 5."

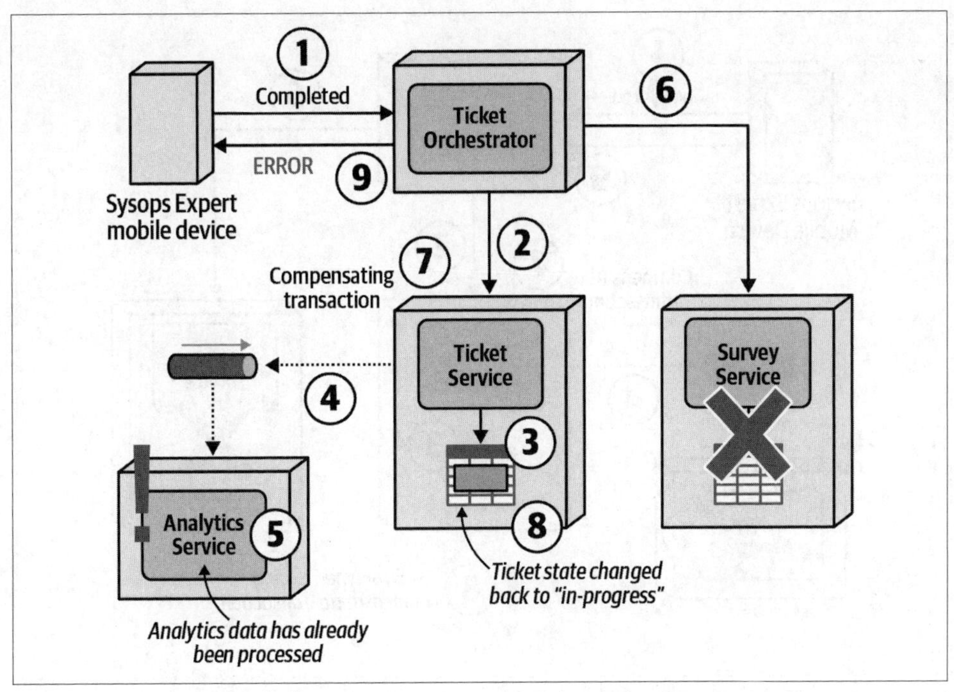

Figure 12-23. Epic Saga(sao) requires compensation, but side effects can occur

"We call this a *side effect* within distributed architectures. By reversing the transaction in the Ticket Service, actions performed by other services using data from the prior update may have already taken place and might not be able to be reversed. This scenario points to the importance of *isolation* within a transaction, something that distributed transactions do not support. To address this issue, the Ticket Service could send another request through the data pump to the Analytics Service, telling that service to ignore the prior ticket information, but just imagine the amount of complex code and timing logic that would be required in the Analytics Service to address this compensating change. Furthermore, there may have been additional downstream actions taken on the analytical data already processed by the Analytics Service, further complicating the chain of events to reverse and correct. With distributed architectures and distributed transactions, it really is sometimes *turtles all the way down* (*https://oreil.ly/zP8dK*)."

Logan paused for a moment, then continued, "Another issue—"

Austen interrupted, "Another issue?"

Logan smiled. "Another issue regarding compensating updates is compensation failures. Keeping with the same Epic Saga example for completing a ticket, notice in Figure 12-24 that in step 7 a compensating update is issued to the Ticket Service to change the state from *completed* back to *in-progress*. However, in this case, the Ticket Service generates an error when trying to change the state of the ticket (step 8)."

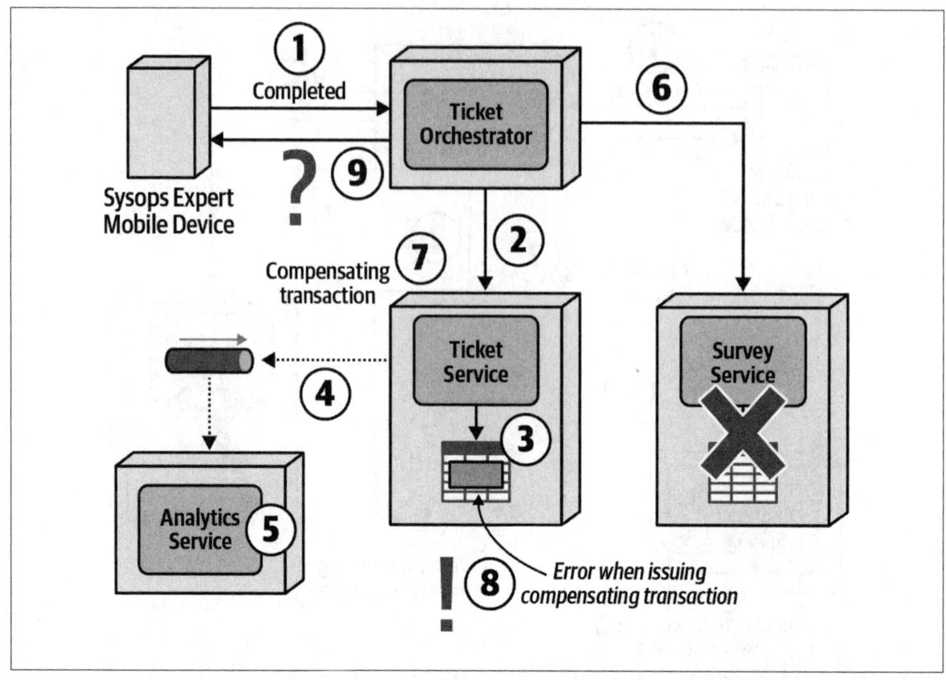

Figure 12-24. Compensating updates within an Epic Saga can fail, leading to inconsistency and confusion about what action to take in the event of a compensation failure

"I've seen that happen! It took forever to track that down," said Addison.

"Architects and developers tend to assume that compensating updates will always work," Logan said. "But sometimes they don't. In this case, as shown in Figure 12-24, there is confusion about what sort of response to send back to the end user (in this case, the Sysops Squad expert). The ticket status is already marked as *complete* because the compensation failed, so attempting the "mark as complete" request again might only lead to yet another error (such as *Ticket already marked as complete*). Talk about confusion on the part of the end user!"

"Yeah, I can imagine the developers coming to us to ask us how to resolve this issue," Addison said.

"Often developers are good checks on incomplete or confusing architecture solutions. If they are confused, there may be a good reason," said Logan. "OK, one more issue. Atomic distributed transactions and corresponding compensating updates also impact responsiveness. If an error occurs, the end user must wait until all corrective action is taken (through compensating updates) before a response is sent telling the user about the error."

"Isn't that where the change to eventual consistency helps, for responsiveness?" asked Austen.

"Yes, while responsiveness can sometimes be resolved by asynchronously issuing compensating updates through eventual consistency (such as with the Parallel Saga and the Anthology Saga pattern), nevertheless most atomic distributed transactions have worse responsiveness when compensating updates are involved."

"OK, that makes sense—atomic coordination will always have overhead," Austen said.

"That's a lot of information. Let's build a table to summarize some of the trade-offs associated with atomic distributed transactions and compensating updates." (See Table 12-12.)

Trade-Offs

Table 12-12. Trade-offs associated with atomic distributed transactions and compensating updates

Advantages	Disadvantages
All data restored to prior state	No transaction isolation
Allows retries and restart	Side effects may occur on compensation
	Compensation may fail
	Poor responsiveness for the end user

Logan said, "While this compensating transaction pattern exists, it also offers a number of challenges. Who wants to name one?"

"I know: *a service cannot perform a rollback*," said Austen. "What if one of the services *cannot* successfully undo the previous operation? The orchestrator must have coordination code to indicate that the transaction wasn't successful."

"Right—what about another?"

"*To lock or not lock participating services?*" said Addison. "When the mediator places a call to a service and it updates a value, the mediator will make calls to subsequent services that are part of the workflow. However, what happens if another request appears for the first service contingent on the outcome of the first request's resolution, either from the same mediator or a different context? This distributed architecture problem becomes worse when the calls are asynchronous rather than synchronous (illustrated in "Phone Tag Saga(sac) Pattern" on page 330). Alternatively, the mediator could insist that other services don't accept calls during the course of a workflow, which guarantees a valid transaction but destroys performance and scalability."

Logan said, "Correct. Let's get philosophical for a moment. Conceptually, transactions force participants to stop their individual worlds and synchronize on a particular value. This is so easy to model with monolithic architectures and relational databases that architects overuse transactions in those systems. Much of the real world isn't transactional, as observed in the famous essay by Gregor Hohpe, "Starbucks Does Not Use Two-Phase Commit" (*https://oreil.ly/feCe1*). Transactional coordination is one of the hardest parts of architecture, and the broader the scope, the worse it becomes."

"Is there an alternative to using an Epic Saga?" Addison asked.

"Yes!" Logan said. "A more realistic approach to the scenario described in Figure 12-24 might be to use either a Fairy Tale Saga or a Parallel Saga pattern. These sagas rely on asynchronous eventual consistency and state management rather than atomic distributed transactions with compensating updates when errors occur. With these types of sagas, the user is less impacted by errors that might occur within the distributed transaction, because the error is addressed behind the scenes, without end-user involvement. Responsiveness is also better with the state management and eventual consistency approach, because the user does not have to wait for corrective action to be taken within the distributed transaction. If we have issues with atomicity, we can investigate those patterns as alternatives."

"Thanks—that's a lot of material, but now I see why the architects made some of the decisions in the new architecture," Addison said.

CHAPTER 13

Contracts

`Friday, April 15, 12:01`

Addison met with Sydney over lunch in the cafeteria to chat about coordination between the Ticket Orchestrator and the services it integrated with for the ticket management workflow.

"Why not just use gRPC for all the communication? I heard it's really fast," said Sydney.

"Well, that's an implementation, not an architecture," Addison said. "We need to decide what types of contracts we want before we choose how to implement them. First, we need to decide between tight or loose contracts. Once we decide on the type, I'll leave it to you to decide how to implement them, as long as they pass our fitness functions."

"What determines what kind of contract we need?" Sydney said.

In Chapter 2, we began discussing the intersection of three important forces—communication, consistency, and coordination—and how to develop trade-offs for them. We modeled the intersectional space of the three forces in a joined three-dimensional space, shown again in Figure 13-1. In Chapter 12, we revisited these three forces with a discussion of the various communication styles and their trade-offs.

However much an architecture can discern a relationship like this one, some forces cut across the conceptual space and affect all of the other dimensions equally. If pursuing the visual three-dimensional metaphor, these cross-cutting forces act as an additional dimension, much as *time* is orthogonal to the three physical dimensions.

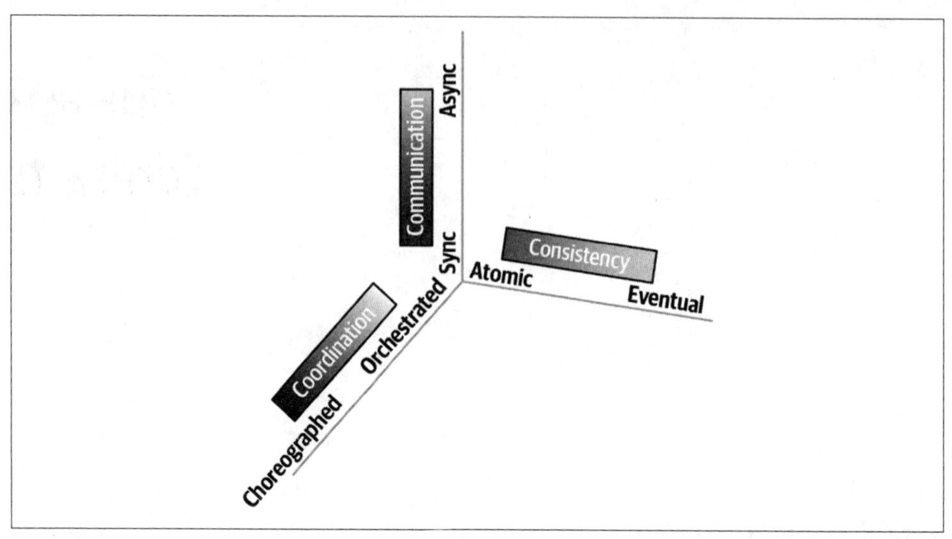

Figure 13-1. Three-dimensional intersecting space for messaging forces in distributed architectures

One constant factor in software architecture that cuts across and affects virtually every aspect of architect decision making is *contracts*, broadly defined as how disparate parts of an architecture connect with one another. The dictionary definition of a contract is as follows:

contract
: A written or spoken agreement, especially one concerning employment, sales, or tenancy, that is intended to be enforceable by law.

In software, we use contracts broadly to describe things like integration points in architecture, and many contract formats are part of the design process of software development: SOAP, REST, gRPC, XMLRPC, and an alphabet soup of other acronyms. However, we broaden that definition and make it more consistent:

hard parts contract
: The format used by parts of an architecture to convey information or dependencies.

This definition of *contract* encompasses all techniques used to "wire together" parts of a system, including transitive dependencies for frameworks and libraries, internal and external integration points, caches, and any other communication among parts.

This chapter illustrates the effects of contracts on many parts of architecture, including static and dynamic quantum coupling, as well as ways to improve (or harm) the effectiveness of workflows.

Strict Versus Loose Contracts

Like many things in software architecture, contracts don't exist within a binary but rather on a broad spectrum, from strict to loose. Figure 13-2 illustrates this spectrum, using example contract types.

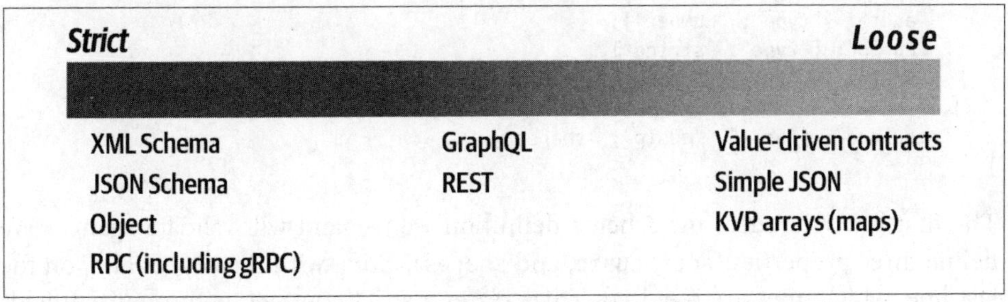

Figure 13-2. The spectrum of contract types, from strict to loose

A strict contract requires adherence to names, types, ordering, and all other details, leaving no ambiguity. An example of the strictest possible contract in software is a remote method call, using a platform mechanism such as RMI in Java. In that case, the remote call mimics an internal method call, matching name, parameters, types, and all other details.

Many strict contract formats mimic the semantics of method calls. For example, developers see a host of protocols that include some variation of the RPC, traditionally an acronym for *Remote Procedure Call*. gRPC (*https://grpc.io*) is an example of a popular remote invocation framework that defaults to strict contracts.

Many architects like strict contracts because they model the identical semantic behavior of internal method calls. However, strict contracts create brittleness in integration architecture—something to avoid. As discussed in Chapter 8, something that is simultaneously changing frequently and used by several distinct architecture parts creates problems in architecture. Contracts fit that description because they form the glue within a distributed architecture: the more frequently they must change, the more rippling problems they cause for other services. However, architects aren't forced to use strict contracts and should do so only when advantageous.

Even an ostensibly loose format such as JSON (*https://www.json.org*) offers ways to selectively add schema information to simple name-value pairs. Example 13-1 shows a strict JSON contract with schema information attached.

Example 13-1. Strict JSON contract

```
{
    "$schema": "http://json-schema.org/draft-04/schema#",
    "properties": {
      "acct": {"type": "number"},
      "cusip": {"type": "string"},
      "shares": {"type": "number", "minimum": 100}
   },
    "required": ["acct", "cusip", "shares"]
}
```

The first line references the schema definition we use and will validate against. We define three properties (`acct`, `cusip`, and `shares`), along with their types and, on the last line, which ones are required. This creates a strict contract, with required fields and types specified.

Examples of looser contracts include formats such as REST (*https://oreil.ly/tzoUg*) and GraphQL (*https://graphql.org*), very different formats but similar in demonstrating looser coupling than RPC-based formats. For REST, the architect models resources rather than method or procedure endpoints, making for less brittle contracts. For example, if an architect builds a RESTful resource that describes parts of an airplane to support queries about seats, that query won't break in the future if someone adds details about engines to the resource—adding more information doesn't break what's there.

Similarly, GraphQL is used by distributed architectures to provide read-only aggregated data rather than perform costly orchestration calls across a wide variety of services. Consider the two GraphQL representations in Examples 13-2 and 13-3, providing two different but capable views of the Profile contract.

Example 13-2. Customer Wishlist `Profile` representation

```
type Profile {
    name: String
}
```

Example 13-3. Customer `Profile` representation

```
type Profile {
    name: String
    addr1: String
    addr2: String
    country: String
    . . .
}
```

The concept of *profile* appears in both examples but with different values. In this scenario, the Customer Wishlist doesn't have internal access to the customer's name, only a unique identifier. Thus, it needs access to a Customer Profile that maps the identifier to the customer name. The Customer Profile, on the other hand, includes a large amount of information about the customer in addition to the name. As far as Wishlist is concerned, the only interesting thing in Profile is the name.

A common anti-pattern that some architects fall victim to is to assume that Wishlist might eventually need all the other parts, so the architects include them in the contract from the outset. This is an example of stamp coupling and an anti-pattern in most cases, because it introduces breaking changes where they aren't needed, making the architecture fragile yet providing little benefit. For example, if the Wishlist cares about only the customer name from Profile, but the contract specifies every field in Profile (just in case), then a change in Profile that Wishlist doesn't care about causes a contract breakage and coordination to fix. Keeping contracts at a "need to know" level strikes a balance between semantic coupling and necessary information without creating needless fragility in integration architecture.

At the far end of the spectrum of contract coupling lie extremely loose contracts, often expressed as name-value pairs in formats like YAML (*https://yaml.org*) or JSON, as illustrated in Example 13-4.

Example 13-4. Name-value pairs in JSON

```
{
  "name": "Mark",
  "status": "active",
  "joined": "2003"
}
```

Nothing but the raw facts in this example! No additional metadata, type information, or anything else, just name-value pairs.

Using such loose contracts allows for extremely decoupled systems, often one of the goals in architectures, such as microservices. However, the looseness of the contract comes with trade-offs such as lack of contract certainty, verification, and increased application logic. We illustrate in "Contracts in Microservices" on page 372 how architects resolve this problem by using contract fitness functions.

Trade-Offs Between Strict and Loose Contracts

When should an architect use strict contracts and when should they use looser ones? Like all the hard parts of architecture, no generic answer exists for this question, so it is important for architects to understand when each is most suitable.

Strict contracts

Stricter contracts have a number of advantages, including these:

Guaranteed contact fidelity
: Building schema verification within contracts ensures exact adherence to the values, types, and other governed metadata. Some problem spaces benefit from tight coupling for contract changes.

Versioned
: Strict contracts generally require a versioning strategy to support two endpoints that accept different values or to manage domain evolution over time. This allows gradual changes to integration points while supporting a selective number of past versions to make integration collaboration easier.

Easier to verify at build time
: Many schema tools provide mechanisms to verify contracts at build time, adding a level of type checking for integration points.

Better documentation
: Distinct parameters and types provide excellent documentation with no ambiguity.

Strict contracts also have a few disadvantages:

Tight coupling
: By our general definition of coupling, strict contracts create tight coupling points. If two services share a strict contract and the contract changes, both services must change.

Versioned
: This appears in both advantages and disadvantages. While keeping distinct versions allows for precision, it can become an integration nightmare if the team doesn't have a clear deprecation strategy or tries to support too many versions.

The trade-offs for strict contracts are summarized in Table 13-1.

Trade-Offs

Table 13-1. Trade-offs for strict contracts

Advantage	Disadvantage
Guaranteed contract fidelity	Tight coupling
Versioned	Versioned
Easier to verify at build time	
Better documentation	

Loose contracts

Loose contracts, such as name-value pairs, offer the least coupled integration points, but they too have trade-offs, as summarized in Table 13-2.

These are some advantages of loose contracts:

Highly decoupled
: Many architects have a stated goal for microservices architectures that includes high levels of decoupling, and loose contracts provide the most flexibility.

Easier to evolve
: Because little or no schema information exists, these contracts can evolve more freely. Of course, semantic coupling changes still require coordination across all interested parties—implementation cannot reduce semantic coupling—but loose contracts allow easier implementation evolution.

Loose contracts also have a few disadvantages:

Contract management
: Loose contracts by definition don't have strict contract features, which may cause problems such as misspelled names, missing name-value pairs, and other deficiencies that schemas would fix.

Requires fitness functions
: To solve the contract issues just described, many teams use consumer-driven contracts as an architecture fitness function to make sure that loose contracts still contain sufficient information for the contract to function.

Trade-Offs

Table 13-2. Trade-offs for loose contracts

Advantage	Disadvantage
Highly decoupled	Contract management
Easier to evolve	Requires fitness functions

For an example of the common trade-offs encountered by architects, consider the example of contracts in microservice architectures.

Contracts in Microservices

Architects must constantly make decisions about how services interact with one another, what information to pass (the semantics), how to pass it (the implementation), and how tightly to couple the services.

Coupling levels

Consider two microservices with independent transactionality that must share domain information such as *Customer Address*, shown in Figure 13-3.

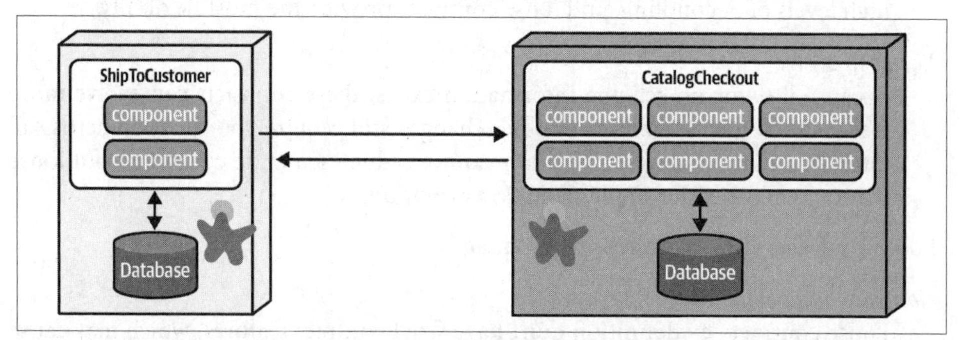

Figure 13-3. Two services that must share domain information about the customer

The architect could implement both services in the same technology stack and use a strictly typed contract, either a platform-specific remote procedure protocol (such as RMI) or an implementation-independent one like gRPC, and pass the customer information from one to another with high confidence of contract fidelity. However, this tight coupling violates one of the aspirational goals of microservices architectures, where architects try to create decoupled services.

Consider the alternative approach, where each service has its own internal representation of Customer, and the integration uses name-value pairs to pass information from one service to another, as illustrated in Figure 13-4.

Here, each service has its own bounded-context definition of Customer. When passing information, the architect utilizes name-value pairs in JSON to pass the relevant information in a loose contract.

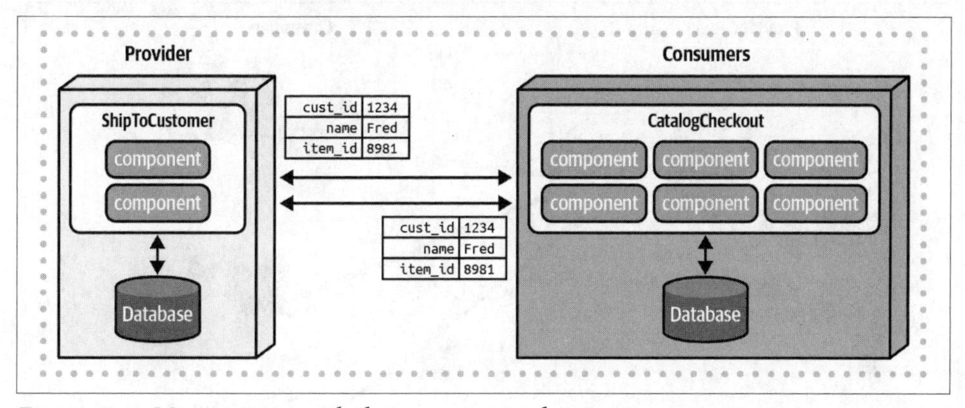

Figure 13-4. Microservices with their own internal semantic representation can pass values in simple messages

This loose coupling satisfies many of the overarching goals of microservices. First, it creates highly decoupled services modeled after bounded contexts, allowing each team to evolve internal representations as aggressively as needed. Second, it creates implementation decoupling. If both services start in the same technology stack, but the team in the second decides to move to another platform, it likely won't affect the first service at all. All platforms in common use can produce and consume name-value pairs, making them the lingua franca of integration architecture.

The biggest downside of loose contracts is contract fidelity—as an architect, how can I know that developers pass the correct number and type of parameters for integration calls? Some protocols, such as JSON, include schema tools to allow architects to overlay loose contracts with more metadata. Architects can also use a style of architect fitness function called a *consumer-driven contract*.

Consumer-driven contracts

A common problem in microservices architectures is the seemingly contradictory goals of loose coupling yet contract fidelity. One innovative approach that utilizes advances in software development is a *consumer-driven contract*, common in microservices architectures.

In many architecture integration scenarios, a service decides what information to emit to other integration partners (a *push* model—the service provider pushes a contract to consumers). The concept of a consumer-driven contract inverses that relationship into a *pull* model; here, the consumer puts together a contract for the items they need from the provider, and passes the contract to the provider, who includes it in their build and keeps the contract test green at all times. The contract encapsulates the information the consumer needs from the provider. This may work for a network of interlocking requests that the Provider must honor, as illustrated in Figure 13-5.

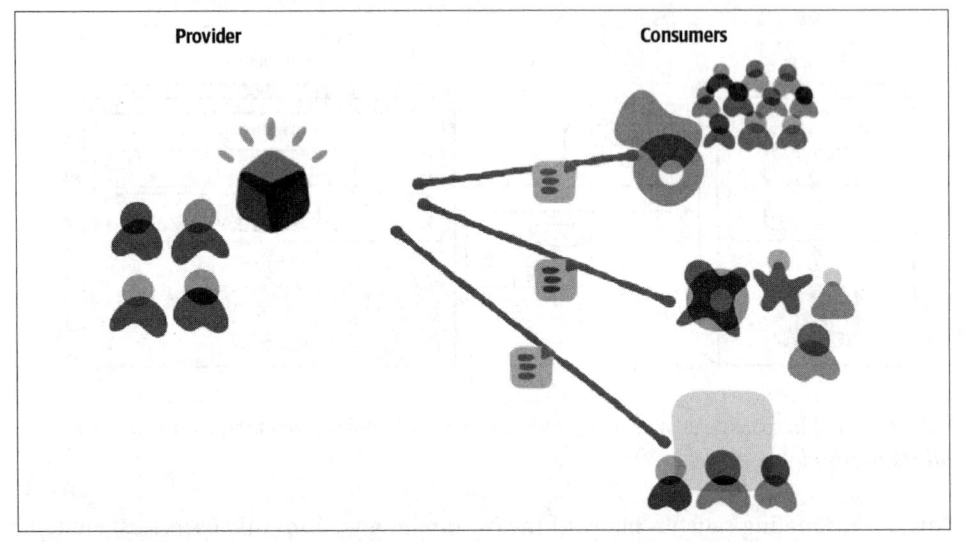

Figure 13-5. Consumer-driven contracts allow the provider and consumers to stay in sync via automated architectural governance

In this example, the team on the left provides bits of (likely) overlapping information to each of the consumer teams on the right. Each consumer creates a contract specifying required information and passes it to the provider, who includes their tests as part of a continuous integration or deployment pipeline. This allows each team to specify the contract as strictly or loosely as needed while guaranteeing contract fidelity as part of the build process. Many consumer-driven contract testing tools provide facilities to automate build-time checks of contracts, providing another layer of benefit similar to stricter contracts.

Consumer-driven contracts are quite common in microservices architecture because they allow architects to solve the dual problems of loose coupling and governed integration. Trade-offs of consumer-driven contracts are shown in Table 13-3.

Advantages of consumer-driven contracts are as follows:

Allow loose contract coupling between services
: Using name-value pairs is the loosest possible coupling between two services, allowing implementation changes with the least chance of breakage.

Allow variability in strictness

If teams use architecture fitness functions, architects can build stricter verifications than typically offered by schemas or other type-additive tools. For example, most schemas allow architects to specify things like numeric type but not acceptable ranges of values. Building fitness functions allows architects to build as much specificity as they like.

Evolvable

Loose coupling implies evolvability. Using simple name-value pairs allows integration points to change implementation details without breaking the semantics of the information passed between services.

These are disadvantages of consumer-driven contracts:

Require engineering maturity

Architecture fitness functions are a great example of a capability that really works well only when well-disciplined teams have good practices and don't skip steps. For example, if all teams run continuous integration that includes contract tests, then fitness functions provide a good verification mechanism. On the other hand, if many teams ignore failed tests or are not timely in running contract tests, integration points may be broken in architecture longer than desired.

Two interlocking mechanisms rather than one

Architects often look for a single mechanism to solve problems, and many of the schema tools have elaborate capabilities to create end-to-end connectivity. However, sometimes two simple interlocking mechanisms can solve the problem more simply. Thus, many architects use the combination of name-value pairs and consumer-driven contracts to validate contracts. However, this means that teams require two mechanisms rather than one.

The architect's best solution for this trade-off comes down to team maturity and decoupling with loose contracts versus complexity plus certainty with stricter contracts.

Trade-Offs

Table 13-3. Trade-offs for consumer-driven contracts

Advantage	Disadvantage
Allows loose contract coupling between services	Requires engineering maturity
Allows variability in strictness	Two interlocking mechanisms rather than one
Evolvable	

Stamp Coupling

A common pattern and sometimes anti-pattern in distributed architectures is stamp coupling (*https://oreil.ly/Jau2N*), which describes passing a large data structure between services, but each service interacts with only a small part of the data structure. Consider the example of four services shown in Figure 13-6.

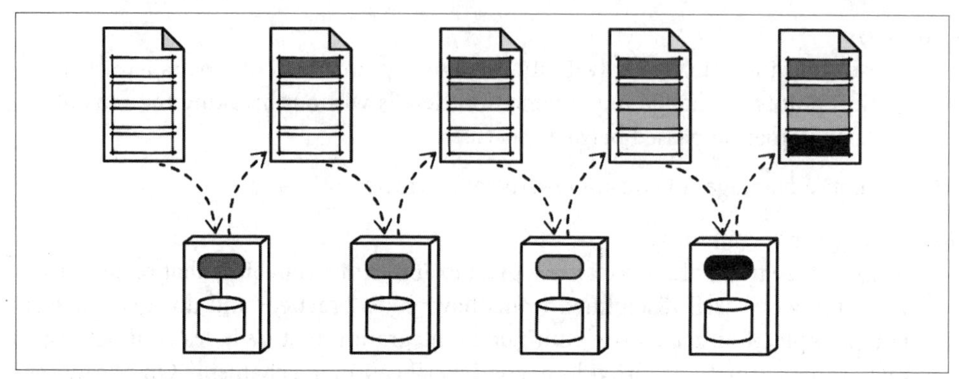

Figure 13-6. Stamp coupling between four services

Each service accesses (either reads, writes, or both) only a small portion of the data structure passed between each service. This pattern is common when an industry-standard document format exists, typically in XML. For example, the travel industry has a global standard XML document format that specifies details about travel itineraries. Several systems that work with travel-related services pass the entire document around, updating only their relevant sections.

Stamp coupling, however, is often an accidental anti-pattern, where an architect has over-specified the details in a contract that aren't needed or accidentally consumes far too much bandwidth for mundane calls.

Over-Coupling via Stamp Coupling

Going back to our Wishlist and Profile Services, consider tying the two together with a strict contract combined with stamp coupling, as illustrated in Figure 13-7.

In this example, even though the Wishlist Service needs only the name (accessed via a unique ID), the architect has coupled Profile's entire data structure as the contract, perhaps in a misguided effort for future proofing. However, the negative side effect of too much coupling in contracts is brittleness. If Profile changes a field that Wishlist doesn't care about, such as `state`, it still breaks the contract.

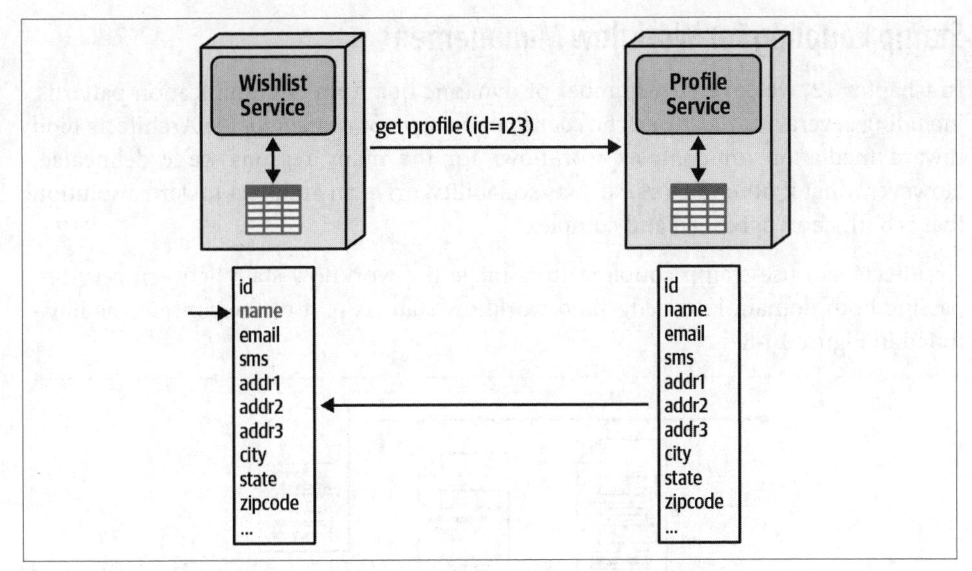

Figure 13-7. The Wishlist Service is stamp coupled to the Profile Service

Over-specifying details in contracts is generally an anti-pattern but easy to fall into when also using stamp coupling for legitimate concerns, including uses such as workflow management (see "Stamp Coupling for Workflow Management" on page 378).

Bandwidth

The other inadvertent anti-pattern that some architects fall into is one of the famous fallacies of distributed computing: *bandwidth is infinite*. Architects and developers rarely have to consider the cumulative size of the number of method calls they make within a monolith because natural barriers exist. However, many of those barriers disappear in distributed architectures, inadvertently creating problems.

Consider the previous example for 2,000 requests per second. If each payload is 500 KB, then the bandwidth required for this single request equals 1,000,000 KB per second! This is obviously an egregious use of bandwidth for no good reason. Alternatively, if the coupling between Wishlist and Profile contained only the necessary information, `name`, the overhead changes to 200 *bytes* per second, for a perfectly reasonable 400 KB.

Stamp coupling can create problems when overused, including issues caused by coupling too tightly to bandwidth. However, like all things in architecture, it has beneficial uses as well.

Stamp Coupling for Workflow Management

In Chapter 12, we covered a number of dynamic quantum communication patterns, including several that featured the *coordination* style of *choreography*. Architects tend toward mediation for complex workflows for the many reasons we've delineated. However, what if other factors, such as scalability, drive an architect toward a solution that is both choreographed and complex?

Architects can use stamp coupling to manage the workflow state between services, passing both domain knowledge and workflow state as part of the contract, as illustrated in Figure 13-8.

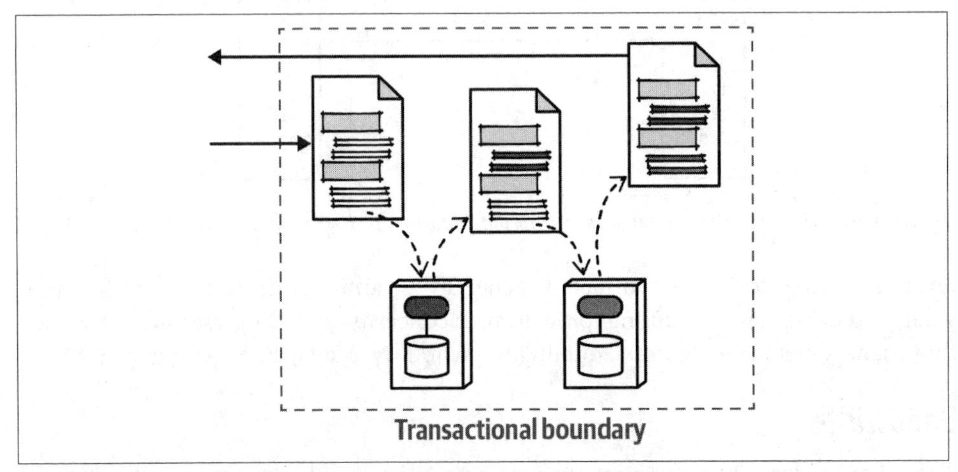

Figure 13-8. Using stamp coupling for workflow management

In this example, an architect designs the contract to include workflow information: status of the workflow, transactional state, and so on. As each domain service accepts the contract, it updates its portion of the contract and state for the workflow, then passes it along. At the end of the workflow, the receiver can query the contract to determine success or failure, along with status and information such as error messages. If the system needs to implement transactional consistency throughout, then domain services should rebroadcast the contract to previously visited services to restore atomic consistency.

Using stamp coupling to manage workflow does create higher coupling between services than nominal, but the semantic coupling must go somewhere—remember, an architect cannot reduce semantic coupling via implementation. However, in many cases, switching to choreography can improve throughput and scalability, making the choice of stamp coupling over mediation an attractive one. Table 13-4 shows the trade-offs for stamp coupling.

Trade-Offs

Table 13-4. Trade-offs for stamp coupling

Advantage	Disadvantage
Allows complex workflows within choreographed solutions	Creates (sometimes artificially) high coupling between collaborators
	Can create bandwidth issues at high scale

Sysops Squad Saga: Managing Ticketing Contracts

`Tuesday, May 10, 10:10`

Sydney and Addison met again in the cafeteria over coffee to discuss the contracts in the ticket management workflow.

Addison said, "Let's look at the workflow under discussion, the ticket management workflow. I've sketched out the types of contracts we should use, and wanted to run it by you to make sure I wasn't missing anything. It's illustrated in Figure 13-9."

"The contracts between the orchestrator and the two ticket services, Ticket Management and Ticket Assignment, are tight; that information is highly semantically coupled and likely to change together," Addison said. "For example, if we add new types of things to manage, the assignment must sync up. The Notification and Survey Service can be much looser—the information changes more slowly, and doesn't benefit from brittle coupling."

Sydney said, "All those decisions make sense—but what about the contract between the orchestrator and the Sysops Squad expert application? It seems that would need as tight a contract as assignment."

"Good catch—nominally, we would like the contract with the mobile application to match ticket assignment. However, we deploy the mobile application through a public app store, and their approval process sometimes takes a long time. If we keep the contracts looser, we gain flexibility and slower rate of change."

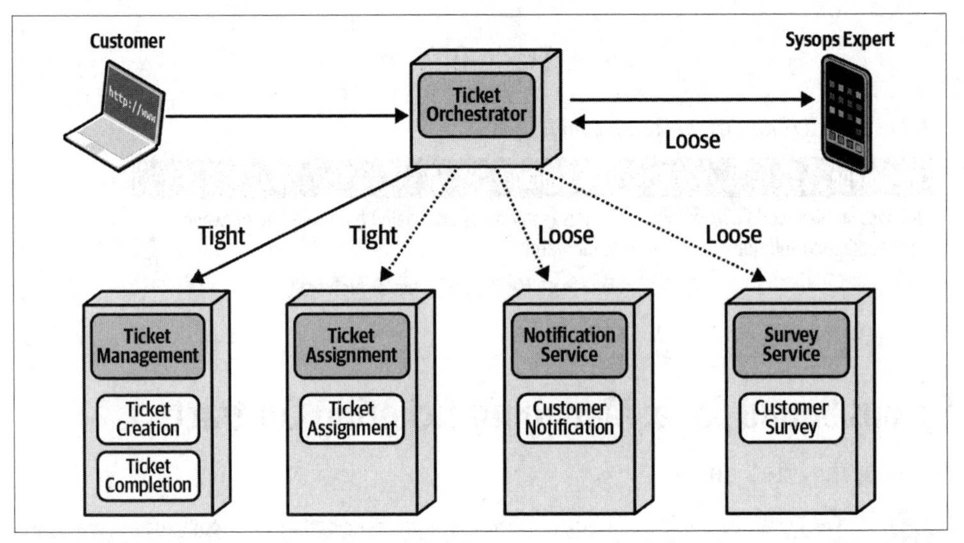

Figure 13-9. Types of contracts between collaborators in the ticket management workflow

They both wrote an ADR for this:

> *ADR: Loose Contract for Sysops Squad Expert Mobile Application*
>
> *Context*
>
> The mobile application used by Sysops Squad experts must be deployed through the public app store, imposing delays on the ability to update contracts.
>
> *Decision*
>
> We will use a loose, name-value pair contract to pass information to and from the orchestrator and the mobile application.
>
> We will build an extension mechanism to allow temporary extensions for short-term flexibility.
>
> *Consequences*
>
> The decision should be revisited if the app store policy allows for faster (or continuous) deployment.
>
> More logic to validate contracts must reside in the orchestrator and mobile application.

CHAPTER 14

Managing Analytical Data

Tuesday, May 31, 13:23

Logan and Dana (the data architect) were standing outside the big conference room, chatting after the weekly status meeting.

"How are we going to handle analytical data in this new architecture?" asked Dana. "We're splitting the databases into small parts, but we're going to have to glue all that data back together for reporting and analytics. One of the improvements we're trying to implement is better predictive planning, which means we are using more data science and statistics to make more strategic decisions. We now have a team that thinks about analytical data, and we need a part of the system to handle this need. Are we going to have a data warehouse?"

Logan said, "We looked into creating a data warehouse, and while it solved the consolidation problem, it had a bunch of issues for us."

Much of this book has been concerned with how to analyze trade-offs within existing architectural styles such as microservices. However, the techniques we highlight can also be used to understand brand-new capabilities as they appear in the software development ecosystem; *data mesh* is an excellent example.

Analytical and operational data have widely different purposes in modern architectures (see "The Importance of Data in Architecture" on page 4); much of this book has dealt with the difficult trade-offs associated with operational data. When client/server systems became popular and powerful enough for large enterprises, architects and database administrators looked for a solution that would allow specialized queries.

Previous Approaches

The split between operational and analytical data is hardly a new problem—the fundamental different uses of data have existed as long as data. As architecture styles have emerged and evolved, approaches for how to handle data have changed and evolved similarly.

The Data Warehouse

Back in earlier eras of software development (for example, mainframe computers or early personal computers), applications were monolithic, including code and data on the same physical system. Not surprisingly, given the context we've covered up until this point, transaction coordination across different physical systems became challenging. As data requirements became more ambitious, coupled with the advent of local area networks in offices, this led to the rise of *client/server* applications, where a powerful database server runs on the network and desktop applications run on local computers, accessing data over the network. The separation of application and data processing allowed better transactional management, coordination, and numerous other benefits, including the ability to start utilizing historical data for new purposes, such as analytics.

Architects made an early attempt to provide queriable analytical data with the *Data Warehouse* pattern. The basic problem they tried to address goes to the core of the separation of operational and analytical data: the formats and schemas of one don't necessarily fit (or even allow the use of) the other. For example, many analytical problems require aggregations and calculations, which are expensive operations on relational databases, especially those already operating under heavy transactional load.

The Data Warehouse patterns that evolved had slight variations, mostly based on vendor offerings and capabilities. However, the pattern had many common characteristics. The basic assumption was that operational data was stored in relational databases directly accessible via the network. Here are the main characteristics of the Data Warehouse pattern:

Data extracted from many sources
: As the operational data resided in individual databases, part of this pattern specified a mechanism for extracting the data into another (massive) data store, the "warehouse" part of the pattern. It wasn't practical to query across all the various databases in the organization to build reports, so the data was extracted into the warehouse solely for analytical purposes.

Transformed to single schema

Often, operational schemas don't match the ones needed for reporting. For example, an operational system needs to structure schemas and behavior around transactions, whereas an analytical system is rarely OLTP data (see Chapter 1) but typically deals with large amounts of data, for reporting, aggregations, and so on. Thus, most data warehouses utilized a *Star Schema* to implement dimensional modelling, transforming data from operational systems in differing formats into the warehouse schema. To facilitate speed and simplicity, warehouse designers denormalize the data to facilitate performance and simpler queries.

Loaded into warehouse

Because the operational data resides in individual systems, the warehouse must build mechanisms to regularly extract the data, transform it, and place it in the warehouse. Designers either used built-in relational database mechanisms like replication or specialized tools to build translators from the original schema to the warehouse schema. Of course, any changes to operational systems schemas must be replicated in the transformed schema, making change coordination difficult.

Analysis done on the warehouse

Because the data "lives" in the warehouse, all analysis is done there. This is desirable from an operational standpoint: the data warehouse machinery typically featured massively capable storage and compute, offloading the heavy requirements into its own ecosystem.

Used by data analysts

The data warehouse utilized data analysts, whose job included building reports and other business intelligence assets. However, building useful reports requires domain understanding, meaning that domain expertise must reside in both the operational data system and the analytical systems, where query designers must use the same data in a transformed schema to build meaningful reports and business intelligence.

BI reports and dashboards

The output of the data warehouse included business intelligence reports, dashboards that provide analytical data, reports, and any other information to allow the company to make better decisions.

SQL-ish interface

To make it easier for DBAs to use, most data warehouse query tools provided familiar affordances, such as a SQL-like language for forming queries. One of the reasons for the data transformation step mentioned previously was to provide users with a simpler way to query complex aggregations and other intelligence.

The Star Schema

The *Star Schema pattern* was popular with data marts and warehouses. It separates the data semantics into *facts*, which hold the organization's quantifiable data, and *dimensions*; hence they are also known as *dimensional models*, which include descriptive attributes of the fact data.

Examples of fact data for the Sysops Squad might include hourly rate, time to repair, distance to client, and other concretely measurable things. Dimensions might include squad member specialties, squad person names, store locations, and other metadata.

Most significantly, the Star Schema is purposely denormalized to facilitate simpler queries, simplified business logic (in other words, fewer complex joins), faster queries and aggregations, complex analytics such as data cubes, and the ability to form muti-dimensional queries. Most Star Schemas become incredibly complex.

The Data Warehouse pattern provides a good example of *technical partitioning* in software architecture: warehouse designers transform the data into a schema that facilitates queries and analysis but loses any domain partitioning, which must be re-created in queries where required. Thus, highly trained specialists were required to understand how to construct queries in this architecture.

However, the major failings of the Data Warehouse pattern included integration brittleness, extreme partitioning of domain knowledge, complexity, and limited functionality for intended purpose:

Integration brittleness
: The requirement built into this pattern to transform the data during the injection phase creates crippling brittleness in systems. A database schema for a particular problem domain is highly coupled to the semantics of that problem; changes to the domain require schema changes, which in turn require data import logic changes.

Extreme partitioning of domain knowledge
: Building complex business workflows requires domain knowledge. Building complex reports and business intelligence also requires domain knowledge, coupled with specialized analytics techniques. Thus, the Venn diagrams of domain expertise overlap, but only partially. Architects, developers, DBAs, and data scientists must all coordinate on data changes and evolution, forcing tight coupling between vastly different parts of the ecosystem.

Complexity
: Building an alternate schema to allow advanced analytics adds complexity to the system, along with the ongoing mechanisms required to injest and transform data. A data warehouse is a separate project outside the normal operational systems for an organization, so must be maintained as a wholly separate ecosystem, yet highly coupled to the domains embedded inside the operational systems. All these factors contribute to complexity.

Limited functionality for intended purpose
: Ultimately, most data warehouses failed because they didn't deliver business value commensurate to the effort required to create and maintain the warehouse. Because this pattern was common long before cloud environments, the physical investment in infrastructure was huge, along with the ongoing development and maintenance. Often, data consumers would request a certain type of report that the warehouse couldn't provide. Thus, such an ongoing investment for ultimately limited functionality doomed most of these projects.

Synchronization creates bottlenecks
: The need in a data warehouse to synchronize data across a wide variety of operational systems creates both operational and organizational bottlenecks—a location where multiple and otherwise independent data streams must converge. A common side effect of the data warehouse is the synchronization process impacting operational systems despite the desire for decoupling.

Operational versus analytical contract differences
: Systems of record have specific contract needs (discussed in Chapter 13). Analytical systems also have contractual needs that often differ from the operational ones. In a data warehouse, the pipelines often handle the transformation as well as ingestion, introducing contractual brittleness in the transformation process.

Table 14-1 shows the trade-offs for the data warehouse pattern.

Trade-Offs

Table 14-1. Trade-offs for the Data Warehouse pattern

Advantage	Disadvantage
Centralized consolidation of data	Extreme partitioning of domain knowledge
Dedicated analytics silo provides isolation	Integration brittleness
	Complexity
	Limited functionality for intended purpose

`Tuesday, May 31, 13:33`

"We looked at creating a data warehouse, but realized that it fit better with older, monolithic kinds of architectures than modern distributed ones," said Logan. "Plus, we have a ton more machine learning cases now that we need to support."

"What about the data lake idea I've been hearing about?" asked Dana. "I read a blog post on Martin Fowler's site.[1] It seems like it addresses a bunch of the issues with the data warehouse, and it is more suitable for ML use cases."

"Oh, yes, I read that post when it came out," Logan said. "His site is a treasure trove of good information, and that post came out right after the topic of microservices became hot. In fact, I first read about microservices on that same site in 2014, and one of the big questions at the time was, *How do we manage reporting in architectures like that?* The data lake was one of the early answers, mostly as a counter to the data warehouse, which definitely won't work in something like microservices."

"Why not?" Dana asked.

The Data Lake

As in many reactionary responses to the complexity, expense, and failures of the data warehouse, the design pendulum swung to the opposite pole, exemplified by the *Data Lake* pattern, intentionally the inverse of the Data Warehouse pattern. While it keeps the centralized model and pipelines, it inverts the "transform and load" model of the data warehouse to a "load and transform" one. Rather than do the immense work of transformation, the philosophy of the Data Lake pattern holds that, rather than do useless transformations that may never be used, do no transformations, allowing business users access to analytical data in its natural format, which typically required transformation and massaging for their purpose. Thus, the burden of work was made *reactive* rather than *proactive*—rather than do work that might not be needed, do transformation work only on demand.

The basic observation that many architects made was that the prebuilt schemas in data warehouses were frequently not suited to the type of report or inquiry required by users, requiring extra work to understand the warehouse schema enough to craft a solution. Additionally, many machine learning models work better with data "closer" to the semi-raw format rather than a transformed version. For domain experts who already understood the domain, this presented an excruciating ordeal, where data was stripped of domain separation and context to be transformed into the data

1 Martin Fowler posted an influential message about the Data Lake pattern on his blog in 2015 at *https://martinfowler.com/bliki/DataLake.html*.

warehouse, only to require domain knowledge to craft queries that weren't natural fits of the new schema!

Characteristics of the Data Lake pattern are as follows:

Data extracted from many sources
: Operational data is still extracted in this pattern, but less transformation into another schema takes place—rather, the data is often stored in its "raw," or native, form. Some transformation may still occur in this pattern. For example, an upstream system might dump formatted files into a lake that are organized based on a column-based snapshots.

Loaded into the lake
: The lake, often deployed in cloud environments, consists of regular data dumps from the operational systems.

Used by data scientists
: Data scientists and other consumers of analytical data discover the data in the lake and perform whatever aggregations, compositions, and other transformations necessary to answer specific questions.

The Data Lake pattern, while an improvement in many ways to the Data Warehouse pattern, still suffered many limitations.

This pattern still takes a *centralized* view of data, where data is extracted from operational systems' databases and replicated into a more or less free-form lake. The burden was on the consumer to discover how to connect disparate data sets together, which often happened in the data warehouse despite the level of planning. The logic followed that, if we're going to have to do pre-work for some analytics, let's do it for all, and skip the massive up-front investment.

While the Data Lake pattern avoided the transformation-induced problems from the Data Warehouse pattern, it also either didn't address or created new problems.

Difficulty in discovery of proper assets
: Much of the understanding of data relationships within a domain evaporates as data flows into the unstructured lake. Thus, domain experts must still involve themselves in crafting analysis.

PII and other sensitive data
: Concern around PII has risen in concert with the capabilities of the data scientist to take disparate pieces of information and learn privacy-invading knowledge. Many countries now restrict not just private information, but also information that can be combined to learn and identify, for ad targeting or other less savory purposes. Dumping unstructured data into a lake often risks exposing information that can be stitched together to violate privacy. Unfortunately, just as in the

discovery process, domain experts have the knowledge necessary to avoid accidental exposures, forcing them to reanalyze data in the lake.

Still technically, not domain, partitioned
The current trend in software architecture shifts focus from partitioning a system based on technical capabilities into ones based on domains, whereas both the Data Warehouse and Data Lake patterns focus on technical partitioning. Generally, architects design each of those solutions with distinct ingestion, transformation, loading, and serving partitions, each focused on a technical capability. Modern architecture patterns favor domain partitioning, encapsulating technical implementation details. For example, the microservices architecture attempts to separate services by domain rather than technical capabilities, encapsulating domain knowledge, including data, inside the service boundary. However, both the Data Warehouse and Data Lake patterns try to separate data as a separate entity, losing or obscuring important domain perspectives (such as PII data) in the process.

The last point is critical—increasingly, architects design around *domain* rather than *technical* partitioning in architecture, and both previous approaches exemplify separating data from its context. What architects and data scientists need is a technique that preserves the appropriate kind of macro-level partitioning, yet supports a clean separation of analytical from operational data. Table 14-2 lists the trade-offs for the Data Lake pattern.

Trade-Offs

Table 14-2. Trade-offs for the Data Lake pattern

Advantage	Disadvantage
Less structured than data warehouse	Sometimes difficult to understand relationships
Less up-front transformation	Requires ad hoc transformations
Better suited to distributed architectures	

The disadvantages around brittleness and pathological coupling of pipelines remain. Although they do less transformation in the Data Lake pattern, it is still common, as well as data cleansing.

The Data Lake pattern pushes data integrity testing, data quality, and other quality issues to downstream lake pipelines, which can create some of the same operational bottlenecks that manifest in the Data Warehouse pattern.

Because of both technical partitioning and the batch-like nature, solutions may suffer from data staleness. Without careful coordination, architects either ignore the changes in upstream systems, resulting in stale data, or allow the coupled pipelines to break.

`Tuesday, May 31, 14:43`

"OK, so we can't use the data lake either!" exclaimed Dana. "What now?"

"Fortunately, some recent research has found a way to solve the problem of analytical data with distributed architectures like microservices," replied Logan. "It adheres to the domain boundaries we're trying to achieve, but also allows us to project analytical data in a way that the data scientists can use. And, it eliminates the PII problems our lawyers are worried about."

"Great!" Dana replied. "How does it work?"

The Data Mesh

Observing other trends in distributed architectures, Zhamak Dehghani and several other innovators derived the core idea of the Data Mesh pattern from domain-oriented decoupling of microservices, service mesh, and sidecars (see "Sidecars and Service Mesh" on page 234), and applied it to analytical data, with modifications. As we mentioned in Chapter 8, the *Sidecar Pattern* provides a nonentangling way to organize orthogonal coupling (see "Orthogonal Coupling" on page 238); the separation between operational and analytical data is another excellent example of just such a coupling, but with more complexity than simple operational coupling.

Definition of Data Mesh

Data mesh is a sociotechnical approach to sharing, accessing, and managing analytical data in a decentralized fashion. It satisfies a wide range of analytical use cases, such as reporting, ML model training, and generating insights. Contrary to the previous architecture, it does so by aligning the architecture and ownership of the data with the business domains and enabling a peer-to-peer consumption of data.

Data mesh is founded on four principles:

Domain ownership of data

Data is owned and shared by the domains that are most intimately familiar with the data: the domains that either are originating the data, or are the first-class consumers of the data. This architecture allows for distributed sharing and accessing the data from multiple domains and in a peer-to-peer fashion without any intermediary and centralized lake or warehouse, and without a dedicated data team.

Data as a product
: To prevent siloing of data and encourage domains to share their data, data mesh introduces the concept of data served as a product. It puts in place the organizational roles and success metrics necessary to ensure that domains provide their data in a way that delights the experience of data consumers across the organization. This principle leads to the introduction of a new architectural quantum called *data product quantum*, to maintain and serve discoverable, understandable, timely, secure, and high-quality data to the consumers. This chapter introduces the architectural aspect of the data product quantum.

Self-serve data platform
: To empower the domain teams to build and maintain their data products, data mesh introduces a new set of self-serve platform capabilities. The capabilities focus on improving the experience of data product developers and consumers. It includes features such as declarative creation of data products, discoverability of data products across the mesh through search and browsing, and managing the emergence of other intelligent graphs, such as lineage of data and knowledge graphs.

Computational federated governance
: This principle assures that despite decentralized ownership of the data, organization-wide governance requirements—such as compliance, security, privacy, and quality of data, as well as interoperability of data products—are met consistently across all domains. Data mesh introduces a federated decision-making model composed of domain data product owners. The policies they formulate are automated and embedded as code in each and every data product. The architectural implication of this approach to governance is a platform-supplied embedded sidecar in each data product quantum to store and execute the policies at the point of access: data read or write.

Data mesh is a wide-ranging topic, fully covered in the book *Data Mesh* by Zhamak Dehghani (O'Reilly). In this chapter, we focus on the core architectural element, the *data product quantum*.

Data Product Quantum

The core tenet of the data mesh overlays modern distributed architectures such as microservices. Just as in the *service mesh*, teams build a *data product quantum* (DPQ) adjacent but coupled to their service, as illustrated in Figure 14-1.

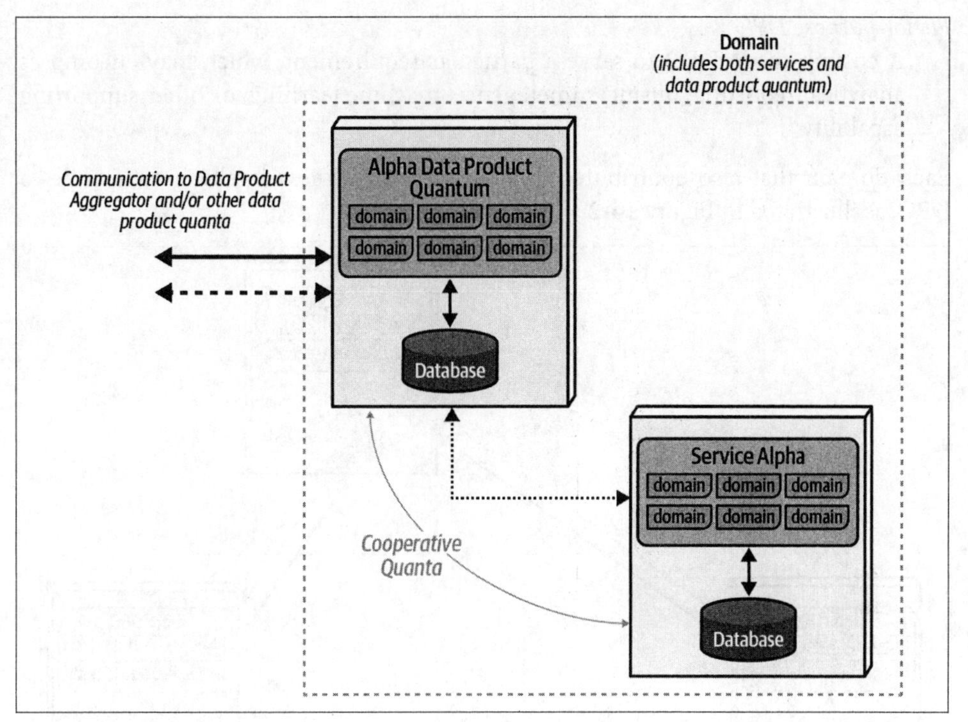

Figure 14-1. Structure of a data product quantum

In this example, the service *Alpha* contains both behavior and transactional (operational) data. The domain includes a *data product quantum*, which also contains code and data, and which acts as an interface to the overall analytical and reporting portion of the system. The DPQ acts as an operationally independent but highly coupled set of behaviors and data.

Several types of DPQs commonly exist in modern architectures:

Source-aligned (native) DPQ
: Provides analytical data on behalf of the collaborating architecture quantum, typically a microservice, acting as a cooperative quantum.

Aggregate DQP
: Aggregates data from multiple inputs, either synchronously or asynchronously. For example, for some aggregations, an asynchronous request may be sufficient; for others, the aggregator DPQ may need to perform synchronous queries for a source-aligned DPQ.

Fit-for-purpose DPQ
: A custom-made DPQ to serve a particular requirement, which may encompass analytical reporting, business intelligence, machine learning, or other supporting capability.

Each domain that also contributes to analysis and business intelligence includes a DPQ, as illustrated in Figure 14-2.

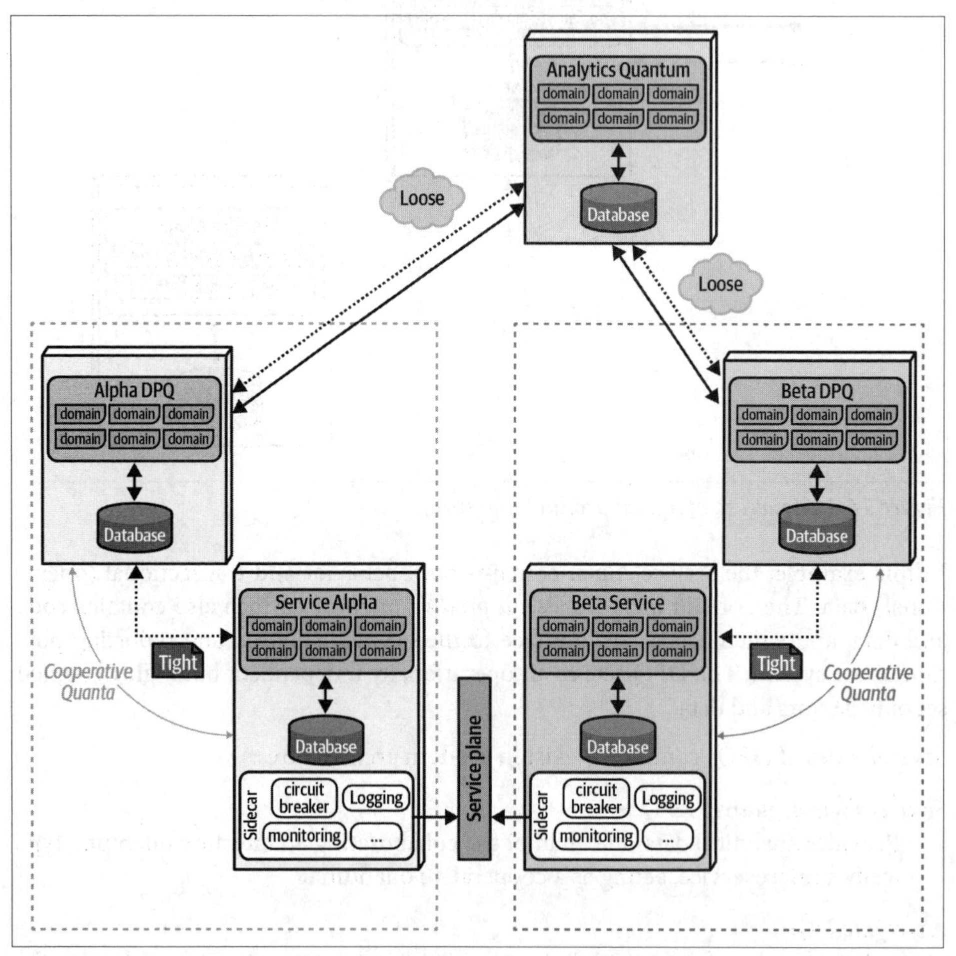

Figure 14-2. The data product quantum acts as a separate but highly coupled adjunct to a service

Here, the DPQ represents a component owned by the domain team responsible for implementing the service. It overlaps information stored in the database, and may have interactions with some of the domain behavior asynchronously. The data product quantum also likely has behavior as well as data for the purposes of analytics and business intelligence.

Each data product quantum acts as a cooperative quantum for the service itself:

Cooperative quantum
: An operationally separate quantum that communicates with its cooperator via asynchronous communication and eventual consistency, yet features tight contract coupling with its cooperator and generally looser contract coupling to the analytics quantum, the service responsible for reports, analysis, business intelligence, and so on. While the two cooperating quanta are operationally independent, they represent two sides of data: operational data in the quantum and analytical data in the data product quantum.

Some portion of the system will carry the responsibility for analytics and business intelligence, which will form its own domain and quantum. To operate, this analytical quantum has static quantum coupling to the individual data product quanta it needs for information. This service may make either synchronous or asynchronous calls to the DPQ, depending on the type of request. For example, some DPQs will feature a SQL interface to the analytical DPQ, allowing synchronous queries. Other requirements may aggregate information across multiple DPQs.

Data Mesh, Coupling, and Architecture Quantum

Because analytical reporting is probably a required feature of a solution, the DPQ and its communication implementation belong to the static coupling of an architecture quantum. For example, in a microservices architecture, the service plane must be available, just as a message broker must be available if the design calls for messaging. However, like the Sidecar pattern in a service mesh, the DPQ should be orthogonal to implementation changes within the service, and maintain a separate contract with the data plane.

From a dynamic quantum coupling standpoint, the data sidecar should always implement one of the communication patterns that features both eventual consistency and asynchronicity: either the "Parallel Saga(aeo) Pattern" on page 346 or "Anthology Saga(aec) Pattern" on page 349. In other words, a data sidecar should never include a transactional requirement to keep operational and analytical data in sync, which would defeat the purpose of using a DPQ for orthogonal decoupling. Similarly, communication to the data plane should genearlly be asynchronous, so as to have minimal impact on the operational architecture characteristics of the domain service.

When to Use Data Mesh

Like all things in architecture, this pattern has trade-offs associated with it, as shown in Table 14-3.

Trade-Offs

Table 14-3. Trade-offs for the Data Mesh pattern

Advantage	Disadvantage
Highly suitable for microservices architectures	Requires contract coordination with data product quantum
Follows modern architecture principles and engineering practices	Requires asynchronous communication and eventual consistency
Allows excellent decoupling between analytical and operational data	
Carefully formed contracts allow loosely coupled evolution of analytical capabilities	

It is most suitable in modern distributed architectures such as microservices with well-contained transactionality and good isolation between services. It allows domain teams to determine the amount, cadence, quality, and transparency of the data consumed by other quanta.

It is more difficult in architectures where analytical and operational data must stay in sync at all times, which presents a daunting challenge in distributed architectures. Finding ways to support eventual consistency, perhaps with very strict contracts, allows many patterns that don't impose other difficulties.

Data mesh is an outstanding example of the constant incremental evolution that occurs in the software development ecosystem; new capabilities create new perspectives, which in turn help address some persistent headaches from the past, such as the artificial separation of domain from data, both operational and analytical.

Sysops Squad Saga: Data Mesh

`Friday, June 10, 09:55`

Logan, Dana, and Addison met in the big conference room, which often had leftover snacks (or, this early in the day, breakfast) from previous meetings.

"I just returned from a meeting with our data scientists, and they are trying to figure out a way we can solve a long-term problem for us—we need to become data-driven in expert supply planning, for skill sets demand for different geographical locations at different points in time. That capability will help recruitment, training, and other supply-related functions," said Logan.

"I haven't been involved in much of the data mesh implementation—how far along are we?" asked Addison.

"Each new service we've implemented includes a DPQ. The domain team is responsible for running and maintaining the DQP cooperative quantum for their service. We've only just started. We're gradually building out the capabilities as we identify the needs. I have a picture of the Ticket Management Domain in Figure 14-3."

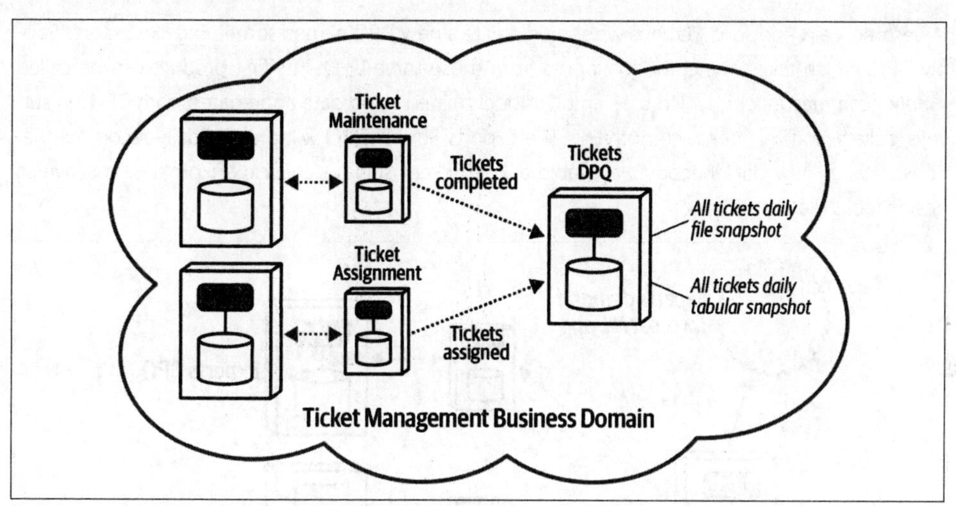

Figure 14-3. Ticket Management Domain, including two services with their own DPQs, with a Tickets DPQ

Logan said, "Tickets DPQ is its own architecture quantum, and acts as an aggregation point for a couple of different ticket views that other systems care about."

"How much does each team have to build versus already supplied?" Addison asked.

"I can answer that," said Dana. "The data mesh platform team is supplying the data users and data product developers with a set of self-serve capabilities. That allows any team that wants to build a new analytical use case to search and find the data products of choice within existing architecture quanta, directly connect to them, and start using them. The platform also supports domains that want to create new data products. The platform continuously monitors the mesh for any data product downtimes, or incompatibility with the governance policies and informs the domain teams to take actions."

Logan said, "The domain data product owners in collaboration with security, legal, risk, and compliance SMEs, as well as the platform product owners, have formed a global federated governance group, which decides on aspects of the DPQs that must be standardized, such as their data-sharing contracts, modes of asynchronous transport of data, access control, and so on. The platform team, over a span of time, enriches the DPQ's sidecar with new policy execution capabilities and upgrades the sidecars uniformly across the mesh."

"Wow, we're further along that I thought," said Dana. "What data do we need in order to supply the information for the expert supply problem?"

Logan replied, "In collaboration with the data scientists, we have determined what information we need to aggregate. It looks like we have the correct information: the Tickets DPQ serves the long-term view of all tickets raised and resolved, the User Maintenance DPQ provides daily snapshots for all expert profiles, and the Survey DPQ provides a log of all survey results from customers."

"Awesome," said Addison. "Perhaps we should create a new DPQ named something like Experts Supply DPQ, which takes asynchronous inputs from those three DPQs? Its first product can be called *supply recommendations*, which uses an ML model trained using data aggregated from DPQs in surveys, tickets, and maintenance domains. The Experts Supply DPQ will provide daily recommendations data, as new data becomes available about tickets, surveys and expert profiles. The overall design looks like Figure 14-4."

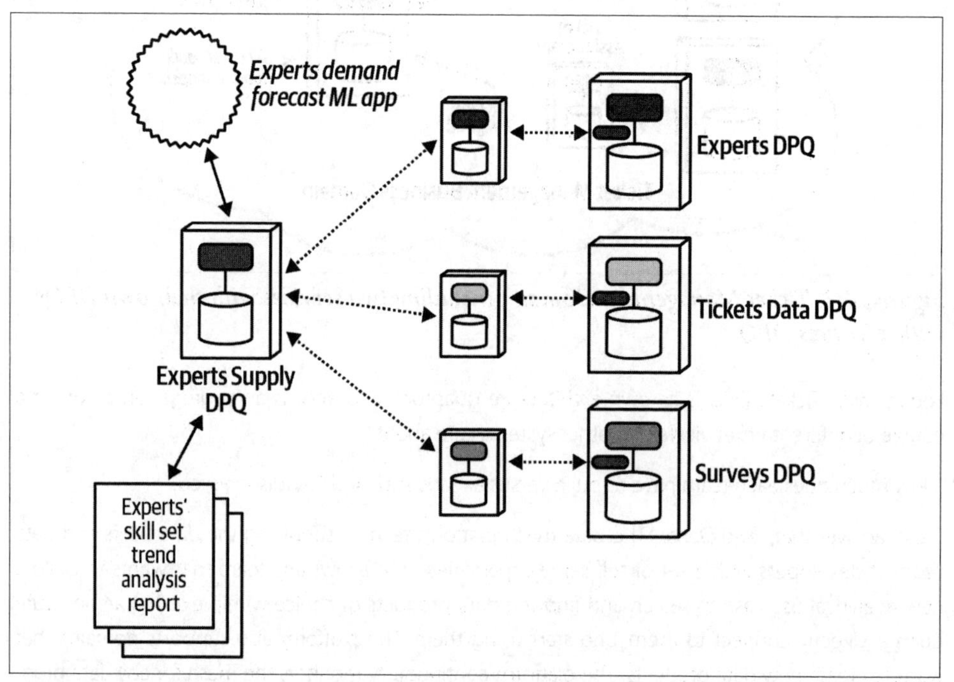

Figure 14-4. Implementing the Experts Supply DPQ

"OK, that looks perfectly reasonable," said Dana. "The services are already done; we just have to make sure the specific endpoints exist in each of the source DPQs, and implement the new Experts Supply DPQ."

"That's right," said Logan. "One thing we need to worry about, though—trend analysis depends on reliable data. What happens if one of the feeder source systems returns incomplete information for a chunk of time? Won't that throw off the trend analysis?"

"That's correct—no data for a time period is better than incomplete data, which makes it seem like there was less traffic than there was," Dana said. "We can just exempt an empty day, as long as it doesn't happen much."

"OK, Addison, you know what than means, right?" Logan said.

"Yes, I certainly do—an ADR that specifies complete information or none, and a fitness function to make sure we get complete data."

> *ADR: Ensure that Expert Supply DPQ Sources Supply an Entire Day's Data or None*
>
> *Context*
> The Expert Supply DPQ performs trend analysis over specified time periods. Incomplete data for a particular day will skew trend results and should be avoided.
>
> *Decision*
> We will ensure that each data source for the Expert Supply DPQ receives complete snapshots for daily trends or no data for that day, allowing data scientists to exempt that day.
>
> The contracts between source feeds and the Expert Supply DPQ should be loosely coupled to prevent brittleness.
>
> *Consequences*
> If too many days become exempt because of availability or other problems, accuracy of trends will be negatively impacted.
>
> *Fitness functions*:
>
> *Complete daily snapshot*. Check timestamps on messages as they arrive. Given typical message volume, any gap of more than one minute indicates a gap in processing, marking that day as exempt.
>
> *Consumer-driven contract fitness function for Ticket DPQ and Expert Supply DPQ*. To ensure that internal evolution of the Ticket Domain doesn't break the Experts Supply DPQ.

CHAPTER 15

Build Your Own Trade-Off Analysis

Monday, June 10, 10:01

The conference room somehow seemed more brightly lit than it did on that fateful day in September when the business sponsors of the Sysops Squad were about to pull the plug on the entire support contract business line. People in the conference room were chatting with each other before the meeting started, creating an energy not seen in the conference room for a long, long time.

"Well," said Bailey, the main business sponsor and head of the Sysops Squad ticketing application, "I suppose we should get things started. As you know, the purpose of this meeting is to discuss how the IT department was able to turn things around and repair what was nine months ago a train wreck."

"We call that a retrospective," said Addison. "And it's really useful for discovering how to do things better in the future, and to also discuss things that seemed to work well."

"So then, tell us, what worked really well? How did you turn this business line around from a technical standpoint?" asked Bailey.

"It really wasn't one single thing," said Austen, "but rather a combination of a lot of things. First of all, we in IT learned a valuable lesson about looking at the business drivers as a way to address problems and create solutions. Before, we always used to focus only on the technical aspects of a problem, and as a result never saw the big picture."

"That was one part of it," said Dana, "but one of the things that turned things around for me and the database team was starting to work together more with the application teams to solve problems. You see, before, those of us on the database side of things did our own thing, and the application development teams did their own thing. We never would have gotten to where we are now without collaborating and working together to migrate the Sysops Squad application."

"For me it was learning how to properly analyze trade-offs," said Addison. "If it weren't for Logan's guidance, insights, and knowledge, we wouldn't be in the shape we're in now. It was because of Logan that we were able to justify our solutions from a business perspective."

"About that," said Bailey, "I think I speak for everyone here when I say that your initial business justifications were what prompted us to give you one last shot at repairing the mess we were in. That was something we weren't accustomed to, and, well, quite frankly it took us by surprise—in a good way."

"OK," said Parker, "so now that we all agree things seem to be going well, how do we keep this pace going? How do we encourage other departments and divisions within the company from getting into the same mess we were in before?"

"Discipline," said Logan. "We continue our new habit of creating trade-off tables for all our decisions, continue documenting and communicating our decisions through architecture decision records, and continue collaborating with other teams on problems and solutions."

"But isn't that just adding a lot of extra process and procedures to the mix?" asked Morgan, head of the marketing department.

"No," said Logan. "That's *architecture*. And as you can see, it works."

Throughout this book, the unifying example illustrates how to generically perform trade-off analysis in distributed architectures. However, generic solutions rarely exist in architecture and, if they do, are generally incomplete for highly specific architectures and the unique problems they bring. Thus, we don't think that the communication analysis covered in Chapter 2 is exhaustive, but rather a starting point for you to add more columns for the unique elements entangled with your problem space.

To that end, this chapter provides some advice on how to build your own trade-off analysis, using many of the same techniques we used to derive the conclusions presented in this book.

Our three-step process for modern trade-off analysis, which we introduced in Chapter 2 is as follows:

- Find what parts are entangled together.
- Analyze how they are coupled to one another.
- Assess trade-offs by determining the impact of change to interdependent systems.

We discuss some techniques and considerations for each step next.

Finding Entangled Dimensions

An architect's first step in this process is to discover what dimensions are entangled, or braided, together. This is unique within a particular architecture but discoverable by experienced developers, architects, operations folks, and other roles familiar with the existing overall ecosystem and its capabilities and constraints.

Coupling

The first part of the analysis answers this question for an architect: how are parts within an architecture coupled to one another? The software development world has a wide variety of definitions of coupling, but we use the simplest, most intuitive version for this exercise: if someone changes *X*, will it possibly force *Y* to change?

In Chapter 2, we described the concept of the static coupling between architecture quanta, which provides a comprehensive structural diagram of technical coupling. No generic tool exists to build this because each architecture is unique. However, within an organization, a development team can build a static coupling diagram, either manually or via automation.

For example, to create a static coupling diagram for a microservice within an architecture, an architect needs to gather the following details:

- Operating systems/container dependencies
- Dependencies delivered via transitive dependency management (frameworks, libraries, etc.)
- Persistence dependencies on databases, search engines, cloud environments, etc.
- Architecture integration points required for the service to bootstrap itself
- Messaging infrastructure (such as a message broker) required to enable communication to other quanta

The static coupling diagram does not consider other quanta whose only coupling point is workflow communication with this quantum. For example, if an AssignTicket Service cooperates with the `ManageTicket` within a workflow but has no other coupling points, they are statically independent (but dynamically coupled during the actual workflow).

Teams that already have most of their environments built via automation can build into that generative mechanism an extra capability to document the coupling points as the system builds.

For this book, our goal was to measure the trade-offs in distributed architecture coupling and communication. To determine what became our three dimensions of dynamical quantum coupling, we looked at hundreds of examples of distributed architectures (both microservices and others) to determine the common coupling points. In other words, all the examples we looked at were sensitive to changes to the dimensions of *communication*, *consistency*, and *coordination*.

This process highlights the importance of iterative design in architecture. No architect is so brilliant that their first draft is always perfect. Building sample topologies for workflows (much as we do in this book) allows an architect or team to build a matrix view of trade-offs, allowing quicker and more thorough analysis than ad hoc approaches.

Analyze Coupling Points

Once an architect or team has identified the coupling points they want to analyze, the next step is to model the possible combinations in a lightweight way. Some of the combinations may not be feasible, allowing the architect to skip modeling those combinations. The goal of the analysis is to determine what forces the architect needs to study—in other words, which forces require trade-off analysis? For example, for our architecture quantum dynamic coupling analysis, we chose coupling, complexity, responsiveness/availability, and scale/elasticity as our primary trade-off concerns, in addition to analyzing the three forces of communication, consistency, and coordination, as shown in the ratings table for the "Parallel Saga(aeo) Pattern" on page 346, appearing again in Table 15-1.

Table 15-1. Ratings for the Parallel Saga pattern

Parallel Saga	Ratings
Communication	Asynchronous
Consistency	Eventual
Coordination	Centralized
Coupling	Low
Complexity	Low
Responsiveness/availability	High
Scale/elasticity	High

When building these ratings lists, we considered each design solution (our named patterns) in isolation, combining them only at the end to see the differences, shown in Table 15-2.

Trade-Offs

Table 15-2. Consolidated comparison of dynamic coupling patterns

Pattern	Coupling level	Complexity	Responsiveness/availability	Scale/elasticity
Epic Saga	Very high	Low	Low	Very Low
Phone Tag Saga	High	High	Low	Low
Fairy Tale Saga	High	Very low	Medium	High
Time Travel Saga	Medium	Low	Medium	High
Fantasy Fiction Saga	High	High	Low	Low
Horror Story	Medium	Very high	Low	Medium
Parallel Saga	Low	Low	High	High
Anthology Saga	Very low	High	High	Very high

Once we had analyzed each pattern independently, we created a matrix to compare the characteristics, leading to interesting observations. First, notice the direct inverse correlation between *coupling level* and *scale/elasticity*: the more coupling present in the pattern, the worse its scalability. This intuitively makes sense; the more services involved in a workflow, the more difficult for an architect to design for scale.

Second, we made a similar observation around *responsiveness/availability* and *coupling level*, which is not quite as direct as the preceding correlation but also significant: higher coupling leads to less responsiveness and availability because the more services involved in a workflow, the more likely the entire workflow will fail based on a service failure.

This analysis technique exemplifies *iterative* architecture. No architect, regardless of their cleverness, can instantly understand the nuances of a truly unique situation—and these nuances constantly present themselves in complex architectures. Building a matrix of possibilities informs the modeling exercises an architect might want to do in order to study the implications of permutating one or more dimensions to see the resulting effect.

Assess Trade-Offs

Once you have built a platform that allows iterative "what if" scenarios, focus on the fundamental trade-offs for a given situation. For example, we focused on synchronous versus asynchronous communication, a choice that creates a host of possibilities and restrictions—everything in software architecture is a trade-off. Thus, choosing a fundamental dimension like synchronicity first limits future choices. With that dimension now fixed, perform the same kind of iterative analysis on subsequent decisions encouraged or forced by the first. An architect team can iterate on this process until they have solved the difficult decisions—in other words, decisions with entangled dimensions. What's left is design.

Trade-Off Techniques

Over time, the authors have created a number of trade-off analyses and have built up some advice on how to approach them.

Qualitative Versus Quantative Analysis

You may have noticed that virtually none of our trade-off tables are *quantitative*—based on numbers—but are rather *qualitative*—measuring the quality of something rather than the quantity, which is necessary because two architectures will always differ enough to prevent true quantitative comparisons. However, using statistical analysis over a large data set allows reasonable qualitative analysis.

For example, when comparing the scalability of patterns, we looked at multiple different *implementations* of communication, consistency, and coordination combinations, assessing scalability in each case, and allowing us to build the comparative scale shown in Table 15-2.

Similarly, architects within a particular organization can carry out the same exercise, building a dimensional matrix of coupled concerns, and look at representative examples (either within the existing organization or localized spikes to test theories).

We recommend you hone the skill of performing qualitative analysis, as few opportunities for true quantitative analysis exist in architecture.

MECE Lists

It is important for architects to be sure they are comparing the same things rather than wildly different ones. For example, it's not a valid comparison to compare a simple message queue to an enterprise service bus, which contains a message queue but dozens of other components as well.

A useful concept borrowed from the technology strategy world to help architects get the correct match of things to compare is a *MECE list*, an acronym for *mutually exclusive, collectively exhaustive*:

Mutually exclusive

None of the capabilities can overlap between the compared items. As in the preceding example, it is invalid to compare a message queue to an entire ESB because they aren't really the same category of thing. If you want to compare just the messaging capabilities absent the other parts, that reduces the comparison to two mutually comparable things.

Collectively exhaustive
: This suggests that you've covered all the possibilities in the decision space, and that you haven't left out any obvious capabilities. For example, if a team of architects is evaluating high-performance message queues and consider only an ESB and simple message queue but not Kafka, they haven't considered all the possibilities in the space.

The goal of a MECE list is to cover a category space completely, with no holes or overlaps, as shown pictorially in Figure 15-1.

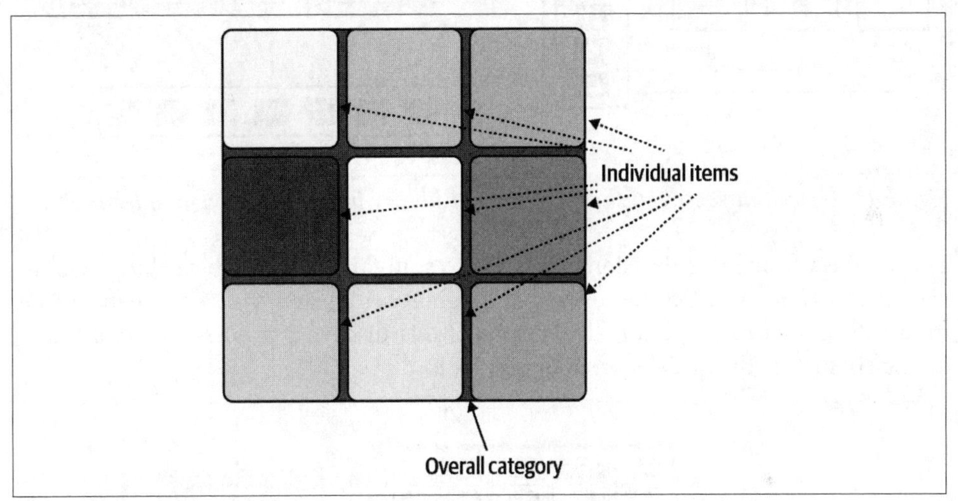

Figure 15-1. A MECE list is mutually exclusive and collectively exhaustive

The software development ecosystem constantly evolves, uncovering new capabilities along the way. When making a decision with long-term implications, an architect should make sure a new capability hasn't just arrived that changes the criteria. Ensuring that comparison criteria is collectively exhaustive encourages that exploration.

The "Out-of-Context" Trap

When assessing trade-offs, architects must make sure to keep the decision in context; otherwise, external factors will unduly affect their analysis. Often, a solution has many beneficial aspects, but lacks critical capabilities that prevent success. Architects need to make sure they balance the *correct* set of trade-offs, not all available ones.

For example, perhaps an architect is trying to decide whether to use a shared service or shared library for common functionality within a distributed architecture, as illustrated in Figure 15-2.

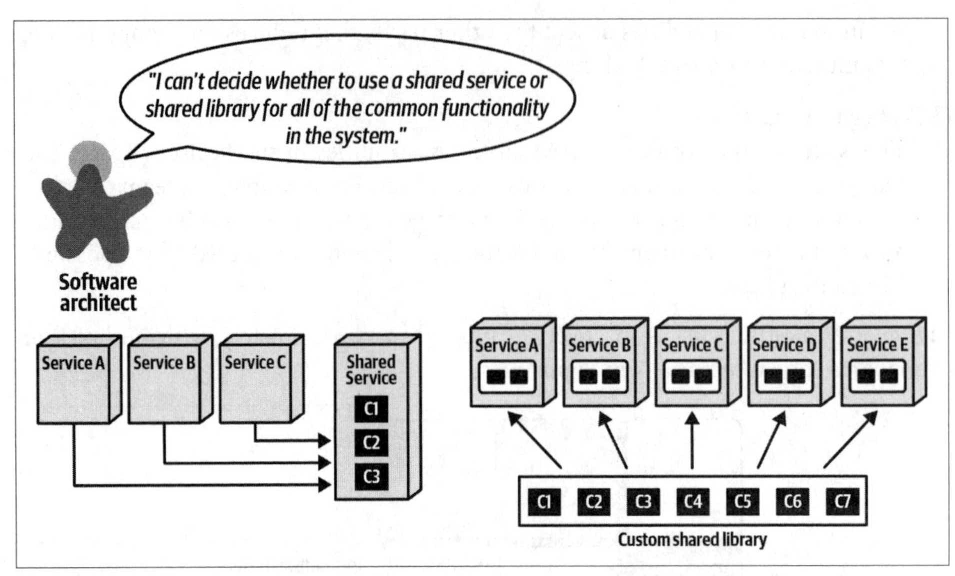

Figure 15-2. Deciding between shared service or library in a distributed architecture

The architect facing this decision will begin to study the two possible solutions, both via general characteristics discovered through research and via experimental data from within their organization. The results of that discovery process lead to a trade-off matrix such as the one shown in Figure 15-3.

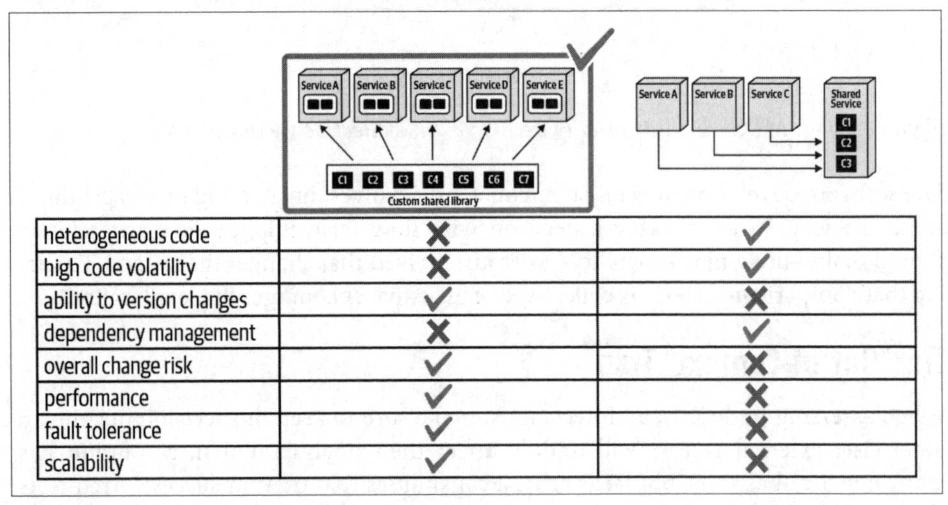

	Custom shared library	Shared service
heterogeneous code	✗	✓
high code volatility	✗	✓
ability to version changes	✓	✗
dependency management	✗	✓
overall change risk	✓	✗
performance	✓	✗
fault tolerance	✓	✗
scalability	✓	✗

Figure 15-3. Trade-off analysis for two solutions

The architect seems justified in choosing the shared library approach, as the matrix clearly favors that solution...*overall.* However, this decision exemplifies the *out-of-context* problem—when the extra context for the problem becomes clear, the decision criteria changes, as illustrated in Figure 15-4.

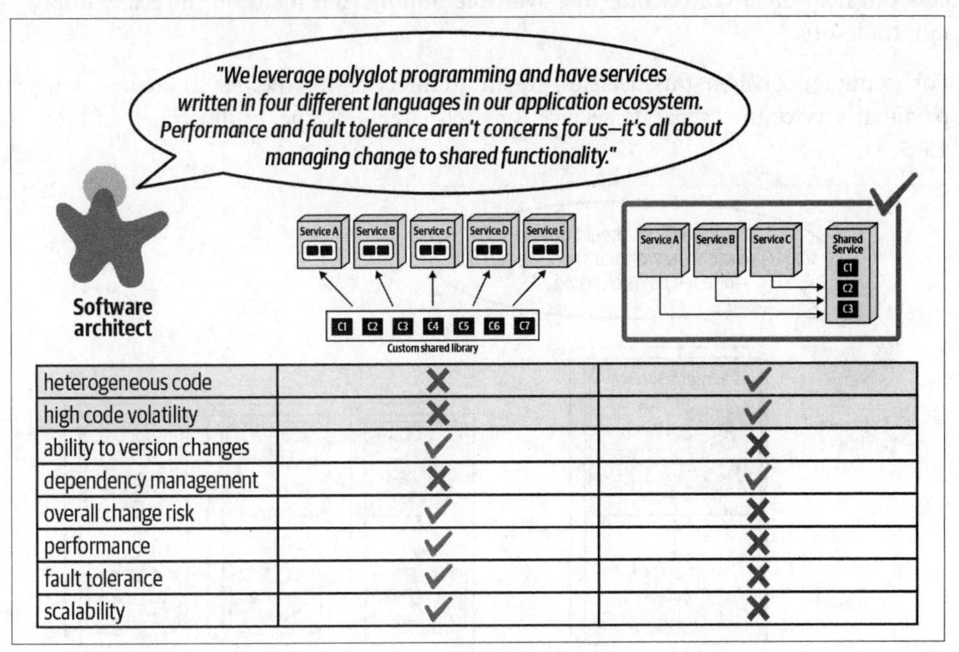

heterogeneous code	✗	✓
high code volatility	✗	✓
ability to version changes	✓	✗
dependency management	✗	✓
overall change risk	✓	✗
performance	✓	✗
fault tolerance	✓	✗
scalability	✓	✗

Figure 15-4. Shifting decision based on additional context

The architect continued to research not only the generic problem of service versus library, but the actual context that applies in this situation. Remember, generic solutions are rarely useful in real-world architectures without applying additional situation-specific context.

This process emphasizes two important observations. First, finding the best context for a decision allows the architect to consider fewer options, greatly simplifying the decision process. One common piece of advice from software sages is "embrace simple designs," without ever explaining how to achieve that goal. Finding the correct *narrow* context for decisions allows architects to think about less, in many cases simplifying design.

Second, it's critical for architects to understand the importance of *iterative* design in architecture, diagramming sample architectural solutions to play qualitative "what-if" games to see how architecture dimensions impact one another. Using iterative design, architects can investigate possible solutions and discover the proper context in which a decision belongs.

Model Relevant Domain Cases

Architects shouldn't make decisions in a vacuum, without relevant drivers that add value to the specific solution. Adding those domain drivers back to the decision process can help the architect filter the available options and focus on the really important trade-offs.

For example, consider this decision by an architect as to whether to create a single payment service or a separate service for each payment type, as illustrated in Figure 15-5.

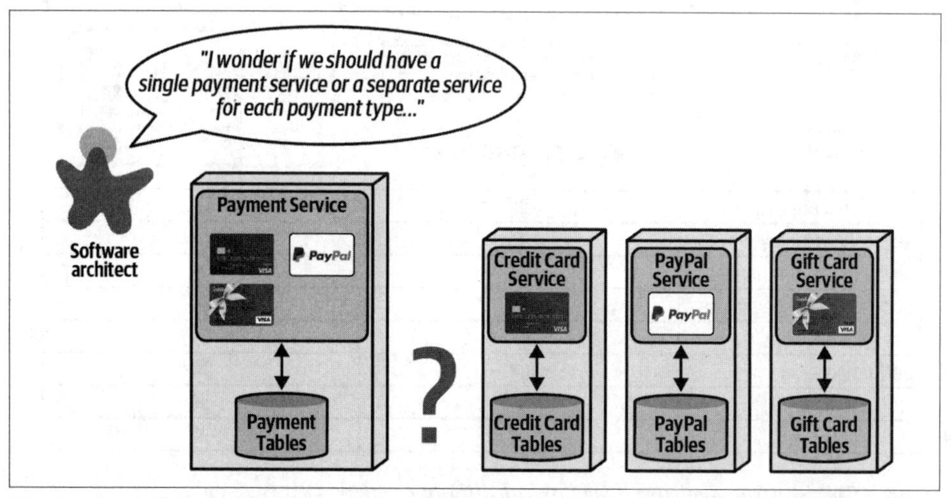

Figure 15-5. Choosing between a single payment service or one per payment type

As we discussed in Chapter 7, architects can choose from a number of integrators and disintegrators to assist this decision. However, those forces are generic—an architect may add more nuance to the decision by modeling some likely scenarios.

For example, consider the first scenario, illustrated in Figure 15-6, to update a credit card processing service.

In this scenario, having separate services provides better *maintainability*, *testability*, and *deployability*, all based on quantum-level isolation of the services. However, the downside of separate services is often duplicated code to prevent static quantum coupling between the services, which damages the benefit of having separate services.

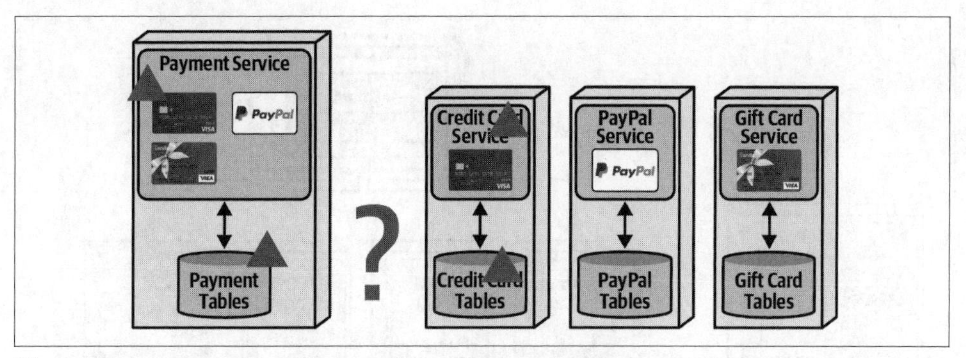

Figure 15-6. Scenario 1: update credit card processing service

In the second scenario, the architect models what happens when the system adds a new payment type, as shown in Figure 15-7.

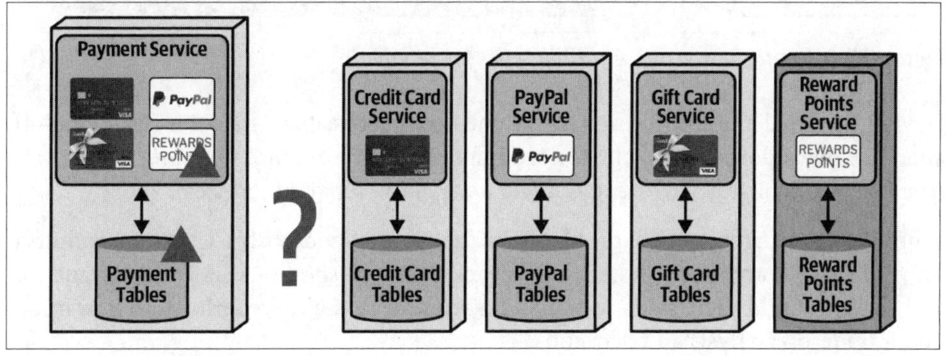

Figure 15-7. Scenario 2: adding a payment type

The architect adds a *reward points* payment type to see what impact it has on the architecture characteristics of interest, highlighting *extensibility* as a benefit of separate services. So far, separate services look appealing.

However, as in many cases, more complex workflows highlight the difficult parts of the architecture, as shown in the third scenario in Figure 15-8.

In this scenario, the architect starts gaining insight into the real trade-offs involved in this decision. Utilizing separate services requires coordination for this workflow, best handled by an orchestrater. However, as we discussed in Chapter 11, moving to an orchestrator likely impacts performance negatively and makes data consistency more of a challenge. The architect could avoid the orchestrator, but the workflow logic must reside somewhere—remember, semantic coupling can only be increased via implementation, never decreased.

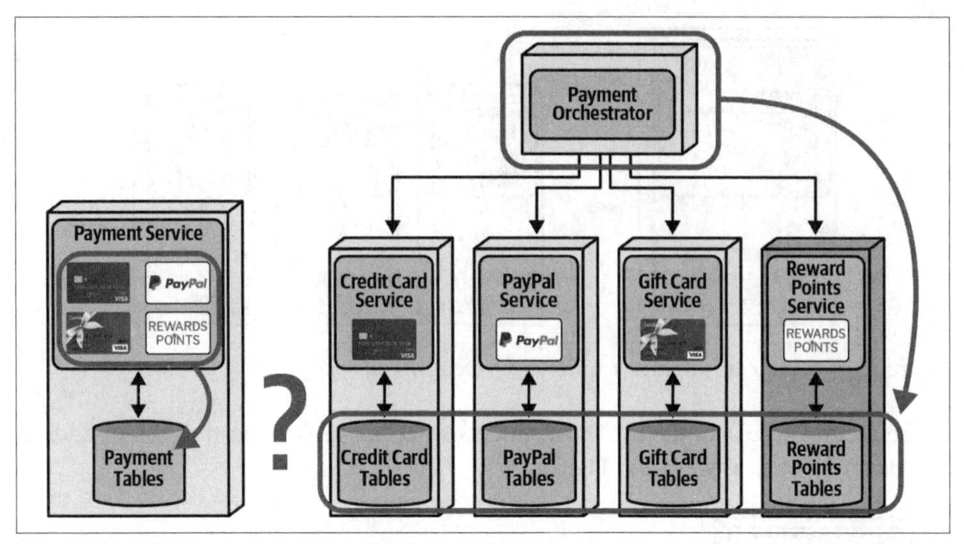

Figure 15-8. Scenario 3: using multiple types for payment

Having modeled these three scenarios, the architect realizes that the real trade-off analysis comes down to which is more important: *performance and data consistency* (a single payment service) or *extensibility and agility* (separate services).

Thinking about architecture problems in the generic and abstract gets an architect only so far. As architecture generally evades generic solutions, it is important for architects to build their skills in modeling relevant domain scenarios to home in on better trade-off analysis and decisions.

Prefer Bottom Line over Overwhelming Evidence

It's easy for architects to build up an enormous amount of information in pursuit of learning all the facets of a particular trade-off analysis. Additionally, anyone who learns something new generally wants to tell others about it, especially if they think the other party will be interested. However, many of the technical details that architects uncover are arcane to nontechnical stakeholders, and the amount of detail may overwhelm their ability to add meaningful insight into the decision.

Rather than show all the information they have gathered, an architect should reduce the trade-off analysis to a few key points, which are sometimes aggregates of individual trade-offs.

Consider the common problem an architect might face in a microservices architecture about the choice of synchronous or asynchronous communication, illustrated in Figure 15-9.

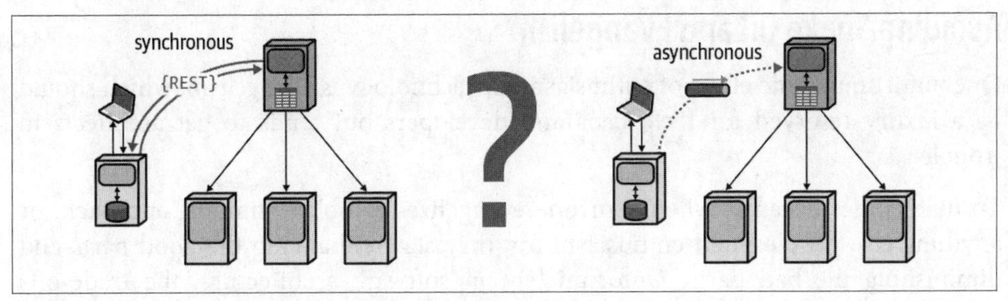

Figure 15-9. Deciding between communication types

The synchronous solution orchestrator makes synchronous REST calls to communicate with workflow collaborators, whereas the asynchronous solution uses message queues to implement asynchronous communication.

After considering the generic factors that point to one versus the other, the architect next thinks about specific domain scenarios of interest to nontechnical stakeholders. To that end, the architect will build a trade-off table that resembles Table 15-3.

Trade-Offs

Table 15-3. Trade-offs between synchronous and asynchronous communication for credit card processing

Synchronous advantage	Synchronous disadvantage	Asynchronous advantage	Asynchronous disadvantage
	Customer must wait for credit card approval process to start	No wait for process to start	
Credit approval is guaranteed to start before customer request ends			No guarantee that the process has started
	Customer application rejected if orchestrator is down	Application submission not dependent on orchestrator	

After modeling these scenarios, the architect can create a bottom-line decision for the stakeholders: which is more important, *a guarantee that the credit approval process starts immediately* or *responsiveness and fault-tolerance*? Eliminating confusing technical details allows the nontechnical domain stakeholders to focus on outcomes rather than design decisions, which help avoids drowning them in a sea of details.

Avoiding Snake Oil and Evangelism

One unfortunate side effect of enthusiasm for technology is evangelism, which should be a luxury reserved for tech leads and developers but tends to get architects in trouble.

Trouble comes because, when someone evangelizes a tool, technique, approach, or anything else people build enthusiasm for, they start enhancing the good parts and diminishing the bad parts. Unfortunately, in software architecture, the trade-offs always eventually return to complicate things.

An architect should also be wary of any tool or technique that promises any shocking new capabilities, which come and go on a regular basis. Always force evangelists for the tool or technique to provide an honest assessment of the good and bad—nothing in software architecture is all good—which allows a more balanced decision.

For example, consider an architect who has had success in the past with a particular approach and becomes an evangelist for it, as illustrated in Figure 15-10.

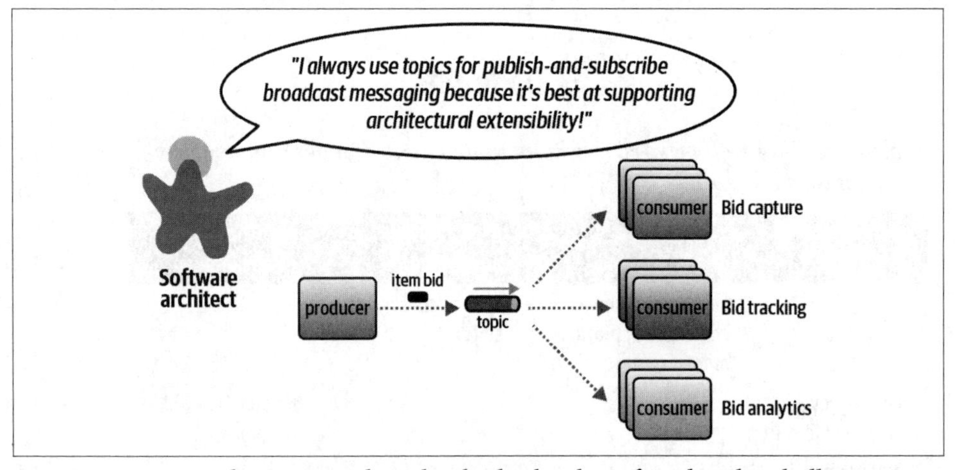

Figure 15-10. An architect evangelist who thinks they have found a silver bullet

This architect has likely worked on problems in the past where extensibility was a key driving architecture characteristic and believes that capability will always drive the decision process. However, solutions in architecture rarely scale outside narrow confines of a particular problem space. On the other hand, anecdotal evidence is often compelling. How do you get to the real trade-off hiding behind the knee-jerk evangelism?

While experience is useful, scenario analysis is one of an architect's most powerful tools to allow iterative design without building whole systems. By modeling likely scenarios, an architect can discover if a particular solution will, in fact, work well.

In the example shown in Figure 15-10, an existing system uses a single topic to broadcast changes. The architect's goal is to add *bid history* to the workflow—should the team keep the existing publish-and-subscribe approach or move to point-to-point messaging for each consumer?

To discover the trade-offs for this specific problem, the architect should model likely domain scenarios using the two topologies. Adding bid history to the existing publish-and-subscribe design appears in Figure 15-11.

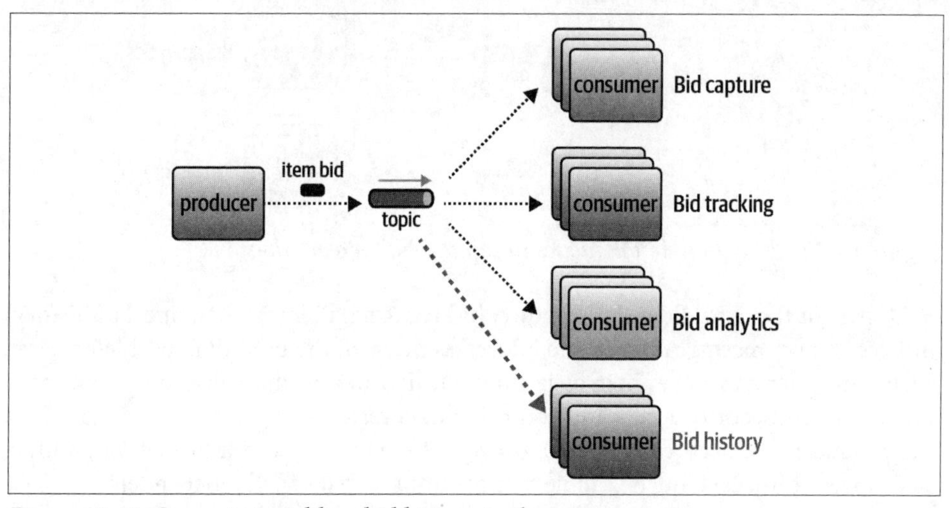

Figure 15-11. Scenario 1: Adding bid history to the existing topic

While this solution works, it has issues. First, what if the teams need different contracts for each consumer? Building a single large contract that encompasses everything implements the "Stamp Coupling for Workflow Management" on page 378 anti-pattern; forcing each team to unify on a single contract creates an accidental coupling point in the architecture—if one team changes its required information, all the teams must coordinate on that change. Second, what about data security? Using a single publish-and-subscribe topic, each consumer has access to all the data, which can create both security problems and PII (Personally Identifiable Information, discussed in Chapter 14) issues as well. Third, the architect should consider the operational architecture characteristic differences between the different consumers. For example, if the operations team wanted to monitor queue depth and use auto-scaling for *bid capture* and *bid tracking* but not for the other two services, using a single topic prevents that capability—the consumers are now operationally coupled together.

To mitigate these shortcomings, the architect should model the alternative solution to see if it addresses the preceding problems (and doesn't introduce new intractable ones). The individual queue version appears in Figure 15-12.

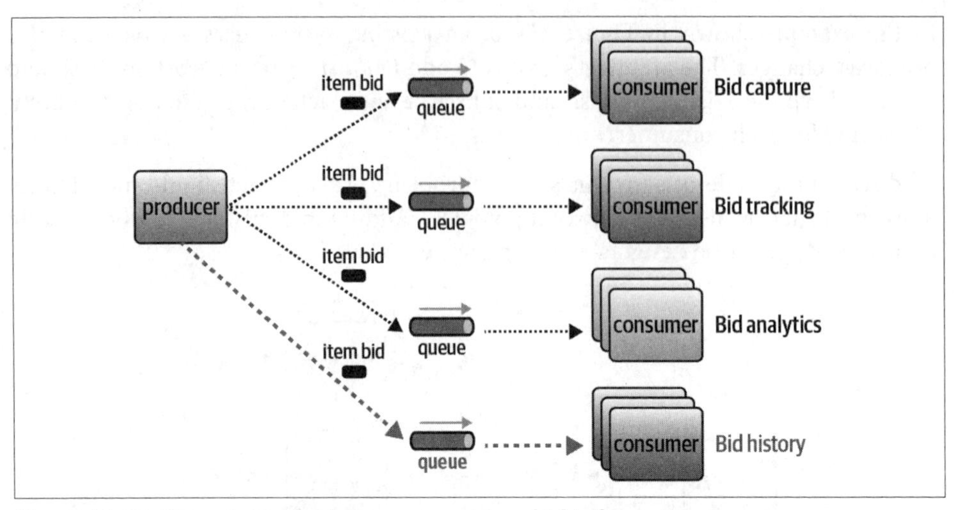

Figure 15-12. Using individual queues to capture bid information

Each part of this workflow (bid capture, bid tracking, bid analytics, and bid history) utilizes its own message queues and addresses many of the preceding problems. First, each consumer can have their own contract, decoupling the consumers from each other. Second, security access and control of data resides within the contract between the producer and each consumer, allowing differences in both information and rate of change. Third, each queue can now be monitored and scaled independently.

Of course, by this point in the book, you should realize that the point-to-point based system isn't perfect either but offers a different set of trade-offs.

Once the architect has modeled both approaches, it seems that the differences boil down to the choices shown in Table 15-4.

Trade-Offs

Table 15-4. Trade-offs between point-to-point versus publish-and-subscribe messaging

Point-to-point	Publish-and-subscribe
Allows heterogeneous contracts	Extensibility (easy to add new consumers)
More granular security access and data control	
Individual operational profiles per consumer	

In the end, the architect should consult with interested parties (operations, enterprise architects, business analysts, and so on) to determine which of these sets of trade-offs is more important.

Sometimes an architect doesn't choose to evangelize something but is rather coerced into playing an opposite foil, particularly for something that has no clear advantage. Technologies develop fans, sometimes fervent ones, who tend to downplay disadvantages and enhance upsides.

For example, recently a tech lead on a project tried to wrangle one of the authors into an argument about monorepo (*https://oreil.ly/PEEBC*) versus trunk-based development (*https://oreil.ly/HCtsh*). Both have good and bad aspects, a classic software architecture decision. The tech lead was a fervent supporter of the monorepo approach, and tried to force the author to take the opposing position—it's not an argument if two sides don't exist.

Instead, the architect pointed out that it was a trade-off, gently explaining that many of the advantages touted by the tech lead required a level of discipline that had never manifested within the team in the past, but will surely improve.

Rather than be forced into taking the opposing position, instead the architect forced a real-world trade-off analysis, not based on generic solutions. The architect agreed to try the Monorepo approach but also gather metrics to make sure that the negative aspects of the solution didn't manifest. For example, one of the damaging antipatterns they wanted to avoid was accidental coupling between two projects because of repository proximity, so the architect and team built a series of fitness functions to ensure that, while technically possible to create a coupling point, the fitness function prevented it.

Don't allow others to force you into evangelizing something—bring it back to trade-offs.

We advise architects to avoid evangelizing and to try to become the objective arbiter of trade-offs. An architect adds real value to an organization not by chasing silver bullet after silver bullet but rather by honing their skills at analyzing the trade-offs as they appear.

Sysops Squad Saga: Epilogue

`Monday, June 20, 16:55`

"OK, I think I finally get it. We can't really rely on generic advice for our architecture—it's too different from all the others. We have to do the hard work of trade-off analysis constantly."

"That's correct. But it's not a disadvantage—it's an advantage. Once we all learn how to isolate dimensions and perform trade-off analysis, we're learning concrete things about *our* architecture. Who cares about other, generic ones? If we can boil the number of trade-offs for a problem down to a small enough number to actually model and test them, we gain invaluable knowledge about our ecosystem. You know, structural engineers have built a ton of math and other predictive tools, but building their stuff is difficult and expensive. Software is a lot...well, softer. I've always said that *testing is the engineering rigor of software development*. While we don't have the kind of math other engineers have, we can incrementally build and test our solutions, allowing much more flexibility and leveraging the advantage of a more flexible medium. Testing with objective outcomes allows our trade-off analyses to go from qualitative to quantitative—from speculation to engineering. The more concrete facts we can learn about our unique ecosystem, the more precise our analysis can become."

"Yeah, that makes sense. Want to go to the after-work gathering to celebrate the big turnaround?"

"Sure."

APPENDIX A

Concept and Term References

In this book, we've made several references to terms or concepts that are explained in detail in our previous book, *Fundamentals of Software Architecture*. The following is a forward reference for those terms and concepts:

Cyclomatic complexity: Chapter 6, page 81

Component coupling: Chapter 7, page 92

Component cohesion: Chapter 7, page 93

Technical versus domain partitioning: Chapter 8, page 103

Layered architecture: Chapter 10, page 135

Service-based architecture: Chapter 13, page 163

Microservices architecture: Chapter 12, page 151

APPENDIX B

Architecture Decision Record References

Each Sysops Squad decision in this book was accompanied by a corresponding Architecture Decision Record. We consolidated all the ADRs here for easy reference:

APPENDIX C

Trade-Off References

The primary focus of this book is trade-off analysis; to that end, we created a number of trade-off tables and figures in Part II to summarize trade-offs around a particular architecture concern. This appendix summarizes all the trade-off tables and figures for easy reference:

Index

D

E

F

G

H

I

J

K

L

M

N

O

P

T

U

V

W

Y

Z

About the Authors

Neal Ford is a director, software architect, and meme wrangler at Thoughtworks, a software company and a community of passionate, purpose-led individuals who think disruptively to deliver technology that addresses the toughest challenges, all while seeking to revolutionize the IT industry and create positive social change. He's an internationally recognized expert on software development and delivery, especially in the intersection of Agile engineering techniques and software architecture. Neal has authored seven books (and counting), a number of magazine articles, and dozens of video presentations and spoken at hundreds of developers conferences worldwide. His topics include software architecture, continuous delivery, functional programming, cutting-edge software innovations, and a business-focused book and video on improving technical presentations. Check out his website, *Nealford.com*.

Mark Richards is an experienced, hands-on software architect involved in the architecture, design, and implementation of microservices architectures, service-oriented architectures, and distributed systems in a variety of technologies. He has been in the software industry since 1983 and has significant experience and expertise in application, integration, and enterprise architecture. Mark is the author of numerous technical books and videos, including the *Fundamentals of Software Architecture*, the "Software Architecture Fundamentals" video series, and several books and videos on microservices as well as enterprise messaging. Mark is also a conference speaker and trainer and has spoken at hundreds of conferences and user groups around the world on a variety of enterprise-related technical topics.

Pramod Sadalage is director of data and DevOps at Thoughtworks. His expertise includes application development, Agile database development, evolutionary database design, algorithm design, and database administration.

Zhamak Dehghani is director of emerging technologies at Thoughtworks. Previously, she worked at Silverbrook Research as a principal software engineer, and Fox Technology as a senior software engineer.

Colophon

The animal on the cover of *Software Architecture: The Hard Parts* is a black-rumped golden flameback woodpecker (*Dinopium benghalense*), a striking species of woodpecker found throughout the plains, foothills, forests, and urban areas of the Indian subcontinent.

This bird's golden back is set atop a black shoulder and tail, the reason for its pyro-inspired name. Adults have red crowns with black-and-white spotted heads and breasts, with a black stripe running from their eyes to the back of their heads. Like other common, small-billed woodpeckers, the black-rumped golden flameback has a straight pointed bill, a stiff tail to provide support against tree trunks, and four-toed feet—two toes pointing forward and two backward. As if its markings weren't distinctive enough, the black-rumped golden flameback woodpecker is often detected by its call of "ki-ki-ki-ki-ki," which steadily increases in pace.

This woodpecker feeds on insects, such as red ant and beetle larvae, underneath tree bark using its pointed bill and long tongue. They have been observed visiting termite mounds and even feeding on the nectar of flowers. The golden flameback also adapts well to urban habitats, subsisting on readily available fallen fruit and food scraps.

Considered relatively common in India, this bird's current conservation status is listed as being of "least concern." Many of the animals on O'Reilly covers are endangered; all of them are important to the world.

The cover image is a color illustration by Karen Montgomery, based on a black and white engraving from *Shaw's Zoology*. The cover fonts are URW Typewriter and Guardian Sans. The text fonts are Adobe Minion Pro and Myriad Pro; the heading font is Adobe Myriad Condensed; and the code font is Dalton Maag's Ubuntu Mono.

O'Reilly Media, Inc.介绍

O'Reilly以“分享创新知识、改变世界”为己任。40多年来我们一直向企业、个人提供成功必需之技能及思想，激励他们创新并做得更好。

O'Reilly业务的核心是独特的专家及创新者网络，他们通过我们分享知识。我们的在线学习（Online Learning）平台提供独家的直播培训、图书及视频，使客户更容易获取业务成功所需的专业知识。几十年来O'Reilly图书一直被视为学习开创未来之技术的权威资料。我们全年举办的诸多会议是活跃的技术聚会场所，来自各领域的专业人士在此建立联系，讨论最佳实践并发现可能影响技术行业未来的新趋势。

我们的客户渴望作出推动世界前进的创新，我们能祝您一臂之力。

业界评论

“O'Reilly Radar博客有口皆碑。”

——Wired

“O'Reilly凭借一系列（真希望当初我也想到了）非凡想法建立了数百万美元的业务。”

——Business 2.0

“O'Reilly Conference是聚集关键思想领袖的绝对典范。”

——CRN

“一本O'Reilly的书就代表一个有用、有前途、需要学习的主题。”

——Irish Times

“Tim是位特立独行的商人，他不光放眼于最长远、最广阔的视野并且切实地按照Yogi Berra的建议去做了：‘如果你在路上遇到岔路口，走小路（岔路）。’回顾过去Tim似乎每一次都选择了小路，而且有几次都是一闪即逝的机会，尽管大路也不错。”

——Linux Journal